To Duane Olson

Flemming Harmuth

Propagation of Nonsinusoidal Electromagnetic Waves

Advances in Electronics and Electron Physics

Edited by

PETER W. HAWKES

Laboratoire d'Optique Electronique
du Centre National de la Recherche Scientifique
Toulouse, France

Propagation of Nonsinusoidal Electromagnetic Waves

HENNING F. HARMUTH

DEPARTMENT OF ELECTRICAL ENGINEERING
THE CATHOLIC UNIVERSITY OF AMERICA
WASHINGTON, D.C.

1986

ACADEMIC PRESS, INC.
Harcourt Brace Jovanovich, Publishers

Orlando San Diego New York Austin
Boston London Sydney Tokyo Toronto

ACADEMIC PRESS, INC.
Orlando, Florida 32887

United Kingdom Edition published by
ACADEMIC PRESS INC. (LONDON) LTD.
24–28 Oval Road, London NW1 7DX

LIBRARY OF CONGRESS CATALOG CARD NUMBER: 63-12814

ISBN 0–12–014580–4

PRINTED IN THE UNITED STATES OF AMERICA

86 87 88 89 9 8 7 6 5 4 3 2 1

To Chang Tong

Professor in the Department of Automation

Tsinghua University

Beijing, China

Contents*

* Equations are numbered consecutively within each of Sections 1.1 to 6.1. Reference to an equation in a different section is made by writing the number of the section in front of the number of the equation, e.g., Eq. (2.1 – 50) for Eq. (50) in Section 2.1.

Illustrations and tables are numbered consecutively within each section, with the number of the section given first, e.g., Fig. 1.2-1, Table 4.2-1.

References are characterized by the name of the author(s), the year of publication, and a lowercase Latin letter if more than one reference by the same author(s) is listed for that year.

Preface

Electromagnetic wave theory has been based on the concept of infinitely extended periodic sinusoidal waves ever since Maxwell published his theory a century ago. On the practical level this worked very well, but on the theoretical level we had always an indication that something was amiss. There was never a satisfactory concept of the propagation velocity of signals within the framework of Maxwell's theory. The often-mentioned group velocity fails on two accounts, one being that it is almost always larger than the velocity of light in radio transmission through the atmosphere.

Beyond the velocity of propagation, we search the literature in vain for a solution of Maxwell's equations for a wave with beginning and end that could represent a signal propagating in a lossy medium. One might think the reason is the practical difficulty of obtaining solutions, but this is only partly correct. The solutions that we will derive are indeed mathematically complex, and they can be made useful only by means of computer plots. However, computers have been with us for 40 years, and for at least half this time they were sufficiently sophisticated and accessible to do the required computations. There was also plenty of incentive to study such solutions. In radar one would like to know the wave produced by the reflection or scattering of a (sinusoidal) pulse rather than a periodic sinusoidal wave, and in the stealth technology one would like to study the absorption of pulses rather than that of infinitely extended periodic waves. Since the enormous efforts devoted by military scientists to these problems have not yielded any satisfactory results, it is clear that something more than mathematical and computational difficulties must be the cause.

A closer study shows that the fault lies with Maxwell's equations rather than with their solution. In general, there can be no solution for signals propagating in lossy media. Expressed more scientifically, Maxwell's equations fail for waves with nonnegligible relative frequency bandwidth propagating in a medium with nonnegligible losses.

The failure is remedied by the addition of a magnetic current density to Maxwell's equations. But the remedy is even more surprising than the failure, since it is generally agreed that magnetic currents have not been observed and it is known from the study of magnetic monopoles that a magnetic current density can be eliminated or created by means of a so-called duality transformation. Both riddles are explained by singularities encountered in the course of calculation. If one chooses the current density zero before reaching the last singularity, one obtains no solution; if one does so after reaching the last singularity, one gets a solution.

It is always a pleasure as well as a duty to say thanks to those who have helped. For more than a decade, Richard B. Schulz, editor of the *IEEE Transactions on Electromagnetic Compatibility,* has helped fight the notion that any attempt to progress beyond the infinitely extended periodic sinusoidal wave is an unsinusoidal activity and thus a heresy. Trevor T. Cole of the School of Electrical Engineering and Jenifer S. Seberry of the Department of Applied Mathematics, University of Sydney, Australia, helped obtain a W. G. Watson Traveling Scholarship in Electrical Engineering; a good part of this book was written while the author enjoyed the opportunity of unfettered research provided by this scholarship. Malek G. M. Hussain of the Department of Electrical Engineering and Computer Science at Kuwait University and Raouf N. Boules of the Department of Applied Mathematics and Physics, Faculty of Engineering, University of Alexandria, Egypt, helped by providing the computer plots for certain functions defined by complicated integrals.

List of Symbols

In order to further international uniformity, the notation of the widely used reference book *Table of Integrals, Series, and Products* by Gradshteyn and Ryzhik (1980) is used. Hyperbolic sine and cosine functions are written sh x and ch x. The inverse functions of sin x and sh x are arcsin x and Arsh x, while $\sin^{-1} x$ and $\mathrm{sh}^{-1} x$ mean $1/\sin x$ and $1/\mathrm{sh}\, x$. The tangent functions are written tg x and th x rather than tan x and tanh x. Unit vectors are denoted $\mathbf{e}_x$, $\mathbf{e}_y$, $\mathbf{e}_z$ for cartesian and $\mathbf{e}_r$, $\mathbf{e}_\vartheta$, $\mathbf{e}_\varphi$ for spherical coordinates. The magnitude of a vector $\mathbf{A}$ is denoted A.

$a = 2^{-1}c(Z\sigma + s/Z) = 2^{-1}c^2(\mu\sigma + \epsilon s) = (\sigma/\epsilon + s/\mu)/2 = \alpha + \alpha'$ $[\mathrm{s}^{-1}]$, Eq. (2.1-50)

$b^2 = (2\pi\kappa)^2 + s\sigma = \beta^2 + s\sigma = \beta^2 + 1/L^2$ $[\mathrm{m}^{-2}]$, Eq. (2.1-50)

$c = (\mu\epsilon)^{-1/2}$ [m/s], velocity of light

d, see Eq. (2.4-20) and following text

j, imaginary unit

$k = j\gamma''$ $[\mathrm{m}^{-1}]$, complex circular wave number, Eq. (2.2-4)

$K = (2\pi)^{-1}(a^2/c^2 - s\sigma)^{1/2}$ $[\mathrm{m}^{-1}]$, Eq. (2.1-63)

$L = (s\sigma)^{-1/2}$ [m]

p_s, p'_s, see Eq. (2.6-13)

q_0, q'_0, see Eq. (2.5-4)

q_e, q'_e, see Eq. (2.9-6)

q_s, q'_s, see Eq. (2.4-6)

$Q = (1 - e^{-\Delta T/\tau})^{-1}$, Eq. (2.3-1)

$r(t)$, exponential ramp function, Fig. 1.2-4b

s, magnetic conductivity [V/A m]

$S(t)$, step function, Fig. 1.2-3

$t' = t - xc^{-1} \sin\vartheta$, Eq. (3.2-33)

$Z = (\mu/\epsilon)^{1/2}$ [V/A], wave impedance

Z_s, wave impedance for sinusoidal waves, Eqs. (2.2-15) and (4.2-4)

$\alpha = Zc\sigma/2 = \sigma/2\epsilon$ $[\mathrm{s}^{-1}]$, Eq. (2.1-66)

$\alpha' = s/2\mu$ $[\mathrm{s}^{-1}]$, Eq. (3.5-5)

α'', attenuation constant for sinusoidal waves, [m^{-1}], Eq. (2.2-6)

$\beta = 2\pi\kappa$ [m^{-1}], circular wave number or phase constant of sinusoidal waves

γ, inverse time constant [s^{-1}], Eq. (2.1-50)

γ'', propagation constant for sinusoidal waves [m^{-1}], Eq. (2.2-3)

$\delta = 2d/Z\sigma \ll 1$, Eq. (2.4-48)

ϵ, permittivity [A s/V m]

$\eta = \beta c/\alpha$, Eq. (2.1-78)

ϑ, angle

ϑ_{i}, ϑ_{r}, ϑ_{t}, angles of incidence, reflection, and transmission

$\theta = \alpha t$, Eq. (2.1-78)

$\theta' = \alpha t'$, Eq. (3.2-50)

κ, wave number [m^{-1}]

μ, permeability [V s/A m]

$\xi = \alpha y/c$, Eq. (2.1-78)

$\xi' = (\alpha y/c) \cos \vartheta$, Eq. (3.2-50)

ρ, electric charge density [A s m^{-3}]

ρ_s, reflection coefficient or function for perpendicularly polarized waves

ρ_p, reflection coefficient or function for parallel polarized waves

$\tau = 1/2a$ [s^{-1}], Eq. (2.3-15); $\tau = 1/2\alpha$, for $s = 0$

τ_s, transmission coefficient for perpendicularly polarized waves

τ_p, transmission coefficient for parallel polarized waves

τ_E, τ_{Es}, τ_{Ep}, transmission function for electric field strength

τ_H, τ_{Hs}, τ_{Hp}, transmission function for magnetic field strength

φ, angle or function $\varphi(y)$

ϕ, scalar potential [V]

$\omega = 2\pi f$, circular frequency [s^{-1}]

1 Introduction

1.1 Solution of Partial Differential Equations

The theoretical investigation of the propagation of electromagnetic waves starts with Maxwell's equations. Assuming that permeability, permittivity, and conductivity of the medium in which we want to study wave propagation are constants, we are dealing with a set of linear, partial differential equations with constant coefficients.

The classical methods for the solution of such equations are d'Alembert's[1] *method of characteristics* and Fourier's[2] *method of standing waves*.[3] d'Alembert's method works well with the one-dimensional wave equation or the equation of the vibrating string. Fourier's method is more readily applicable to many other partial differential equations, but the solutions are generally represented by integrals that could not be evaluated numerically before the development of the electronic computer.

The difficulty of finding general solutions of partial differential equations made scientists usually settle for particular solutions. The first step in finding such particular solutions is almost always the separation of variables by Bernoulli's product method. Consider equations with three space variables and the time variable. If cartesian coordinates are used, the separation of variables will lead to sine and cosine functions of the space variables; cylinder coordinates will lead to cylinder functions, spherical coordinates to spherical functions, etc. Time is always treated like one axis of a cartesian coordinate system, and the time variation is thus always of the form $e^{st} \sin \omega t$ or $e^{st} \cos \omega t$. We see here one of the most important reasons for the preeminence of sinusoidal functions in electrical communications and physics.

In electrical communications one is by definition interested in the transmission of signals and information. In physics, signals and information transmission became of fundamental importance with the arrival of the

---

[1] Jean Le Ronde d'Alembert, mathematician, 1717–1783; born in Paris, France.

[2] Jean Baptiste Joseph de Fourier, mathematician, 1768–1830; born in Auxerre, France.

[3] This method was actually first used by Daniel Bernoulli (mathematician, 1700–1782; born in Groningen, The Netherlands) in work on hydrodynamics (Bernoulli, 1738). His contribution is acknowledged in the often-used expression "separation of variables by Bernoulli's product method."

special theory of relativity, which is based on the concept of a finite propagation velocity of signals.

Solutions of Maxwell's equations with sinusoidal time variation cannot tell us anything about the propagation of signals, since signals can only be represented by *non*sinusoidal waves or functions. Waves that are either periodic or analytic functions of time transmit information at the rate zero, and sinusoidal functions are periodic as well as analytic. The reason for this zero rate of information transmission is that periodic time functions are known for all times if one period is known, and analytic time functions are known for all times if they are known in an infinitesimal neighborhood of a point in time. Since sinusoidal functions and waves are periodic and analytic in the whole interval $-\infty < t < +\infty$, we obtain no information by receiving them after a time $t = -T$, where T is arbitrarily large but finite.

A signal has always a beginning and an end, a rectangular pulse[4] being the most usual representation of a signal. One can approximate a pulse by sinusoidal functions and the Fourier transform. However, this approximation does not tell us how an electromagnetic wave with the initial time variation of a rectangular pulse changes when propagating in a lossy medium.

To obtain a general solution of Maxwell's equations we must find a function that meets three requirements. First, it must satisfy Maxwell's equations; sinusoidal time functions and their superpositions will do that. Second, it must satisfy the boundary conditions; we will see that this can be done by means of a Fourier transform. Third, it must satisfy the initial conditions; this requirement calls for something beyond the Fourier transform.

As far as the author is aware, a general method for satisfying initial conditions was introduced into electrical communications by Krylov (1929), when he calculated currents and voltages along a transmission line from the telegrapher's equation. This method became widely known by the books of Smirnov, which are available in several languages (1964, Vol. 2, Ch. VII).

Having this method to satisfy the initial conditions and having electronic computers that can evaluate numerically the integrals obtained by Fourier's method, one might expect that many solutions of Maxwell's equations representing signals—or nonsinusoidal solutions—should have become known during the last quarter of a century. But this is not the case.

The reason is that Maxwell's equations have solutions with sinusoidal or

[4] Signals in the special theory of relativity are usually represented by "light flashes," which one may represent by rectangular pulses or sinusoidal pulses, if one wants to be more specific about their time variation. For a photon interpreted as a particle, a rectangular pulse is a proper representation, while the same photon interpreted as a wave makes a sinusoidal pulse a good representation.

periodic time variation, but they have generally no solutions for signals that necessarily have a nonsinusoidal and nonperiodic time variation. So far, the only known exception are signals propagating in a loss-free medium, which leads in the case of one space variable to the one-dimensional wave equation and its general solution by d'Alembert.

To give an example for the failure of Maxwell's equations consider an electromagnet. A current is applied at the time $t = 0$ and increases from zero in an unspecified way to a constant value I in a finite time. We know that a constant magnetic field strength but no electric field strength is observed at a sufficiently large time. However, while the current increases from 0 to I we must have an electromagnetic wave that is excited by the electromagnet and propagates away from it. Maxwell's equations give a solution for waves propagating in a loss-free medium, but the solution does not correspond with observation; the electric field strength does not vanish. The assumption of a loss-free medium is apparently too unrealistic. No solution exists for a lossy medium. Hence, this simple and perfectly practical problem is outside the scope of Maxwell's equations.

Speaking more generally, Maxwell's equations fail for waves with nonnegligible relative frequency bandwidth[5] propagating in a medium with non-negligible losses.

Our first task must thus be to modify Maxwell's equations so that solutions can be obtained for waves with arbitrary time variation propagating in a lossy medium. Once we have these modified equations and their solution for excitation by electric or magnetic field strengths with the time variation of a rectangular pulse or a step function, we can investigate the propagation of signals. One of the most fundamental scientific problems we may address—both for electrical communications and the special theory of relativity—will be the velocity of propagation; we will obtain results that differ qualitatively from the group velocity as well as the more advanced concept of propagation velocity found by Sommerfeld (1914) and Brillouin (1914) for light pulses.

Among the technical problems, we may calculate the radar signature of pulses reflected at boundaries of lossy media like ground and seawater, or the effect of absorbing materials on radar pulses; these are basic tasks for stealth technology as well as anti-stealth-radar technology. Maxwell's equations permit such solutions only for periodic sinusoidal waves rather than the sinusoidal pulses actually used by radar. One does not have to go to the extreme of carrier-free pulses to make the solutions for periodic sinusoidal

---

[5] The relative frequency bandwidth of a wave is defined as the ratio $(f_H - f_L)/(f_H + f_L)$, where f_H is the highest and f_L the lowest frequency of interest in the Fourier representation. The term "nonsinusoidal waves" is an abbreviation of the term "waves with non-negligible relative frequency bandwidth."

waves inapplicable; the solutions would not apply to radar pulses with 10 rather than the usual 100 to 1000 sinusoidal cycles per pulse either. A better theory of electromagnetic waves is thus a basic requirement for the development of radar beyond the conventional small-relative-bandwidth radar.

1.2 Functions Modeling the Causality Law

One of the most basic laws of an experimental science is the causality law, which is usually stated in the form "any effect requires a sufficient cause." In communications it is often used in the somewhat less demanding form "an effect cannot precede its cause." The causality law is not a mathematical axiom, and a mathematical calculation will satisfy the causality law only if it is specifically introduced into the calculation.

For an example of how this obvious requirement is violated in communications consider the following problem that may be found in many textbooks (e.g., Close, 1966, p. 479). An idealized low-pass filter is characterized by the transfer function

$$\begin{aligned} h(f) = h(\omega) &= 1 \quad \text{for} \quad 0 \leqq f \leqq \Delta f \\ &= 0 \quad \text{for} \quad f \geqq \Delta f \end{aligned} \tag{1}$$

as shown in Fig. 1.2-1b. In order to calculate an output voltage $v_o(t)$ produced by an input voltage $v_i(t)$ one calculates the Fourier transform $g(f)$ of $v_i(t)$, multiplies it with the transfer function $h(f)$, and calculates the inverse Fourier transform of the product $g(f)h(f)$. A typical input voltage is the step function $V_0S(t)$,

$$\begin{aligned} S(t) &= 0 \quad \text{for} \quad t < 0 \\ &= 1 \quad \text{for} \quad t \geqq 0 \end{aligned} \tag{2}$$

which, unfortunately, has no Fourier transform. However, the function $S(t)e^{-t/\tau}$ shown in Fig. 1.2-1c has a Fourier transform, and this function approaches the step function for $\tau \to \infty$.

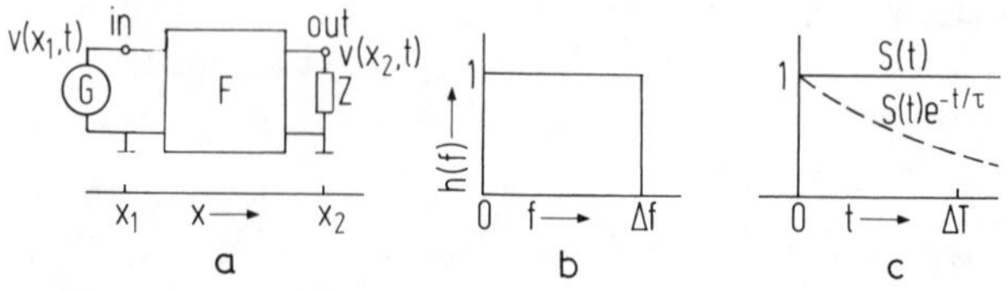

FIG. 1.2-1. (a) Filter with input terminal at location x_1 and output terminal at location x_2. (b) Characterization of the filter by a transfer function $h(f)$. (c) Typical input voltages for the filter.

We write $V_0 S(t)e^{-t/\tau}$ as the sum of the skew-symmetric function

$$V_0[S(t)-\tfrac{1}{2}]e^{-|t/\tau|} = \tfrac{1}{2}V_0 e^{-t/\tau} \qquad \text{for} \quad t>0$$
$$= -\tfrac{1}{2}V_0 e^{-|t/\tau|} \qquad \text{for} \quad t<0$$

and the symmetric function $(V_0/2)e^{-|t/\tau|}$, $-\infty < t < +\infty$. The sine transform of the skew-symmetric function and the cosine transform of the symmetric function yield:

$$g_s(f) = 2\frac{V_0}{2}\int_0^\infty e^{-t/\tau}\sin 2\pi ft\, dt = V_0\frac{2\pi f}{\tau^{-2}+(2\pi f)^2}$$
$$g_c(f) = 2\frac{V_0}{2}\int_0^\infty e^{-t/\tau}\cos 2\pi ft\, dt = V_0\frac{\tau^{-1}}{\tau^{-2}+(2\pi f)^2} \qquad (3)$$

For $\tau \rightarrow \infty$ the sine transform yields $V_0/2\pi f$, while the cosine transform yields zero for all values of f except for $f=0$. From $(V_0/2)e^{-|t/\tau|}$ follows $g_c(0) = V_0/2$ for $\tau \rightarrow \infty$. The inverse transform of $g_s(f)h(f)$ can be performed for $\tau \rightarrow \infty$,

$$2\int_0^\infty g_s(f)h(f)\sin 2\pi ft\, df = 2V_0\int_0^{\Delta f}\frac{\sin 2\pi ft}{2\pi f}\,df$$
$$= \frac{V_0}{\pi}\int_0^{2\pi\Delta f t}\frac{\sin 2\pi ft}{2\pi ft}\,d(2\pi ft) = \frac{V_0}{\pi}\operatorname{Si}(2\pi\,\Delta f\, t)$$

where $\operatorname{Si}(2\pi\,\Delta f\,t)$ is the sine integral[1].

For the inverse transform of $h(f)g_c(f)$ we draw the logical conclusion that a dc component with amplitude $V_0/2$ in the frequency domain yields a constant voltage $V_0/2$ in the time domain. Hence, we get

$$v_o(t) = V_0[\tfrac{1}{2} + \tfrac{1}{\pi}\operatorname{Si}(2\pi\,\Delta f\,t)] \qquad (4)$$

The input voltage $v_i(t)$ is plotted in Fig. 1.2-2a, the output voltage $v_o(t)$ in Fig. 1.2-2b.

At this point it is usual to remark that the output voltage[2] "is not zero for $t<0$, but that it begins before the input is applied. This violation of the normal cause-and-effect relationship means that the ideal low-pass filter is

---

[1] It is sometimes stated that Eq. (3) is the Fourier transform of the step function for $\tau \rightarrow \infty$. This is not quite correct. If the step function had a Fourier transform in the usual sense, it should not depend on the method of its calculation. For instance, the Fourier transform of a rectangular pulse of duration τ should yield the same result $V_0/2\pi f$ and $V_0/2$ for $\tau \rightarrow \infty$, but one may readily verify that this is not so.

[2] Close (1966), p. 479.

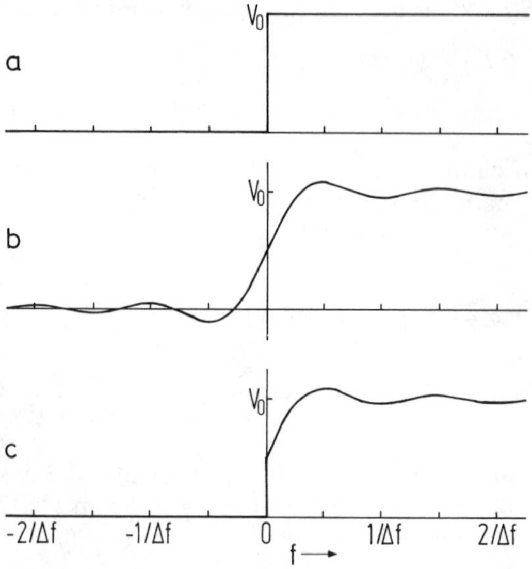

FIG. 1.2-2. (a) Input function to the filter of Fig. 1.2-1. (b) Output function if causality is not imposed as a condition. (c) Output function if causality but no other physical laws are imposed as a condition.

physically unrealizable." There is, of course, no reason why the causality law should be satisfied, since it was nowhere introduced as a condition. We took a Fourier transform of the step function $S(t)$, multiplied it with a function $h(f)$ in the transform domain, and made an inverse Fourier transform. These are all mathematical operations without any physical content, and we obtained a mathematical result without any physical content. The use of terms like input and output voltage creates the misleading impression that one deals with something more than abstract mathematics.

In order to put the causality law into our filter example consider the physical filter F shown in Fig. 1.2-1a. A time variable input voltage $v_i(t) = v(x_1, t)$ is applied to a terminal with spatial coordinate x_1, and a time variable output voltage $v_o(t) = v(x_2, t)$ is obtained at a terminal with the coordinate x_2. We are clearly dealing with a problem containing the time variable t and a space variable x, even though x is degenerated so that it has two values x_1 and x_2 only. The relationship between input and output voltage is determined by Maxwell's equations or an equation derived from them. Since these are linear, partial differential equations with constant coefficients,[3] they are satisfied by functions with sinusoidal time variation.

[3] We assume that the material constants μ, ϵ, and σ of Maxwell's equations are indeed constants, which is not always so.

We assume that the sinusoidal functions at input and output terminal are connected by the transfer function of Eq. (1), without worrying for the time being whether Maxwell's equations could ever yield such a relationship.

In addition to satisfying Maxwell's equations we must satisfy boundary and initial conditions. If we want to apply an input voltage $V_0S(t)$ to the terminal at x_1 in Fig. 1.2-1a, we must satisfy the boundary condition of Eq. (2) at the boundary $x = x_1$:

$$\begin{aligned} v(x_1, t) &= 0 \qquad \text{for} \quad t < 0 \\ &= V_0 \qquad \text{for} \quad t \geqq 0 \end{aligned} \tag{5}$$

We can do this by representing the function $S(t)e^{-t/\tau}$ of Fig. 1.2-1c by a Fourier transform and then approach the limit $\tau \to \infty$.

The initial condition provides the opportunity to introduce the causality law. We must demand that the output voltage $v(x_2, t)$ for $t < 0$ does not depend in any way on the input voltage $v(x_1, t)$ for $t \geqq 0$. We may choose any function we want for $v(x_2, t)$, $t < 0$, but the simplest choice is to assume the filter contains no power source or stored energy and the output voltage is zero for $t < 0$:

$$v(x_2, 0) = 0 \qquad \text{for} \quad t < 0 \tag{6}$$

We obtain now an output voltage with the time variation shown in Fig. 1.2-2c. This function is clearly more acceptable than the one of Fig. 1.2-2b, but not necessarily satisfactory. If we want to be sure to get a physically realizable output voltage, we must require that the box F in Fig. 1.2-1 not only satisfies the causality law but also Maxwell's equations.

In the foregoing example we had to introduce the causality law by means of an initial condition since sinusoidal functions do not model it by themselves. There are other functions that inherently model the causality law.

Consider the step function $S(t)$ of Fig. 1.2-3a. Let it pass through a filter[4] that produces an output with the time variation of the function of Fig. 1.2-3b. Let now the rectangular pulse of Fig. 1.2-3c pass through the same filter. We can represent the rectangular pulse by the superposition of the step functions $S(t)$ and $-S(t - \Delta T)$ of lines *a* and *d*. The output produced by $-S(t - \Delta T)$ is shown in line *e*. The superposition of lines *b* and *e* gives the output in line *f* produced by the rectangular function of line *c*.

In this example we did not have to introduce the causality law to obtain an output satisfying it, when we superimposed the two filtered step functions, in contrast to our experience with a superposition of sinusoidal functions. The reason is that sinusoidal functions are both periodic and analytic. Such functions do not model the causality law. The step function $S(t)$, on the other

[4] Such a filter can be implemented by a sliding correlator.

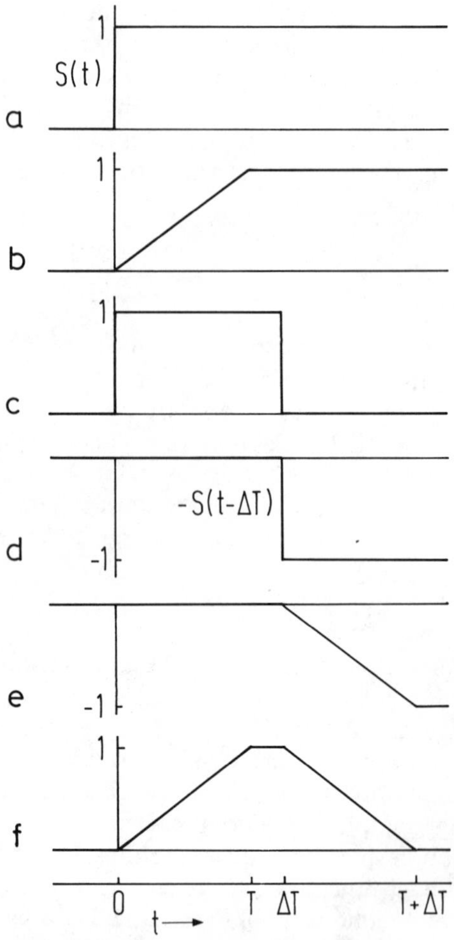

FIG. 1.2-3. The step function $S(t)$ as an example of a causality modeling function.

hand, is nonanalytic and nonperiodic. A superposition

$$\sum_{i=0}^{n} A(i)S(t - i\,\Delta T)$$

of such functions is always zero for $t < 0$, regardless of the amplitudes $A(i)$, and it is not surprising that the superposition of such functions yields zero for $t < 0$. A superposition of sinusoidal functions representing the function $S(t)e^{-t/\tau}$ in Fig. 1.2-1c will yield zero for $t < 0$ only if the amplitudes—or the amplitude densities—of the sinusoidal functions are uniquely chosen. Any

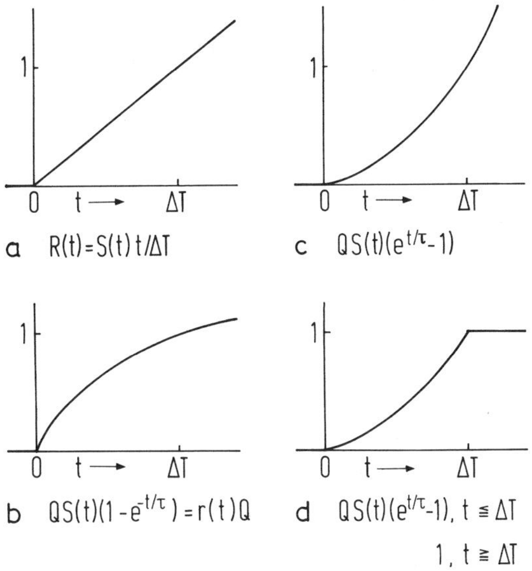

FIG. 1.2-4. (a) Linear ramp function $R(t)$; (b), (c) exponential ramp functions; (d) exponential step function; $\tau = \Delta T/\ln 5$, $Q = 0.25$.

filtering modifies this unique choice and usually[5] produces nonzero values for $t < 0$.

Functions like the step function $S(t)$ that model the causality law are for the study of the transients of electromagnetic waves of a significance comparable to that of sinusoidal waves for the study of the steady state of electromagnetic waves. The step function is the simplest of this class of functions, but it is not the only one. We review a few more that will be of interest for general transient solutions of Maxwell's equations.

The *linear ramp function* $R(t) = S(t)t/\Delta T$ shown in Fig. 1.2-4a avoids the discontinuity of the step function $S(t)$, and is thus physically more desirable. But it causes problems for a different reason. Linear partial differential equations with constant coefficients favor exponential time functions if the separation of variables by Bernoulli's product method is used for their solution. The linear ramp function is not so distinguished and leads to either unnecessary or unsurmountable difficulties.

The distinction of exponential functions does not mean that only sinusoi-

[5] Paley and Wiener (1934) derived a theorem about the representation of quasianalytic functions by a Fourier transform (Theorem XII, p. 16). This is strictly a theorem of abstract mathematics without any physical content, despite many claims to the contrary (e.g., Wallman, 1948).

dal functions are useful, since $e^{t/\tau}$ and $e^{-t/\tau}$ are as much exponential functions as $e^{j\omega t}$ and $e^{-j\omega t}$. The two functions $QS(t)(1-e^{-t/\tau})$ and $QS(t)(e^{t/\tau}-1)$ shown in Fig. 1.2-4b and c are of particular interest, since they replace the discontinuity of $S(t)$ at $t=0$ in Fig. 1.2-3 by a continuous rise. Let us observe that $QS(t)$ is the step function with amplitude Q. Hence, we are using a superposition of the exponential functions $e^{t/\tau}$ and $e^{-t/\tau}$ with the step function. The step function in turn may be viewed as the limit

$$\lim_{\tau\to\infty} e^{-t/\tau}=1$$

of the exponential function $e^{-t/\tau}$ for $t\geqq 0$, which is the time interval of interest here.

The *exponential ramp functions* $QS(t)(1-e^{-t/\tau})$ and $QS(t)(e^{t/\tau}-1)$ share with the linear ramp function the feature that the sum of two delayed functions yields a constant value. Consider the relation

$$(e^{t/\tau}-1)-e^{\Delta T/\tau}(e^{(t-\Delta T)/\tau}-1)=e^{\Delta T/\tau}-1 \tag{7}$$

and the choice of the constant Q:

$$Q(e^{\Delta T/\tau}-1)=1, \qquad \tau=\Delta T/\ln(1+1/Q) \tag{8}$$

The sum of two delayed exponential ramp functions then becomes

$$\begin{aligned} &Q[S(t)(e^{t/\tau}-1)\\ &\quad -S(t-\Delta T)e^{\Delta T/\tau}(e^{(t-\Delta T)/\tau}-1)]=0 && \text{for}\quad t\leqq 0\\ &\qquad\qquad =Q(e^{t/\tau}-1) && \text{for}\quad 0\leqq t\leqq \Delta T\\ &\qquad\qquad =1 && \text{for}\quad t\geqq \Delta T \end{aligned} \tag{9}$$

This *exponential step function* is shown in Fig. 1.2-4d for $Q=0.25$.

For the function $QS(t)(1-e^{-t/\tau})$ we obtain with

$$Q=(1-e^{-\Delta T/\tau})^{-1}, \qquad \tau=-\Delta T/\ln(1-1/Q) \tag{10}$$

the corresponding relation

$$\begin{aligned} &Q[S(t)(1-e^{-t/\tau})\\ &\quad -S(t-\Delta T)e^{-\Delta T/\tau}(1-e^{-(t-\Delta T)/\tau})]=0 && \text{for } t\leqq 0\\ &\qquad\qquad =Q(1-e^{-t/\tau}) && \text{for } 0\leqq t\leqq \Delta T\\ &\qquad\qquad =1 && \text{for } t\geqq \Delta T \end{aligned} \tag{11}$$

The linear ramp function $R(t)=S(t)t/\Delta T$ of Fig. 1.2-4a yields a *linear step function* by subtracting the time-shifted function $R(t-\Delta T)$ from $R(t)$ as

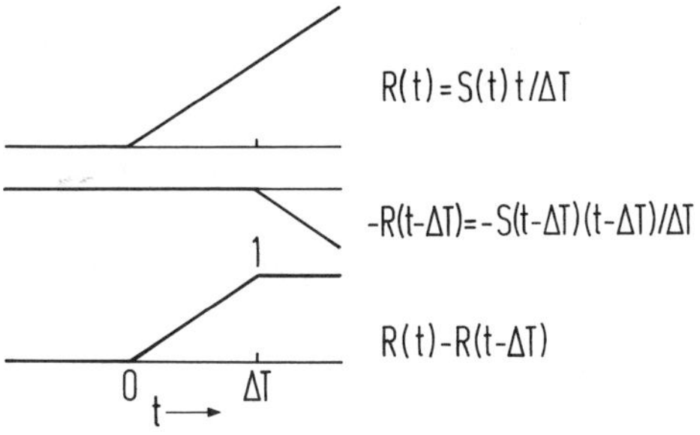

FIG. 1.2-5. Sum of two delayed linear ramp functions produces a linear step function.

shown in Fig. 1.2-5:

$$\begin{aligned} R(t) - R(t-\Delta T) \\ = [S(t)t - S(t-\Delta T)(t-\Delta T)]/\Delta T &= 0 && \text{for} \quad t \leqq 0 \\ &= t/\Delta T && \text{for} \quad 0 \leqq t \leqq \Delta T \\ &= 1 && \text{for} \quad t \geqq \Delta T \end{aligned} \tag{12}$$

Using the approximations

$$1 - e^{-t/\tau} \doteq t/\tau, \qquad 1 - e^{-\Delta T/\tau} \doteq \Delta T/\tau \qquad \text{for} \quad t/\tau, \qquad \Delta T/\tau \ll 1 \tag{13}$$

we may write an exponential step function in the form

$$\begin{aligned} (1 - e^{-\Delta T/\tau})^{-1}[S(t)(1 - e^{-t/\tau}) \\ - S(t-\Delta T)(1 - e^{-(t-\Delta T)/\tau})] &= 0 && \text{for} \quad t \leqq 0 \\ &= t/\Delta T && \text{for} \quad 0 \leqq t \leqq \Delta T \\ &= 1 && \text{for} \quad t \geqq \Delta T \end{aligned} \tag{14}$$

Hence, the exponential step function is approximated by the linear step function, if the conditions of Eq. (13) are satisfied.

1.3 Representation of Transients

Transient solutions of Maxwell's equations are easy to obtain if the transients are due to boundary conditions with the time variation of the step function $S(t)$ or the exponential ramp function $S(t)(1 - e^{-t/\tau})$ shown in Figs. 1.2-3a and 1.2-4b. However, we will need solutions for boundary conditions with arbitrary time variation. They may be obtained by means of series expansions in time-shifted step functions $S(t - i\Delta T)$ and exponential ramp functions $S(t - i\Delta T)(1 - e^{-(t-i\Delta T)/\tau})$. These series expansions do for transient solutions what the Fourier series does for periodic solutions.

We have already shown in Fig. 1.2-3 how a rectangular pulse in the interval $0 \leqq t \leqq \Delta T$ can be represented by the superposition $S(t) - S(t - \Delta T)$ of two step functions. If the step function is changed to $S(t)g(t)$ during propagation through a linear[1], distorting medium, the rectangular pulse will be changed to $S(t)g(t) - S(t - \Delta T)g(t - \Delta T)$.

One might object that the step function $S(t)$ as well as the exponential ramp function of Fig. 1.2-4c are not quadratically integrable, while we require this feature for signals that are going to be represented by a superposition of such functions. This is a valid concern, but experience has shown that we can find solutions of Maxwell's equations with the step function or the exponential ramp function as boundary condition. One could, of course, use the rectangular pulse as basic function for the representation of signals. However, the rectangular pulse is represented by the sum of two step functions, and one ends up with solutions that are twice as long to write as for step functions.

The situation becomes even more severe for exponential ramp functions. One requires two exponential ramp functions to produce the exponential step function of Fig. 1.2-4d, and four exponential ramp functions to produce the pulse of Fig. 1.3-1 with exponential transients.

Figure 1.3-2 shows the representation of a discrete function $F(t)$ by a superposition of step functions. In the general case of a function with n discontinuities at $t_0, t_1, \ldots, t_{n-1}$—where n is finite[2]—and constant values $F(t_0), F(t_1), \ldots, F(t_{n-1})$ to the right of the discontinuities one

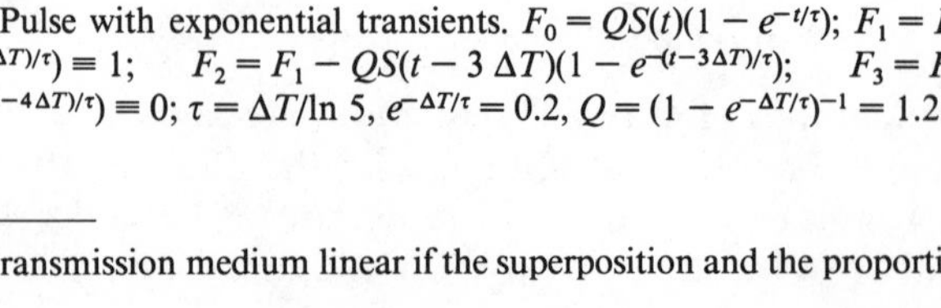

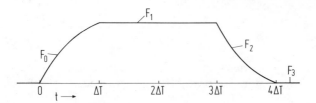

FIG. 1.3-1. Pulse with exponential transients. $F_0 = QS(t)(1 - e^{-t/\tau})$; $F_1 = F_0 - 0.2QS(t - \Delta T)(1 - e^{-(t-\Delta T)/\tau}) \equiv 1$; $F_2 = F_1 - QS(t - 3\,\Delta T)(1 - e^{-(t-3\Delta T)/\tau})$; $F_3 = F_2 + 0.2QS(t - 4\,\Delta T)(1 - e^{-(t-4\Delta T)/\tau}) \equiv 0$; $\tau = \Delta T/\ln 5$, $e^{-\Delta T/\tau} = 0.2$, $Q = (1 - e^{-\Delta T/\tau})^{-1} = 1.25$.

[1] We call a transmission medium linear if the superposition and the proportionality law are satisfied.

[2] A larger number n can never be observed in an experimental science. The development of the theory for denumerable or nondenumerable values of n can be justified in an experimental science only if this brings some simplification.

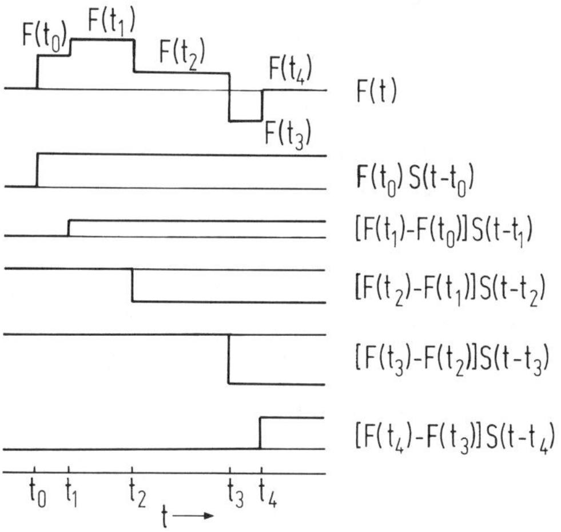

FIG. 1.3-2. Representation of a discontinuous function by a superposition of time-shifted step functions $Q_i S(t - t_i)$.

obtains the sum of weighted and delayed step functions

$$\begin{aligned} F(t) &= \sum_{i=0}^{k} [F(t_i) - F(t_{i-1})] S(t - t_i), \qquad t_k \leqq t < t_{k+1} \\ &= \sum_{i=0}^{n-1} [F(t_i) - F(t_{i-1})] S(t - t_i), \qquad t_{n-1} \leqq t \end{aligned} \tag{1}$$

$$F(t_{-1}) = 0$$

Let a wave with time variation $F(t)$ at the surface of a linear medium propagate through this medium. If the medium changes $S(t)$ to $S(t)g(t)$, the function $F(t)$ will be changed to $F_c(t)$:

$$\begin{aligned} F_c(t) &= \sum_{i=0}^{k} [F(t_i) - F(t_{i-1})] S(t - t_i) g(t - t_i), \qquad t_k \leqq t < t_{k+1} \\ &= \sum_{i=0}^{n-1} [F(t_i) - F(t_{i-1})] S(t - t_i) g(t - t_i), \qquad t_{n-1} \leqq t \end{aligned} \tag{2}$$

If we have found $g(t)$, which usually requires an analytical solution followed by a numerical evaluation of $g(t)$ by computer, we may use Eq. (2) to find $F_c(t)$ for more complicated functions by shifting $S(t)g(t)$, multiplying it, and summing the product. These operations are easy to perform with a computer.

Although the step function $S(t)$ in conjunction with Eqs. (1) and (2) is

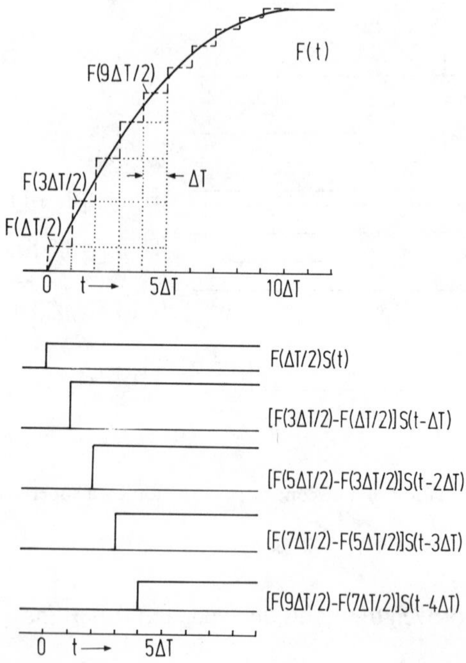

FIG. 1.3-3. Representation of a continuous function by a superposition of time-shifted step functions $Q_i S(t - i\,\Delta T)$.

exclusively intended for computer evaluation, we need the abstraction[3] of Eqs. (1) and (2) for continuously varying functions $F(t)$, since most of our calculations require analytical solutions followed by numerical evaluations. Refer to Fig. 1.3-3 for the representation of a function that is zero for $t < 0$ and varies continuously for $t \geqq 0$, by a superposition of step functions. We divide the interval $0 \leqq t < \infty$ into equal subintervals of width ΔT. The function is then approximated in the time interval $k\Delta T \leqq t < (k+1)\Delta T$ by the step function

$$F(t) = \sum_{i=0}^{k} \{F[(i+\tfrac{1}{2})\Delta T] - F[(i-\tfrac{1}{2})\Delta T]\}S(t - i\,\Delta T) \tag{3}$$

We substitute

$$t'' = i\,\Delta T, \qquad i = 0, 1\,\ldots, k \tag{4}$$

[3] We use the word abstraction rather than generalization, since continuous functions exist only in abstract mathematics; they can never be observed in an experimental science.

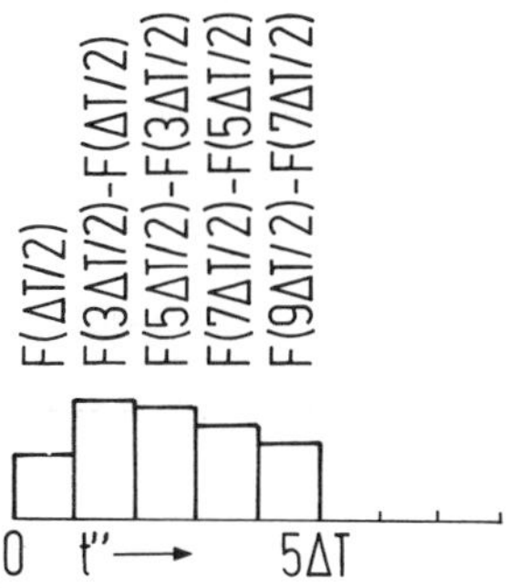

FIG. 1.3-4. Transition from a sum to an integral for the representation of a continuous function by time-shifted step functions.

and multiply on both sides with ΔT:

$$F(t)\Delta T = \sum_{t''=0}^{k\Delta T} [F(t'' + \Delta T/2) - F(t'' - \Delta T/2)]S(t - t'')\Delta T,$$
$$k\,\Delta T \leqq t < (k+1)\Delta T \tag{5}$$

The sum of Eq. (5) is shown in Fig. 1.3-4 for the example of Fig. 1.3-3 in the time interval $4\,\Delta T \leqq t < 5\,\Delta T$. A comparison of Figs. 1.3-3 and 1.3-4 shows that the area under the step function in Fig. 1.3-4 equals $F(9\,\Delta T/2)\Delta T$ in accordance with Eq. (3).

Equation (5) may be rewritten:

$$F(t) = \sum_{t''=0}^{k\Delta t} \frac{F(t'' + \Delta T/2) - F(t'' - \Delta T/2)}{\Delta T} S(t - t'')\Delta T \tag{6}$$

Consider now the limit $\Delta T \to dt''$, $k\Delta T \to t$. The area under the step function in Fig. 1.3-4 may then be represented by an integral rather than a sum[4]:

$$F(t) = \int_0^t \frac{dF(t'')}{dt''} S(t - t'')dt'' \tag{7}$$

A medium that changes a wave with the time variation of a step function $S(t)$ into one with the time variation $S(t)g(t)$ will change a wave with time variation $F(t)$ into

$$F_c(t) = \int_0^t \frac{dF(t'')}{dt''} S(t - t'')g(t - t'')dt'' \tag{8}$$

[4] The integrals of Eqs. (7) and (8) are sometimes referred to as DuHamel, Carson, superposition, or convolution integrals (Close, 1966, p. 155).

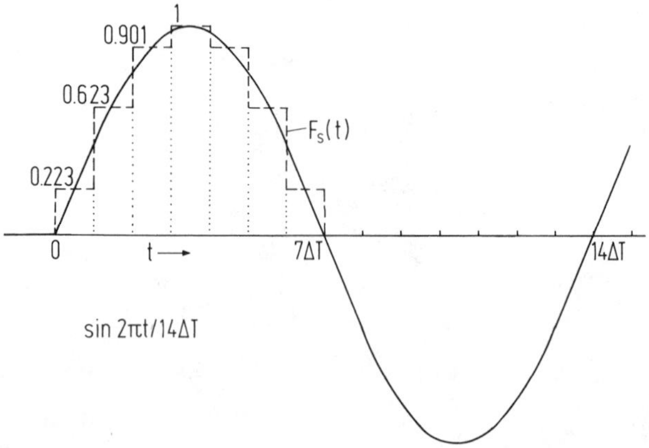

FIG. 1.3-5. Representation of a sinusoidal pulse by a sum of time-shifted step functions.

For numerical calculations one must use the sum of Eq. (3) rather than the integral of Eq. (7) to obtain the changed step function $F_c(t)$:

$$F_c(t) = \sum_{i=0}^{k} \{F[(i+\tfrac{1}{2})\Delta T] - F[(i-\tfrac{1}{2})\Delta T]\}S(t - i\,\Delta T)g(t - i\,\Delta T),$$
$$k \leqq t/\Delta T < k+1 \qquad (9)$$

As an example consider a *sinusoidal pulse* as used in radar. Its first cycle is shown in Fig. 1.3-5. A typical radar pulse consists of 100 to 1000 such cycles, and a typical radar *signal* of many pulses. The first half-cycle is approximated by the function $F_s(t)$:

$$\begin{aligned}F_s(t) = {} & 0.223S(t) + 0.400S(t-\Delta T) + 0.278S(t - 2\,\Delta T) \\ & + 0.099S(t - 3\,\Delta T) - [0.099S(t - 4\,\Delta T) + 0.278S(t - 5\,\Delta T) \\ & + 0.400S(t - 6\,\Delta T)]\end{aligned}$$

The sinusoidal pulse $F(t)$ may then be represented by a sum of time-shifted functions $F_s(t)$ with alternating signs:

$$F(t) = F_s(t) - F_s(t - 7\,\Delta T) + F_s(t - 14\,\Delta T) - \cdots \qquad (10)$$

So far our discussion was directed to the representation of a continuous function $F(t)$ of a continuous variable t by a step function of a continuous variable. Very similar is the representation by a sampled function. Figure 1.3-6 shows again the function $F(t)$ of Fig. 1.3-3. It is now represented at the times $t = k\,\Delta T$ by the samples $F(k\Delta T) = F(k)$. The step function $F(t)$ of Eq.

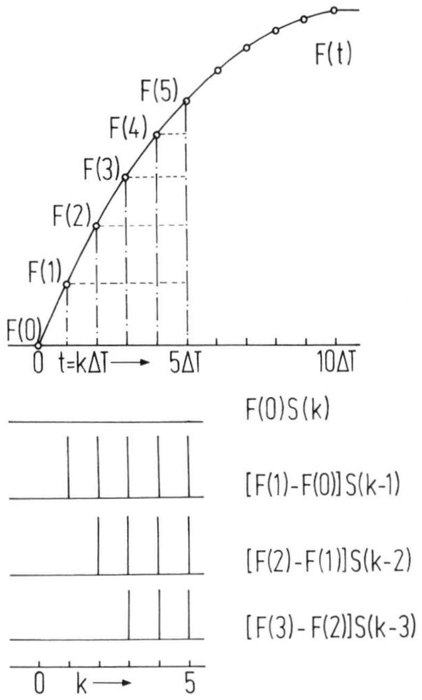

FIG. 1.3-6. Representation of a continuous function $F(t)$ of a continuous variable t by a sampled function defined for $t = k\,\Delta T$.

(3) is replaced by a sampled function $F(k)$:

$$F(k\,\Delta T) = F(k) = \sum_{i=0}^{k} [F(i) - F(i-1)]S(k-i),$$
$$F(-1) = 0 \qquad\qquad (11)$$
$$S(k-i) = 0 \quad \text{for} \quad k - i < 0$$
$$= 1 \quad \text{for} \quad k - i \geqq 0$$

Similarly, the changed step function $F_c(t)$ of Eq. (9) is replaced by the changed sampled function $F_c(k)$:

$$F_c(k) = \sum_{i=0}^{k} [F(i) - F(i-1)]S(k-i)g(k-i),$$
$$F(-1) = 0 \qquad\qquad (12)$$

We turn to the representation of transients by a superposition of linear

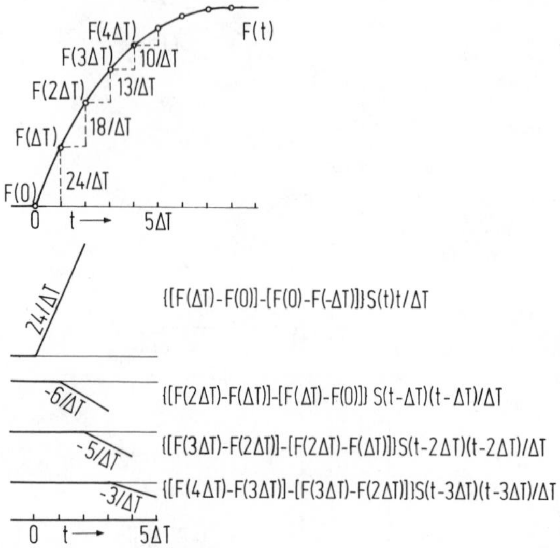

FIG. 1.3-7. Representation of a continuous function by a superposition of time-shifted linear ramp functions $Q_i S(t - i\,\Delta T)(t - i\,\Delta T)/\Delta T$.

ramp functions $R(t) = S(t)t/\Delta T$ as shown in Fig. 1.2-4a. Refer to Fig. 1.3-7, which shows the representation of the function $F(t)$. In the first interval $0 \leqq t \leqq \Delta T$ we have:

$$F(t) = F(\Delta T)S(t)t/\Delta T = [F(\Delta T) - 2F(0) + F(-\Delta T)]S(t)t/\Delta T$$
$$F(0) = F(-\Delta T) = 0 \tag{13}$$

For the second interval $\Delta T \leqq t \leqq 2\,\Delta T$ we get:

$$F(t) = [F(\Delta T) - 2F(0) + F(-\Delta T)]S(t)t/\Delta T$$
$$+ [F(2\Delta T) - 2F(\Delta T) + F(0)]S(t - \Delta T)(t - \Delta T)/\Delta T \tag{14}$$

One may readily recognize the generalization for all further intervals:

$$F(t) = \sum_{i=0}^{k} \{F[(i+1)\Delta T] - 2F(i\,\Delta T)$$
$$+ F[(i-1)\Delta T]\}S(t - i\,\Delta T)(t - i\,\Delta T)/\Delta T,$$
$$k\,\Delta T \leqq t \leqq (k+1)\Delta T, \qquad F(0) = F(-\Delta T) = 0 \tag{15}$$

Substitution of

$$t'' = i\,\Delta T \tag{16}$$

and division as well as multiplication by $(\Delta T)^2$ yields:

$$F(t) = \sum_{t''=0}^{k\Delta T} \frac{F(t''+\Delta T) - 2F(t'') + F(t''-\Delta T)}{(\Delta T)^2} S(t - i\,\Delta T)(t - t'')\Delta T \tag{17}$$

Note that the product of $1/\Delta T$ with $(\Delta T)^2$ yields ΔT.

The first four terms of Eq. (17) are shown in Fig. 1.3-8 for the example of Fig. 1.3-7. The sum represents the area under this step function.

Consider the limit $\Delta T \to dt''$ with $k\Delta T = t$. The first term of the sum of Eq. (17) becomes the second derivative of $F(t'')$. The area under the step function in Fig. 1.3-8 can be represented by an integral rather than a sum:

$$F(t) = \int_0^t \frac{d^2F(t'')}{dt''^2} S(t - t'')(t - t'')dt'' \tag{18}$$

A medium that changes a wave excited by a field strength with the time variation of a linear ramp function

$$R(t) = S(t)t/\Delta T \tag{19}$$

FIG. 1.3-8. Transition from a sum to an integral for the representation of a continuous function by time-shifted linear ramp functions.

into one with the time variation

$$R_c(t) = S(t)h(t/\Delta T) \tag{20}$$

will change a wave excited by a field strength with the time variation $F(t)$ to

$$F_c(t) = \sum_{t''=0}^{k\Delta T} \frac{F(t'' - \Delta T) - 2F(t'') + F(t'' + \Delta T)}{(\Delta T)^2} \times S(t - t'')h[(t - t'')/\Delta T](\Delta T)^2,$$
$$F(0) = F(-\Delta T) = 0 \tag{21}$$

according to Eq. (17). Note that $1/\Delta T$ is no longer multiplied by $(\Delta T)^2$ to yield ΔT as in Eq. (17). In order to turn Eq. (21) into an integral, we define a new function

$$\lim_{\Delta T \to dt} \Delta T\, h[(t - t'')/\Delta T] = \tau\, h_0(t - t'') \tag{22}$$

where τ is a constant with dimension of time, and obtain from Eq. (21) in analogy to Eq. (18):

$$F_c(t) = \tau \int_0^t \frac{d^2F(t'')}{dt''^2} S(t - t'')h_0(t - t'')dt'' \tag{23}$$

For numerical calculations one does not need the division and multiplication by $(\Delta T)^2$ in Eq. (21); furthermore, the notation $i\,\Delta T$ instead of t'' is clearer:

$$F_c(t) = \sum_{i=0}^{k} \{F[(i + 1)\Delta T] - 2F(i\,\Delta T) + F[(i - 1)\Delta T]\} \times S(t - i\Delta T)h(t/\Delta T - i),$$
$$F(0) = F(-\Delta T) = 0, \qquad k\,\Delta T \leqq t \leqq (k + 1)\Delta T \tag{24}$$

For $k = 0$, we get

$$F_c(t) = F(\Delta T)S(t)h(t/\Delta T) \tag{25}$$

In Eq. (15) we had the term $t - i\Delta T$ instead of $h(t/\Delta T - i)$, which insured that $F(0)$ was zero for $k = 0$ and $t = 0$. Since $F_c(t)$ is caused by $F(t)$, and $F(t)$ is zero for $t \leqq 0$, the causality law requires that $F_c(0)$ must be zero, too. We introduce the causality law into Eqs. (24) and (25) by the condition

$$h(0) = 0 \tag{26}$$

Note that in the discrete case of Eqs. (21) and (24) the normalizing time unit is ΔT, which is the time resolution chosen for the computation. In the continuous case of Eq. (23) the normalizing time unit τ must come from

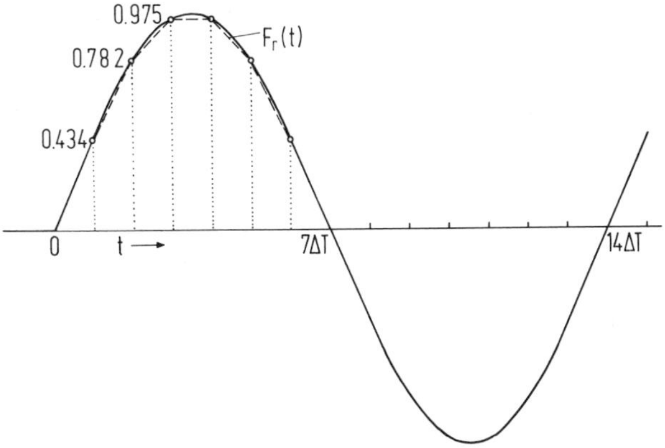

FIG. 1.3-9. Representation of a sinusoidal pulse by a sum of time-shifted linear ramp functions $R(t)$.

somewhere else. We will see later on that τ is determined by the physical content of the problem.[5]

As an example of the representation of a function by linear ramp functions we consider again a sinusoidal pulse as shown in Fig. 1.3-9. The first half-cycle is approximated by the function $F_r(t)$:

$$\begin{aligned} F_r(t) = [&0.434S(t)t - 0.086S(t-\Delta T)(t-\Delta T) \\ &- 0.155S(t-2\,\Delta T)(t-2\,\Delta T) \\ &- 0.193S(t-3\,\Delta T)(t-3\,\Delta T) \\ &- 0.193S(t-4\,\Delta T)(t-4\,\Delta T) \\ &- 0.155S(t-5\,\Delta T)(t-5\,\Delta T) \\ &- 0.086S(t-6\,\Delta T)(t-6\,\Delta T)]/\Delta T \end{aligned} \tag{27}$$

As in the case of Fig. 1.3-5 one may represent a sinusoidal pulse $F(t)$ by a sum of time-shifted functions $F_r(t)$:

$$F(t) = F_r(t) - F_r(t-7\,\Delta T) + F_r(t-14\,\Delta T) - \cdots \tag{28}$$

The approximation of Fig. 1.3-9 is much better than the one of Fig. 1.3-5. Furthermore, the continuous variation of the linear ramp function is inher-

[5] This radical difference between the discrete and the continuous calculation is a minor irritation here, but it is the cause for the basic difference between discrete and continuous topologies for physical space–time, first mentioned by Riemann (1854, p. 285, last paragraph). For a more detailed discussion see Harmuth (1987).

ently better suited than the discontinuous variation of the step function for the study of continuous physical processes.

In Figs. 1.3-7 and 1.3-9 we represented continuous functions by sectionally linear functions. We replace now the sectionally linear function $F(t)$ in Eq. (15) by a sampled function $F(k)$ defined for $t = k\,\Delta T = 0, \Delta T, 2\,\Delta T, \ldots$ only:

$$F(k) = \sum_{i=0}^{k} [F(i+1) - 2F(i) + F(i-1)]S(k-i)(k-i),$$
$$F(-1) = F(0) = 0$$
$$S(k-i) = 0 \quad \text{for} \quad k-i<0$$
$$= 1 \quad \text{for} \quad k-i \geqq 0 \tag{29}$$

Note that the term $i = k$ of the sum of Eq. (29) is always zero due to the factor $k - i$.

For the changed function $F_c(t)$ of Eq. (24) we obtain in analogy to Eq. (29):

$$F_c(k) = \sum_{i=0}^{k} [F(i+1) - 2F(i) + F(i-1)]S(k-i)h(k-i),$$
$$h(0) = 0, \qquad F(-1) = F(0) = 0 \tag{30}$$

Let us note that $h(0) = 0$ will automatically be obtained if the function $h(t)$ is calculated for a physical problem, since the physical problem will introduce the causality law and thus lead to $h(0) = 0$. The introduction of $h(0) = 0$ by Eq. (26) makes it possible to use this result without calculating $h(t)$ in each case.

We turn to the exponential ramp function. Since we can solve Maxwell's equations for transients directly only for the step function $S(t)$ and the exponential ramp function $S(t)Q(1 - e^{-t/\tau})$, but not for the linear ramp function $S(t)t/\Delta T$, we must work out the representation of a general function $F(t)$ by a superposition of time-shifted exponential ramp functions. Refer to Fig. 1.3-10 which shows on top the general transient function $F(t)$. In the interval $0 \leqq t \leqq \Delta T$ it is represented by the function

$$(1 - e^{-\Delta T/\tau})^{-1}[F(\Delta T) - F(0)]S(t)(1 - e^{-t/\tau})$$

which equals $F(0) = 0$ for $t = 0$, $F(\Delta T)$ for $t = \Delta T$, and varies like $1 - e^{-t/\tau}$ as shown in Fig. 1.3-10b. If we subtract for $t \geqq \Delta T$ the function

$$(1 - e^{-\Delta T/\tau})^{-1}[F(\Delta T) - F(0)]e^{-\Delta T/\tau}S(t - \Delta T)(1 - e^{-(t-\Delta T)/\tau})$$

we get according to Eq. (1.2-11) the function of Fig. 1.3-10c, which is constant for $t \geqq \Delta T$. If we furthermore add

$$(1 - e^{-\Delta T/\tau})^{-1}[F(2\,\Delta T) - F(\Delta T)]S(t - \Delta T)(1 - e^{-(t-\Delta T)/\tau})$$

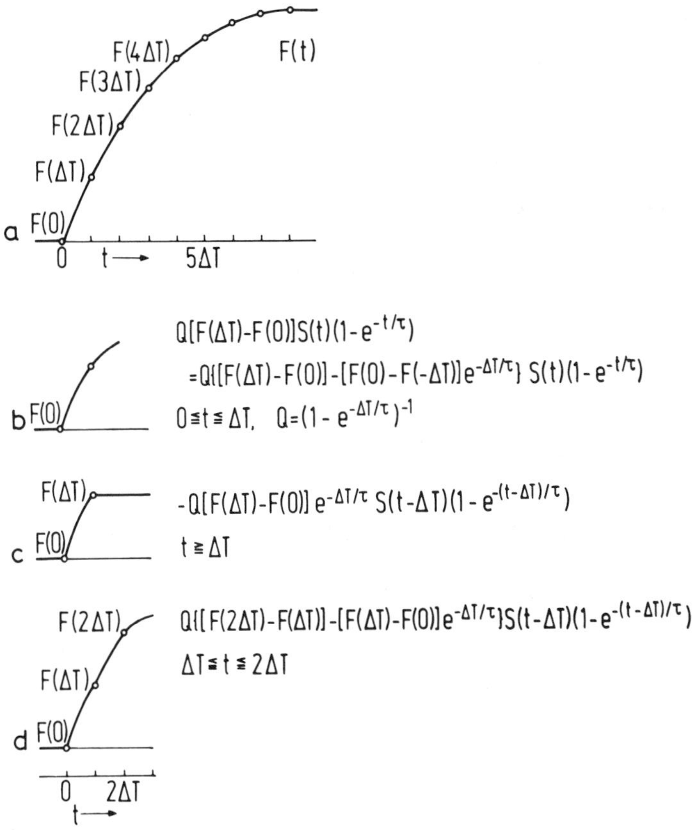

FIG. 1.3-10. Representation of a continuous function by a superposition of time-shifted exponential ramp functions $QS(t)(1 - e^{-t/\tau})$.

we get the function shown in Fig. 1.3-10d, which represents $F(t)$ in the interval $\Delta T \leqq t \leqq 2\,\Delta T$.

It is evident that at the next step one must first subtract

$$(1 - e^{-\Delta T/\tau})^{-1}[F(2\,\Delta T) - F(\Delta T)]e^{-\Delta T/\tau}S(t - 2\,\Delta T)(1 - e^{-(t-2\,\Delta T)/\tau})$$

and then add

$$(1 - e^{-\Delta T/\tau})^{-1}[F(3\,\Delta T) - F(2\,\Delta T)]S(t - 2\,\Delta T)(1 - e^{-(t-2\,\Delta T)/\tau})$$

The process is quite analogous to Fig. 1.3-7, except that the general term for Fig. 1.3-7

$$[\{F[(i+1)\Delta T] - F(i\,\Delta T)\} - \{F(i\,\Delta T) - F[(i-1)\Delta T]\}]$$
$$\times S(t - i\,\Delta T)(t - i\,\Delta T)/\Delta T$$

is replaced by

$$(1 - e^{-\Delta T/\tau})^{-1}[\{F[(i+1)\Delta T] - F(i\,\Delta T)\} - \{F(i\,\Delta T) - F[(i-1)\Delta T]\}e^{-\Delta T/\tau}] \times S(t - i\,\Delta T)(1 - e^{-(t-i\,\Delta T)/\tau})$$

Instead of Eq. (15) we get thus

$$F(t) = (1 - e^{-\Delta T/\tau})^{-1} \sum_{i=0}^{k} \{F[(i+1)\Delta T] - (1 + e^{-\Delta T/\tau})F(i\,\Delta T) + e^{-\Delta T/\tau}F[(i-1)\Delta T]\}S(t - i\,\Delta T)(1 - e^{-(t-i\,\Delta T)/\tau}),$$
$$k\,\Delta T \leqq t \leqq (k+1)\Delta T, \qquad F(0) = F(-\Delta T) = 0 \tag{31}$$

We write t'' for $i\,\Delta T$ and choose $\Delta T/\tau \ll 1$ to obtain an equation similar to Eq. (17):

$$F(t) = \sum_{t''=0}^{k\Delta T} \frac{F(t'' + \Delta T) - 2F(t'') + F(t'' - \Delta T)}{(\Delta T)^2} S(t - t'') \times \tau(1 - e^{-(t-t'')/\tau})\Delta T \tag{32}$$

In analogy to the transition from Eq. (17) to Eq. (18) we now consider the limit $\Delta T \rightarrow dt''$ with $k\,\Delta T = t$:

$$F(t) = \tau \int_0^t \frac{d^2F(t'')}{dt''^2} S(t - t'')(1 - e^{-(t-t'')/\tau})dt'' \tag{33}$$

A medium that changes a wave excited by a field strength with the time variation of an exponential ramp function

$$S(t)(1 - e^{-\Delta T/\tau})^{-1}(1 - e^{-t/\tau})$$

into one with the time variation

$$S(t)h(t/\Delta T)$$

will change a wave excited by a field strength with the time variation $F(t)$ to

$$F_c(t) = \sum_{i=0}^{k} \{F[(i+1)\Delta T] - (1 + e^{-\Delta T/\tau})F(i\,\Delta T) + e^{-\Delta T/\tau}F[(i-1)\Delta T]\} \times S(t - i\,\Delta T)h(t/\Delta T - i),$$
$$F(-1) = F(0) = 0, \qquad h(0) = 0 \tag{34}$$

The condition $h(0) = 0$ is required here as in Eq. (26) by the causality law.

We write again t'' for $i\,\Delta T$ and choose $\Delta T/\tau \ll 1$:

$$F_c(t) = \sum_{t''=0}^{k\Delta T} \frac{F(t''+\Delta T) - 2F(t'') + F(t''-\Delta T)}{(\Delta T)^2} S(t-t'') \times h[(t-t'')/\Delta T](\Delta T)^2 \tag{35}$$

In Eq. (31) we had

$$\lim_{\Delta T/\tau \to 0} (1 - e^{-\Delta T/\tau})^{-1} = \tau/\Delta T \tag{36}$$

and the product of $\tau/\Delta T$ with $(\Delta T)^2$ gave $\tau\,\Delta T$. No such product exists in Eq. (35). In order to turn Eq. (35) into an integral, we define a new function

$$\lim_{\Delta T \to dt} \Delta T\, h[(t-t'')/\Delta T] = \tau h_0(t-t'') \tag{37}$$

where τ is the same constant as in Eqs. (31)–(33), and obtain from Eq. (35) in analogy to Eq. (33):

$$F_c(t) = \tau \int_0^t \frac{d^2F(t'')}{dt''^2} S(t-t'')h_0(t-t'')dt'' \tag{38}$$

In Fig. 1.3-10 we represented a continuous function by sectionally exponential functions. We replace now the sectionally exponential function $F(t)$ in Eq. (31) by a sampled function $F(k)$ defined for $t = k\,\Delta T = 0, \Delta T, 2\Delta T, \ldots$, only:

$$F(k) = (1 - e^{-\Delta T/\tau})^{-1} \sum_{i=1}^{k} [F(i+1) - (1 + e^{-\Delta T/\tau})F(i) + e^{-\Delta T/\tau}F(i-1)] \times S(k-i)(1 - e^{-(k-i)\Delta T/\tau}), \qquad F(-1) = F(0) = 0 \tag{39}$$

Note that the term $i = k$ of the sum of Eq. (39) is always zero due to the exponent $(k-i)\Delta T/\tau$.

For the changed function $F_c(t)$ of Eq. (34) we obtain in analogy:

$$F_c(k) = \sum_{i=0}^{k} [F(i+1) - (1 + e^{-\Delta T/\tau})F(i) + e^{-\Delta T/\tau}F(i-1)] \times S(k-i)h(k-i),$$
$$F(-1) = F(0) = 0 \qquad h(0) = 0 \tag{40}$$

1.4 Inverse Processes

In Section 1.3 we have seen how a general function $F(t)$ is changed into a function $F_c(t)$ if a step function $S(t)$ is changed into $S(t)g(t)$, a linear ramp function $S(t)t/\Delta T$ is changed into $S(t)h(t/\Delta T)$, or an exponential ramp function $S(t)(1-e^{-\Delta T/\tau})^{-1}(1-e^{-t/\tau})$ is changed into $S(t)h(t/\Delta T)$. For the generalization of Fresnel's equations for waves with general time variation we will need the inverse processes. In essence, the changed function $F_c(t)$ will be known and the function $F(t)$ that caused $F_c(t)$ will have to be determined.

We rewrite Eq. (1.3-8) as convolution[1]:

$$F_c(t) = \int_0^t \frac{dF(t'')}{dt''} S(t-t'')g(t-t'')dt'' = \frac{dF(t'')}{dt''} * S(t-t'')g(t-t'') \tag{1}$$

Using the symbol $\circledast$ for deconvolution we obtain formally

$$F_c(t) \circledast S(t-t'')g(t-t'') = \frac{dF(t'')}{dt''} \tag{2}$$

and

$$F(t) = \int_0^t F_c(t) \circledast S(t-t'')g(t-t'')dt'' \tag{3}$$

There is no known method to evaluate Eq. (3) for general continuous functions $F_c(t)$ and $g(t)$. However, one can obtain $F(t)$ for general sampled functions $F_c(k)$ and $g(k)$. These are the only functions of practical interest since all our results will be so complicated that numerical values can be obtained by computer processing only, and sampled functions are the only ones suitable for computer processing; they are also the only ones obtainable by measurements.

The usual deconvolution operation (Cuenod and Durling, 1968, p.50) solves Eq. (2). We deviate somewhat since we want to solve Eq. (3), and we can do this directly rather than by performing first the deconvolution and then the integration.

Equation (1.3-8) is replaced by Eq. (1.3-12) for sampled functions:

$$F_c(k) = \sum_{i=0}^{k} [F(i) - F(i-1)]S(k-i)g(k-i) \tag{4}$$

First we write Eq. (4) for $k = 0$:

$$F_c(0) = F(0)S(0)g(0) = F(0)g(0) \tag{5}$$

Consider the example of Fig. 1.4-1a. The continuous function shown by the

[1] The integration limits 0 and t of Eq. (1.3-8) can be replaced by $-\infty$ and $+\infty$ since $dF(t'')/dt''$ is zero for $t'' < 0$ and $S(t-t'')$ is zero for $t'' > t$.

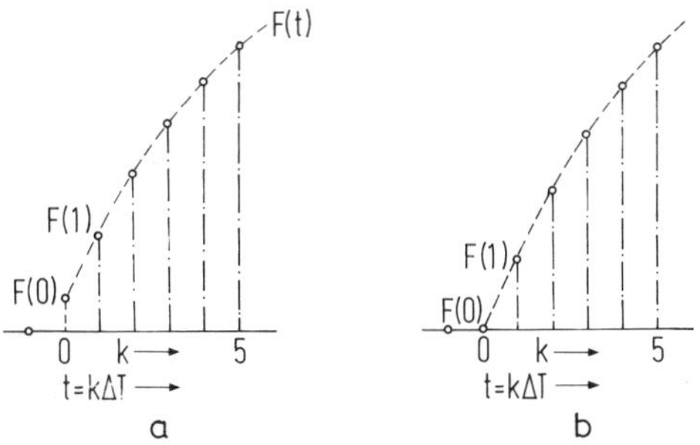

FIG. 1.4-1. A sampled function does not distinguish between a discontinuous rise of a function from zero (a) or a continuous rise (b).

dashed line with a discontinuity at $t = k\,\Delta T = 0$ has the value $F(0) \neq 0$ at $k = 0$. If $F_c(t)$ is obtained from $F(t)$ by a linear transformation[2], we must get from Eq. (5) a value unequal to zero for $F_c(0)$, and $g(0)$ must thus be unequal to zero. However, in the example of Fig. 1.4-1b, where a continuous function rises without discontinuity from 0 at $t = 0$, the first sample unequal to zero is $F(1)$. We then get from Eq. (4):

$$F_c(1) = F(1)g(0) \tag{6}$$

The same argument applies now for $F_c(1)$, $F(1)$, and $g(0)$. We can avoid the distinction between Figs. 1.4-1a and b by always denoting the first sample that is unequal to zero by $F(0)$, which merely implies a shift of the time scale. The important result derived either from Eq. (5) or Eq. (6) is that $g(0)$ is unequal to zero:

$$g(0) \neq 0 \tag{7}$$

We write next Eq. (4) for $k = 1$

$$F_c(1) = F(0)S(1)g(1) + [F(1) - F(0)]S(0)g(0) \tag{8}$$

and solve it for $F(1)$:

$$F(1) = g^{-1}(0)\{F_c(1) - [g(1) - g(0)]F(0)\} \tag{9}$$

For $k = 2$ we get:

$$F_c(2) = F(0)S(2)g(2) + [F(1) - F(0)]S(1)g(1) + [F(2) - F(1)]S(0)g(0) \tag{10}$$

$$F(2) = g^{-1}(0)\{F_c(2) - [g(1) - g(0)]F(1) - [g(2) - g(1)]F(0)\} \tag{11}$$

---

[2] Linear transformation means that the proportionality law and the superposition law hold for the computation of $F_c(t)$ from $F(t)$. We exclude the trivial case where $F_c(t)$ is zero for all values of t.

The pattern is already recognizable from Eqs. (5), (9), and (11). For the general case we get:

$$F(k) = g^{-1}(0)\left(F_c(k) - \sum_{i=1}^{k} [g(i) - g(i-1)]F(k-i)\right) \tag{12}$$

If we want to emphasize that this is a relation between sampled time functions we may write

$$F(k\,\Delta T) = g^{-1}(0)\left(F_c(k\,\Delta T) - \sum_{i=1}^{k} \{g(i\,\Delta T) - g[(i-1)\Delta T]\}F[(k-i)\Delta T]\right) \tag{13}$$

Consider now the inversion of Eq. (1.3-23) holding for a linear ramp function. Formaly we may write:

$$F_c(t) = \tau \int_0^t \frac{d^2F(t'')}{dt''^2} S(t-t'')h_0(t-t'')dt'' = \frac{d^2F(t'')}{dt''^2} * S(t-t'')\tau h_0(t-t'') \tag{14}$$

$$\frac{d^2F(t'')}{dt''^2} = F_c(t) \mathbin{\bar{*}} S(t-t'')\tau h_0(t-t'') \tag{15}$$

$$F(t) = \int_0^t \left(\int_0^{t\dagger} F_c(t) \mathbin{\bar{*}} S(t-t'')\tau h_0(t-t'')dt''\right)dt^\dagger \tag{16}$$

For the practical computation we start from Eq. (1.3-30) and write it first for $k = 1$ since $k = 0$ only yields $F_c(0) = 0$:

$$F_c(1) = F(1)h(1) \tag{17}$$

We must require that $h(1)$ is not zero in order to obtain an unambiguous value of $F(1)$ from $F_c(1)$. We exclude the case of a discontinuity at $t = 0$ as shown in Fig. 1.4-1a and restrict ourselves to the case of a continuous rise from $F(0) = 0$ represented by Fig. 1.4-1b, since the discontinuous case is best handled by step functions rather than by linear or exponential ramp functions. If $F_c(t)$ is obtained from $F(t)$ by a nontrivial linear transformation, we get from Eq. (17) a value $F_c(1) \neq 0$ from $F(1) \neq 0$, and $h(1)$ is thus unequal to zero.

For $k = 2$ we obtain from Eq. (1.3-30):

$$F_c(2) = F(1)h(2) + [F(2) - 2F(1)]h(1) \tag{18}$$

$$F(2) = h^{-1}(1)\{F_c(2) - [h(2) - 2h(1)]F(1)\} \tag{19}$$

For $k = 3$ we get:

$$F_c(3) = F(1)h(3) + [F(2) - 2F(1)]h(2) + [F(3) - 2F(2) + F(1)]h(1) \quad (20)$$

$$F(3) = h^{-1}(1)\{F_c(3) - [h(2) - 2h(1)]F(2) - [h(3) - 2h(2) + h(1)]F(1)\} \quad (21)$$

One more step, $k = 4$, will clearly reveal the pattern:

$$F_c(4) = F(1)h(4) + [F(2) - 2F(1)]h(3) + [F(3) - 2F(2) + F(1)]h(2) + [F(4) - 2F(3) + F(2)]h(1) \quad (22)$$

$$F(4) = h^{-1}\{F_c(4) - [h(2) - 2h(1)]F(3) - [h(3) - 2h(2) + h(1)]F(2) - [h(4) - 2h(3) + h(2)]F(1)\} \quad (23)$$

One may now recognize the pattern of Eqs. (17), (19), (21), and (23). For the general case we get:

$$F(k) = h^{-1}(1)\left(F_c(k) - [h(2) - 2h(1)]F(k-1) - \sum_{i=1}^{k-2} [h(i+2) - 2h(i+1) + h(i)]F(k-i-1)\right) \quad (24)$$

If one wants to emphasize that this equation gives a relation between sampled time functions, one may use the notation of Eq. (13).

The inversion of Eq. (1.3-38) yields again Eq. (16) due to the formal equality of Eqs. (1.3-23) and (1.3-38). For the inversion of Eq. (1.3-40) we write for $k = 3$:

$$F_c(3) = F(1)h(3) + [F(2) - (1 + e^{-\Delta T/\tau})F(1)]h(2) + [F(3) - (1 + e^{-\Delta T/\tau})F(2) + e^{-\Delta T/\tau}F(1)]h(1) \quad (25)$$

$$F(3) = h^{-1}(1)\{F_c(3) - [h(2) - (1 + e^{-\Delta T/\tau})h(1)]F(2) - [h(3) - (1 + e^{-\Delta T/\tau})h(2) + e^{-\Delta T/\tau}h(1)]F(1)\} \quad (26)$$

A comparison of Eqs. (26) and (21) yields the general case with the help of Eq. (24):

$$F(k) = h^{-1}(1)\left(F_c(k) - [h(2) - (1 + e^{-\Delta T/\tau})h(1)]F(k-1) - \sum_{i=1}^{k-2} [h(i+2) - (1 + e^{-\Delta T/\tau})h(i+1) + e^{-\Delta T/\tau}h(i)]F(k-i-1)\right) \quad (27)$$

1.5 Planar Nonsinusoidal Waves

Maxwell's equations for a medium without charges and currents—having a constant permeability μ, permittivity ϵ, and the conductivity $\sigma = 0$—have the following form:

$$\operatorname{curl} \mathbf{H} = \epsilon\, \partial \mathbf{E}/\partial t \tag{1}$$

$$\operatorname{curl} \mathbf{E} = -\mu\, \partial \mathbf{H}/\partial t \tag{2}$$

$$\epsilon \operatorname{div} \mathbf{E} = \mu \operatorname{div} \mathbf{H} = 0 \tag{3}$$

Consider a planar, transverse electromagnetic (TEM) wave propagating in this medium in the direction y. The condition for a TEM wave requires

$$E_y = H_y = 0 \tag{4}$$

while the condition for a planar wave requires

$$\begin{aligned} \partial E_x/\partial x = \partial E_x/\partial z = \partial E_z/\partial x = \partial E_z/\partial z = 0 \\ \partial H_x/\partial x = \partial H_x/\partial z = \partial H_z/\partial x = \partial H_z/\partial z = 0 \end{aligned} \tag{5}$$

Writing the operator curl in cartesian coordinates

$$\begin{aligned} \operatorname{curl} \mathbf{H} = \mathbf{e}_x(\partial H_z/\partial y - \partial H_y/\partial z) + \mathbf{e}_y(\partial H_x/\partial z - \partial H_z/\partial x) \\ + \mathbf{e}_z(\partial H_y/\partial x - \partial H_x/\partial y) \end{aligned} \tag{6}$$

where $\mathbf{e}_x$, $\mathbf{e}_y$, and $\mathbf{e}_z$ are unit vectors in the direction of the positive x, y, and z axes, and introducing Eqs. (4) and (5) into Eqs. (1) and (2), one obtains:

$$-\partial H_x/\partial y = \epsilon\, \partial E_z/\partial t \tag{7}$$

$$\partial H_z/\partial y = \epsilon\, \partial E_x/\partial t \tag{8}$$

$$\partial E_x/\partial y = \mu\, \partial H_z/\partial t \tag{9}$$

$$-\partial E_z/\partial y = \mu\, \partial H_x/\partial t \tag{10}$$

One may make the substitutions

$$E = E_x = E_z, \qquad H = -H_x = H_z \tag{11}$$

to rewrite the two pairs of Eqs. (7) and (10) as well as (8) and (9) as one pair:

$$\partial E/\partial y + \mu\, \partial H/\partial t = 0 \tag{12}$$

$$\partial H/\partial y + \epsilon\, \partial E/\partial t = 0 \tag{13}$$

Differentiation of Eq. (12) with respect to t and Eq. (13) with respect to y,

$$\partial^2 E/\partial y\, \partial t + \mu\, \partial^2 H/\partial t^2 = 0, \qquad \partial^2 H/\partial y^2 + \epsilon\, \partial^2 E/\partial y\, \partial t = 0$$

permits one to eliminate E and to obtain a differential equation for H:

$$\partial^2 H/\partial y^2 - \mu\epsilon\, \partial^2 H/\partial t^2 = 0 \tag{14}$$

The general solution of this wave equation was found by d'Alembert in the eighteenth century:

$$H(y, t) = H_1 f_1(t - y/c) + H_2 f_2(t + y/c), \qquad c = (\mu\epsilon)^{-1/2} \tag{15}$$

The functions f_1 and f_2, as well as the constants H_1 and H_2 with the dimension of a magnetic field strength, must be determined by initial and boundary conditions; they cannot follow from Maxwell's equations. In particular, there is nothing in Maxwell's equations that forces us to assume that f_1 and f_2 are sinusoidal functions.

The electric field strength associated with the magnetic field strength of Eq. (15) follows from either Eq. (12)

$$E(y, t) = -\mu \int \frac{\partial H}{\partial t}\, dy + E_y(t) \tag{16}$$

or from Eq. (13):

$$E(y, t) = -\epsilon^{-1} \int \frac{\partial H}{\partial y}\, dt + E_t(y) \tag{17}$$

Keeping in mind that the differentiation of $f_1(t - y/c)$ and $f_2(t + y/c)$ with respect to y yields $-c^{-1}f_1'$ and $c^{-1}f_2'$ while differentiation with respect to t only yields f_1' and f_2', one obtains by insertion of Eq. (15) into Eqs. (16) and (17):

$$\begin{aligned} E(y, t) &= -\mu \int [H_1 f_1'(t - y/c) + H_2 f_2'(t + y/c)]dy + E_y(t) \\ &= Z[H_1 f_1(t - y/c) - H_2 f_2(t + y/c)] + E_y(t) \end{aligned} \tag{18}$$

$$\begin{aligned} E(y, t) &= -\epsilon^{-1} \int [-c^{-1} H_1 f_1'(t - y/c) \\ &\qquad + c^{-1} H_2 f_2'(t + y/c)]dt + E_t(y) \\ &= Z[H_1 f_1(t - y/c) - H_2 f_2(t + y/c)] + E_t(y) \end{aligned} \tag{19}$$

$$Z = (\mu/\epsilon)^{1/2}$$

Since both equations must yield the same result, we conclude that the two integration constants $E_t(y)$ and $E_y(t)$ must be independent of y and t, and equal a constant E_c.

Electric and magnetic field strength are completely linked by Eqs. (16) and (17). If a magnetic field strength traveling toward increasing values of y has the time variation $f_1(t - y/c)$, the electric field strength will have the same time variation. This relationship is taken for granted in the theory of sinusoidal waves.

Let us return to Eqs. (12) and (13). We differentiate now Eq. (12) with respect to y and Eq. (13) with respect to t,

$$\partial^2 E/\partial y^2 + \mu\, \partial^2 H/\partial y\, \partial t = 0, \qquad \partial^2 H/\partial y\, \partial t + \epsilon\, \partial^2 E/\partial t^2 = 0$$

and eliminate $\partial^2 H/\partial y\, \partial t$:

$$\partial^2 E/\partial y^2 - \mu\epsilon\, \partial^2 E/\partial t^2 = 0 \tag{20}$$

We have again the wave equation, but this time for the electric field strength. Its general solution

$$E(y, t) = E_1 g_1(t - y/c) + E_2 g_2(t + y/c) \tag{21}$$

gives us two functions g_1 and g_2 that do not have to be equal to f_1 and f_2 in Eq. (15) or depend in any way on f_1 and f_2.

The magnetic field strength associated with the electric field strength of Eq. (21) follows from either one of Eqs. (12) or (13):

$$H(y, t) = -\mu^{-1} \int \frac{\partial E}{\partial y} dt + H_t(y) \tag{22}$$

$$H(y, t) = -\epsilon \int \frac{\partial E}{\partial t} dy + H_y(t) \tag{23}$$

Proceeding as in the transition from Eqs. (16) and (17) to Eqs. (18) and (19) we obtain:

$$\begin{aligned} H(y, t) &= -\mu^{-1} \int [-c^{-1} E_1 g_1'(t - y/c) \\ &\qquad + c^{-1} E_2 g_2'(t + y/c)] dt + H_t(y) \\ &= Z^{-1}[E_1 g_1(t - y/c) - E_2 g_2(t + y/c)] + H_t(y) \end{aligned} \tag{24}$$

$$\begin{aligned} H(y, t) &= -\epsilon \int [E_1 g_1'(t - y/c) + E_2 g_2'(t + y/c)] dy + H_y(t) \\ &= Z^{-1}[E_1 g_1(t - y/c) - E_2 g_2(t + y/c)] + H_y(t) \end{aligned} \tag{25}$$

Again we conclude that the integration constants $H_y(t)$ and $H_t(y)$ must be independent of y and t, and equal a constant H_c.

For the general solution of Eqs. (12) and (13) we must add the solutions of Eqs. (15) and (24) as well as those of Eqs. (21) and (18):

$$\begin{aligned} H(y, t) = H_1 f_1(t - y/c) + Z^{-1} E_1 g_1(t - y/c) + H_2 f_2(t + y/c) \\ - Z^{-1} E_2 g_2(t + y/c) + H_c \end{aligned} \tag{26}$$

$$E(y, t) = ZH_1 f_1(t - y/c) + E_1 g_1(t - y/c) - [ZH_2 f_2(t + y/c) - E_2 g_2(t + y/c)] + E_c \tag{27}$$

The wave traveling toward increasing values of y has the time variation

$$f_1(t - y/c) + Z^{-1} H_1^{-1} E_1 g_1(t - y/c)$$

for both the electric and magnetic field strength. A consequence of this result is that one cannot determine the functions $f_1(t - y/c)$ and $g_1(t - y/c)$ from an observation of $E(y, t)$ and $H(y, t)$; only their linear combination can be observed. A corresponding statement holds for the wave traveling toward decreasing values of y.

1.6 Nonplanar Nonsinusoidal Waves

The equality of time variation of electric and magnetic field strengths for planar waves in a nonconducting medium does not apply generally. To see this we analyze Maxwell's equations in their general form:

$$\text{curl } \mathbf{H} = \partial \mathbf{D}/\partial t + \mathbf{g} \tag{1}$$

$$\text{curl } \mathbf{E} = -\partial \mathbf{B}/\partial t \tag{2}$$

$$\text{div } \mathbf{D} = \rho \tag{3}$$

$$\text{div } \mathbf{B} = 0 \tag{4}$$

We assume that the electric and magnetic field strengths $\mathbf{E}$ and $\mathbf{H}$ are connected with the electric and magnetic flux densities $\mathbf{D}$ and $\mathbf{B}$ by constants ϵ and μ:

$$\mathbf{D} = \epsilon \mathbf{E}, \qquad \mathbf{B} = \mu \mathbf{H} \tag{5}$$

Application of the operator div to Eq. (1) and differentiation of Eq. (3) with respect to time yields the continuity equation:

$$\text{div } \mathbf{g} + \partial \rho/\partial t = 0 \tag{6}$$

Equation (4) is satisfied identically by choosing $\mathbf{B}$ or $\mathbf{H}$ equal to the curl of a vector $\mathbf{A}$:

$$\mathbf{H} = \mu^{-1}\mathbf{B} = \mu^{-1} \text{ curl } \mathbf{A} \tag{7}$$

Inserting $\partial \mathbf{B}/\partial t$ into Eq. (2) and observing that the curl of the gradient of a scalar ϕ is zero, yields the general expression for the electric field strength:

$$\mathbf{E} = -\partial \mathbf{A}/\partial t - \text{grad } \phi \tag{8}$$

With the notation div grad $= \nabla^2$ Eqs. (1) and (3) assume the following form if Eqs. (7) and (8) are inserted:

$$\text{curl curl } \mathbf{A} + \mu\epsilon\, \partial^2\mathbf{A}/\partial t^2 + \mu\epsilon \text{ grad}(\partial\phi/\partial t) = \mu\mathbf{g} \quad (9)$$

$$-\epsilon \text{ div}(\partial\mathbf{A}/\partial t) - \epsilon\, \nabla^2\phi = \rho \quad (10)$$

Equation (7) defines curl **A**, not **A** itself. One may add the gradient of a scalar to **A** without changing curl **A**. This makes it possible to introduce an additional condition that simplifies the solution of Eqs. (9) and (10). We choose the so-called Lorentz[1] convention:

$$\text{div } \mathbf{A} + \mu\epsilon\, \partial\phi/\partial t = 0 \quad (11)$$

With the help of the vector relation

$$\text{curl curl } \mathbf{A} = \text{grad div } \mathbf{A} - \nabla^2\mathbf{A} \quad (12)$$

one obtains from Eqs. (9) and (10):

$$\nabla^2\mathbf{A} - \mu\epsilon\, \partial^2\mathbf{A}/\partial t^2 = -\mu\mathbf{g} \quad (13)$$

$$\nabla^2\phi - \mu\epsilon\, \partial^2\phi/\partial t^2 = -\epsilon^{-1}\rho \quad (14)$$

The homogeneous part of these equations has the form of the wave equation. To obtain the general solution one must add a particular solution of the inhomogeneous equations. Such particular solutions are represented by integrals taken over all current densities **g** and charge densities[2] ρ:

$$\mathbf{A}(x, y, z, t) = \frac{\mu}{4\pi}\int\int\int \mathbf{g}(\xi, \eta, \zeta, t - r/c)r^{-1}\, d\xi\, d\eta\, d\zeta \quad (15)$$

$$\phi(x, y, z, t) = \frac{1}{4\pi\epsilon}\int\int\int \rho(\xi, \eta, \zeta, t - r/c)r^{-1}\, d\xi\, d\eta\, d\zeta \quad (16)$$

Here r is the distance between the coordinates ξ, η, ζ of the current and charge densities and the coordinates x, y, z of the potentials **A** and ϕ:

$$r = [(x - \xi)^2 + (y - \eta)^2 + (z - \zeta)^2]^{1/2} \quad (17)$$

The vector potential **A** is due to the currents. The scalar potential ϕ is inherently due to the charges, but it can also be caused by a current flowing in and out of absorbing material whose wave impedance is equal to the wave impedance $Z = (\mu/\epsilon)^{1/2}$ of the surrounding medium. For an explanation

---

[1] Hendrik Antoon Lorentz, physicist, 1853–1928; born in Arnheim, The Netherlands.

[2] The proof may be found in Abraham and Becker (1932), Pt. III, Ch. X, Sec. 10, or in Abraham and Becker (1950), Pt. III. Ch. X, Sec. 9. These solutions for **A** and ϕ are called general solutions in these books, but later editions correct the mistake. A good but condensed presentation is also given by Heitler (1954), p. 4.

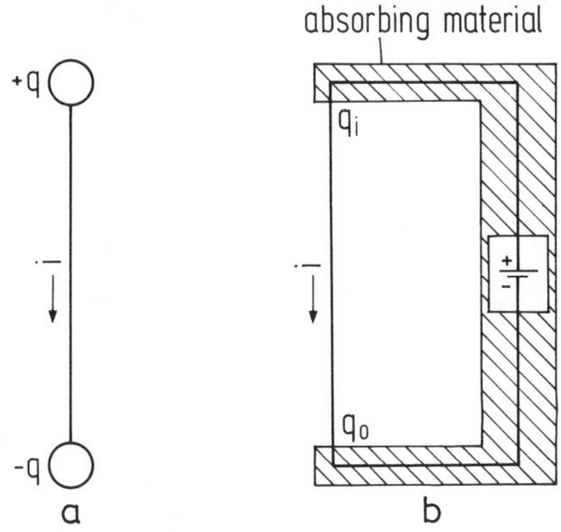

FIG. 1.6-1. (a) Current flowing from a charge $+q$ to a charge $-q$ and (b) from a source q_i to a sink q_o connected by conductors covered with absorbing material to a battery.

refer to Fig. 1.6-1. Two spheres with charges $+q$ and $-q$ are shown on the left. If the two spheres are connected by a wire, a current will flow. The charges produce the scalar potential ϕ and the current the vector potential **A**.

On the right in Fig. 1.6-1 a battery is shown that drives a current through a wire loop. The battery and part of the wire loop are surrounded by absorbing material whose wave impedance is closely matched to that of the surrounding medium.[3] As a result, the current i seems to start at point q_i and end at point q_o, just like the current between the two charged spheres in Fig. 1.6-1a. The entrance and exit points q_i and q_o thus act like they contained infinite charges. Hence, the sources and sinks of currents represented by $+\text{div}\,\mathbf{g}$ and $-\text{div}\,\mathbf{g}$ contribute to the scalar potential just like charges.

Let us return to Eqs. (7) and (8). For a given vector potential **A** there is no assurance that the time variation of curl **A** and $\partial\mathbf{A}/\partial t$ is the same, and the time variation of **H** and **E** is thus not necessarily the same. In addition, the scalar potential ϕ can produce a time variation of **E** without any effect whatsoever on **H**. One may thus draw the theoretically correct conclusion that waves with nonsinusoidal time variation permit one to transmit independent signals with the electric and magnetic field strength. However, before one draws any practical conclusion from this fact, one should consider the previous result that planar waves require equal time variation of **E**

[3] We tacitly assume here that a wave impedance can be defined for nonsinusoidal waves. According to Section 1.5 this is certainly correct for any lossless medium which yields the impedance $Z = (\mu/\epsilon)^{1/2}$. Since an absorbing material necessarily must have losses, it cannot be matched exactly to a medium like empty space or air. However, the matching is often sufficiently good for practical purposes.

and **H**. Since waves in free space become approximately planar at great distances from the source of the waves, and since communications is primarily of interest for such large distances, we have *not* discovered a method to double the number of communication channels.

1.7 Attenuation of Waves by Hysteresis Losses

A major area of application of the theory in this book is the stealth technology,[1] both for the advancement of this technology and the radar that neutralizes it. Since steady-state solutions of Maxwell's equations are well known, they can be used to make targets hard to detect by the conventional radar using a sinusoidal carrier with many cycles per pulse. The logic countermeasure is to study transient solutions and to design radars that rely more on the transient than the steady-state solutions. This in turn forces the developers of the stealth technology to study the effect of their products on transient solutions of Maxwell's equations and modify them so that they remain hard to detect by the *carrier-free radar* or—more generally—the *large-relative-bandwidth radar.*[2] As we will see, the transient solutions require a much better command of mathematics and much more computational effort than the usual steady-state solutions. In particular, we will see that transient solutions cannot be obtained from steady-state solutions by means of the Fourier transform or any other means, since the steady-state solutions are limits of the general transient solutions, and one can derive a limit from the general solution but not the general solution from a limit.

There is, however, one problem of basic interest in the stealth technology that can be solved without any mathematics. It is the use of hysteresis losses for the absorption of electromagnetic waves. The wave impedance of empty space equals $Z_0 = (\mu_0/\epsilon_0)^{1/2}$ and the velocity of propagation of a wave equals $c_0 = (\mu_0\epsilon_0)^{-1/2}$. Let some material have the relative permeability μ_r, relative permittivity ϵ_r, and a very low conductivity σ. The wave impedance of this material equals $Z = (\mu_r\mu_0/\epsilon_r\epsilon_0)^{1/2}$, and the velocity of propagation in this material equals $c = c_0/(\mu_r\epsilon_r)^{1/2}$. For $Z = Z_0$ a wave will pass from empty

---

[1] The term "stealth technology" is the currently highly publicized term for techniques that make targets hard to detect by radar. These techniques are almost as old as radar and were used, e.g., during World War II by the German Navy to mask submarine snorkels (Emerson, 1973).

[2] Although the concept of the carrier-free or large-relative-bandwidth radar is so new that the topic is still controversial for many, one may find some of the basic ideas independently developed in other fields. For instance, a paper by Bates (1971) says in effect that the good ambiguity resolution of echo-location by bats, whales, etc., can be explained by the large relative bandwidth of the signals used. This is the same result as developed by the author for carrier-free radar signals, a few million years after bats and whales incorporated it into their life support systems (Harmuth, 1981, Ch. 6).

space into the material without reflection. This result is usually derived for sinusoidal waves, but we will see later on that it holds generally.

There are many materials with a relative permittivity $\epsilon_r > 1$ regardless of how fast the field strengths of a wave change with time.[3] The relative permeability of ferromagnetic materials may easily be 10,000 for slowly changing field strengths, but only ferrites will yield a relative permeability significantly larger than 1 for waves that change from zero to their maximum value in 1 ns or a still shorter time.[4] With ferrites it is possible to achieve approximately the condition $\mu_r = \epsilon_r$ for the waves generally used by radar, and thus reduce significantly reflection and backscattering. There are two things one can do with such a material. (a) One lets the wave pass from space into the material and again out of the material into space at the opposite side. Such a material is transparent but it is not invisible, since the velocity $c = c_0/(\mu_r\epsilon_r)^{1/2}$ in the material is lower than in space for $\mu_r = \epsilon_r > 1$, which implies a detectable physical effect. (b) One may cover metallic parts with a ferrite material to prevent the strong reflection or backscattering typical for metals. In this case the ferrites not only must closely satisfy the condition $\mu_r = \epsilon_r$, but they must also absorb the incident wave strongly, since otherwise the wave would propagate through the ferrite material to the metallic parts and back out of the ferrite material with no more effect than a delay caused by the ferrite material. There is a variety of effects that cause attenuation of electromagnetic waves in a ferrite. The only ones we can discuss at this time are hysteresis losses and—from Section 2.1 on—ohmic losses.

Let us point out that currently available ferrite materials can match the impedance of space only over a small relative bandwidth[5] and have thus very little effect on radar signals with large relative bandwidth or carrier-free

---

[3] Instead of high and low frequencies of waves one must use the fast and slow change (of field strengths) of waves when discussing waves with general time variation.

[4] Ferrites were first developed in Japan by Yogoro Kato more than 50 years ago. Their desirable features as absorbers for the waves used by radar were understood at the time of World War II, but the war ended before practical use could be made of this knowledge. For more details on ferrites see Johnstone (1984), Stepanov (1968), Wickenden and Howell (1978), Knott (1979), Wartenberg (1968), Birks (1948), Hatakeyama and Inui (1984), as well as the following U.S. patents: Borcherdt (1961), Downen and Eichenberger (1979), Forster and Vanderbilt (1977), Grimes (1972), Grimes *et al.* (1976), Ishino *et al.* (1977, 1978), Klingler (1967), Meinke *et al.* (1972), Naito (1973), Neher (1953), Nielsen (1981), Pratt (1961a,b), Salisbury (1952), Suetake (1973a,b), Tuinila and Bayrd (1969), Wallin (1977), Wesch (1967, 1970, 1973), Wesch and Meinke (1971), Wesch and Ullrich (1969), Wright (1977a,b) and Wright and Wright (1975).

[5] The relative bandwidth is defined by $\eta = (f_H - f_L)/(f_H + f_L)$, where f_H and f_L are the highest and lowest frequency of interest. The conventional small-relative-bandwidth radar has a relative bandwidth in the order of 0.01 or less, while the large-relative-bandwidth radar has a relative bandwidth close to 1.

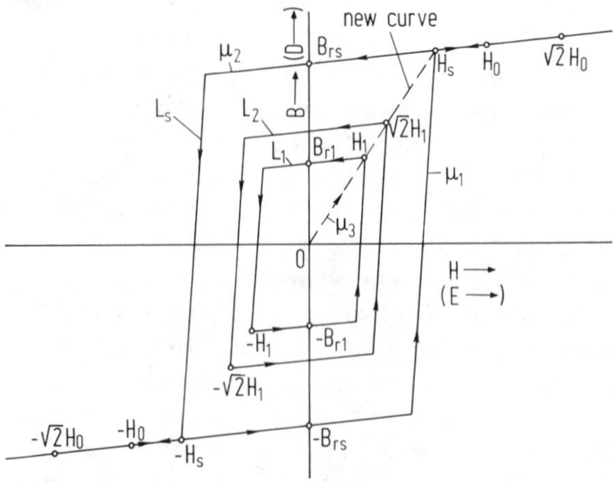

FIG. 1.7-1. Idealized hysteresis loop of a ferrite material.

signals. Two papers by Amin and James (1981a,b) discuss in some detail what can be done at this time. We want to show here mainly the great difference between periodic sinusoidal waves and pulses, and the difficulty of predicting effects for pulses based on theoretical and experimental knowledge for sinusoids.

Refer to Fig. 1.7-1, which shows an idealized hysteresis loop L_s as well as two minor loops L_1 and L_2. For magnetic hysteresis these loops connect the magnitudes B and H of the magnetic flux density and the magnetic field strength; for electric hysteresis they connect the magnitudes D and E of the electric flux density and the electric field strength. An ideal ferrite material with $\mu_r = \epsilon_r$ should have both magnetic and electric hysteresis with equal loops.

Let an electromagnetic wave with the time variation of the sinusoidal pulse of Fig. 1.7-2a arrive at the ferrite material. The material shall be at its unmagnetized and unelectricized point $H = B = 0, E = D = 0$. As the field strength increases from 0 at t_0, the material will be magnetized along the *new curve* in Fig. 1.7-1. If the field strength is sufficiently large, the material will be driven into saturation and the point $\sqrt{2}H_0$ will be reached at the time t_1. The dropping field strength will bring us to the point B_{rs} at the time t_2, to $-\sqrt{2}H_0$ at the time t_3, to $-B_{rs}$ at t_4, and to $\sqrt{2}H_0$ at t_5. Except for the first cycle of the sinusoidal pulse, one gets a complete run around the hysteresis loop for each cycle. Since a typical radar pulse contains between 100 and 1000 sinusoidal cycles—and a typical radar signal many such pulses—we can ignore the deviation during the first cycle.

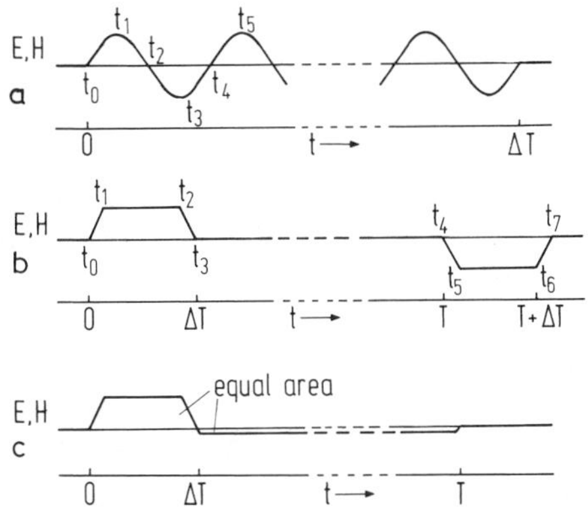

FIG. 1.7-2. Time variation of the electric and magnetic field strength of radar pulses with (a) small relative bandwidth and with (b, c) large relative bandwidth. Note that the duration ΔT of a pulse is the same in all three examples.

It is shown in many textbooks[6] that the energy lost per run around a hysteresis loop is proportionate to the area inside the loop; the proportionality factor is the volume of the material.[7] Denoting the energy lost for one run around the hysteresis loop with W, we get the energy loss nW for n sinusoidal cycles.

Consider now the trapezoidal pulse of Fig. 1.7-2b. Its duration ΔT is the same as that of the sinusoidal pulse. In order to have the same energy, the amplitude of the sinusoidal pulse must be $\sqrt{2}$ times the amplitude of the trapezoidal pulse, if its transition times $t_1 - t_0$ and $t_3 - t_4$ are small compared with its duration ΔT. In Fig. 1.7-1 the magnetization due to field strengths with trapezoidal time variation will again start at time t_0 at $H = 0$, $B = 0$. At the time t_1 the point H_0 rather than $\sqrt{2}H_0$ will be reached and there will be no change until time t_2. At the time t_3 the point B_{rs} will be reached. The absorbed energy is proportionate to the area of the triangle with the points 0, H_s, B_{rs}.

Let now the negative trapezoidal pulse of Fig. 1.7-2b arrive. At time t_4 we are still at point B_{rs} in Fig. 1.7-1; point $-H_0$ is reached at t_5, and point $-B_{rs}$ at t_7. If we continue in this way we see that per trapezoidal pulse energy proportionate to one-half the area of the hysteresis loop L_s is lost after the first pulse, or the energy $W/2$ compared with nW for the sinusoidal pulse.

Let us next consider the case that the electric and magnetic field strengths

---

[6] Kraus and Carver (1973), pp. 232–234, 324–327, or Kraus (1984), pp. 245–249, 338–339.

[7] The dimension of B is V s/m^2, that of H is A/m. The area of the hysteresis loop is proportionate to HB, which has the dimension V A s/m^3. With the volume C measured in m^3 we get thus the dimension of energy for CHB. For electric hysteresis we have the dimension A s/m^2 for D, V/m for E, and V A s/m^3 for ED.

are insufficient to reach saturation. The sinusoidal pulse will only reach the point $\sqrt{2}H_1$ in Fig. 1.7-1 on the new curve and then drive the magnetization along the loop L_2. The trapezoidal pulses of Fig. 1.7-2b will then reach the point H_1 and drive the magnetization along the minor loop L_1. For the idealized hysteresis curve of Fig. 1.7-1 the area of loop L_1 is one-half the area of loop L_2. Hence, the energy loss for the trapezoidal pulses is reduced by another factor $\frac{1}{2}$ compared with sinusoidal pulses if the ferrite material is not driven into saturation.

Consider next the trapezoidal pulse of Fig. 1.7-2c. Starting at the point $H=0, B=0$ the material will be driven along the new curve to the point H_0. At the end of the positive pulse the magnetization will drop to a point slightly to the left of B_{rs}. For all subsequent pulses the magnetization will vary for practical purposes between the points H_0 and B_{rs}. Essentially no energy is lost due to hysteresis, and the relative permeability is always close to 1. In the nonsaturated case the magnetization will vary between point H_1 and essentially B_{r1} along the loop L_1. Again, no energy is lost due to hysteresis, and μ_r is close to 1. The same statements apply to electric hysteresis and ϵ_r if μ_r and ϵ_r are matched to yield $\mu_r/\epsilon_r = 1$. One would thus expect that ferrite materials

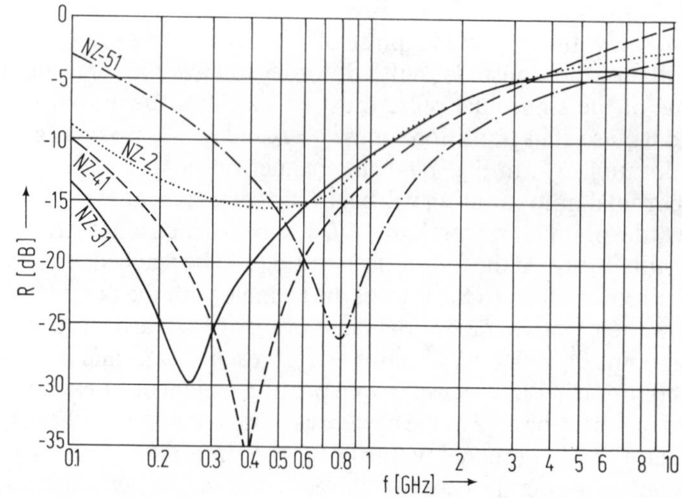

FIG. 1.7-3. Reduction of the power of periodic sinusoidal waves with perpendicular incidence reflected by a metal plate covered with four types of ferrite materials. The ferrite materials consist of flat tiles with the following thicknesses: 6.4 mm (NZ-2), 8 mm (NZ-31), 7 mm (NZ-41), and 5 mm (NZ-51). This illustration is based on data in Technical Bulletin 8-2-17 of Emerson & Cuming, Inc., Canton, Massachusetts.

have very little effect on carrier-free radar pulses. However, experimental results show that the matter is much more complicated. Consider, e.g., the ferrite tiles[8] NZ-31. According to Fig. 1.7-3 they reduce the reflection from a metal plate for perpendicular incidence by about −14 dB at 100 MHz, −30 dB at 250 MHz, and −11 dB at 1 GHz. The peak amplitude of a carrier-free pulse with a duration of about 1 ns was found to be reduced by about −10 to −12 dB under the same circumstances.[9] This does not mean that the energy of the reflected pulse was reduced to 10% or less. We will discuss in Section 4.9 the effect of a metal plate covered with a material having a relative permeability larger than 1 on carrier-free pulses. The data given in Fig. 1.7-3 for periodic sinusoidal waves will turn out to be of no help for this problem.

1.8 Modified Maxwell Equations

The usual form of Maxwell's[1] equations in the international system of units (SI) is

$$\text{curl}\,\mathbf{H} = \partial\mathbf{D}/\partial t + \mathbf{g} \tag{1}$$

$$-\text{curl}\,\mathbf{E} = \partial\mathbf{B}/\partial t \tag{2}$$

$$\text{div}\,\mathbf{D} = \rho, \qquad \text{div}\,\mathbf{B} = 0, \qquad \mathbf{D} = \epsilon\mathbf{E}, \qquad \mathbf{B} = \mu\mathbf{H} \tag{3}$$

where $\mathbf{E}$ and $\mathbf{H}$ are the electric and magnetic field strengths, $\mathbf{D}$ and $\mathbf{B}$ are the electric and magnetic flux densities, $\mathbf{g}$ is the electric current density, ρ the electric charge density, ϵ the permittivity, and μ the permeability.

The question has often been raised whether magnetic currents and charges exist. This matter is usually discussed in the literature under the heading *magnetic monopoles*. Jackson (1975, pp. 251–260) showed that one may indeed write Maxwell's equations in the form

$$\text{curl}\,\mathbf{H} = \partial\mathbf{D}/\partial t + \mathbf{g}_e \tag{4}$$

$$-\text{curl}\,\mathbf{E} = \partial\mathbf{B}/\partial t + \mathbf{g}_m \tag{5}$$

$$\text{div}\,\mathbf{D} = \rho_e, \qquad \text{div}\,\mathbf{B} = \rho_m, \qquad \mathbf{D} = \epsilon\mathbf{E}, \qquad \mathbf{B} = \mu\mathbf{H} \tag{6}$$

where $\mathbf{g}_e$, $\mathbf{g}_m$, ρ_e, and ρ_m stand for electric current density, magnetic current

[8] Made by Emerson & Cuming, Inc., Canton, Massachusetts.

[9] Measurements were made with the *SIR System-4R* produced by Geophysical Survey Systems, Inc., Hudson, New Hampshire.

[1] James Clerk Maxwell, physicist, 1831–1879; born in Edinburgh, Scotland.

density, electric charge density, and magnetic charge density.[2] One may make a *duality transformation* of these equations:

$$
\begin{aligned}
\mathbf{E} &= \mathbf{E}'\cos\lambda + \mathbf{H}'\sin\lambda, & \mathbf{D} &= \mathbf{D}'\cos\lambda + \mathbf{B}'\sin\lambda \\
\mathbf{H} &= -\mathbf{E}'\sin\lambda + \mathbf{H}'\cos\lambda, & \mathbf{B} &= -\mathbf{D}'\sin\lambda + \mathbf{B}'\cos\lambda \\
\rho_{\mathrm{e}} &= \rho'_{\mathrm{e}}\cos\lambda + \rho'_{\mathrm{m}}\sin\lambda, & \mathbf{g}_{\mathrm{e}} &= \mathbf{g}'_{\mathrm{e}}\cos\lambda + \mathbf{g}_{\mathrm{m}}\sin\lambda \\
\rho_{\mathrm{m}} &= -\rho'_{\mathrm{e}}\sin\lambda + \rho'_{\mathrm{m}}\cos\lambda, & \mathbf{g}_{\mathrm{m}} &= -\mathbf{g}'_{\mathrm{e}}\sin\lambda + \mathbf{g}'_{\mathrm{m}}\cos\lambda
\end{aligned}
\tag{7}
$$

The primed quantities again satisfy Eqs. (4)–(6). By choosing the real angle λ so that ρ_{m} and $\mathbf{g}_{\mathrm{m}}$ become zero, one obtains thus again Maxwell's equations. Jackson draws the following conclusion:

> The invariance of the equations of electrodynamics under duality transformations shows that it is a matter of convention to speak of a particle possessing an electric charge, but not magnetic charge. The only meaningful question is whether or not *all* particles have the same ratio of magnetic to electric charge. If they do, then we can make a duality transformation, choosing the angle λ so that $\rho_{\mathrm{m}} = 0$, $\mathbf{g}_{\mathrm{m}} = 0$. We then have Maxwell's equations as they are usually known.

When studying the propagation of waves with general time variation in a medium with ohmic losses, we will find that Maxwell's equations as they are usually known have no solution, but the modified equations of Eqs. (4)–(6) have a solution, even if one makes the transition $\mathbf{g}_{\mathrm{m}} \rightarrow 0$ *in the end.* Maxwell's equations in the form of Eqs. (1)–(3) permit two classes of solutions:

1. *Particular solutions* with sinusoidal time variation and sums of such solutions.
2. Solutions with general time variation of the *specialized equations* with $\mathbf{g} = 0$ almost everywhere, holding for a loss-free transmission medium.

To obtain solutions with general time variation in a lossy medium one must use Eqs. (4)–(6). In addition to the electric Ohm's[3] law

$$\mathbf{g}_{\mathrm{e}} = \sigma\mathbf{E} \tag{8}$$

we introduce the magnetic Ohm's law

$$\mathbf{g}_{\mathrm{m}} = s\mathbf{H}$$

After one has carried out the calculations one may consider the transition

---

[2] The addition of magnetic charges or magnetic monopoles to Maxwell's equations is also discussed by Amaldi (1968). The incentive for adding such terms is the observation by Dirac (1931, 1948) that the existence of a magnetic charge would explain the discreteness of the electric charge. For details on the quantum mechanical argument for magnetic charges see also Carrigan (1965), Sandars (1966), and Schwinger (1969); for experimental searches see Alvarez *et al.* (1970), and Fleischer *et al.* (1969a,b).

[3] Georg Simon Ohm, physicist, 1789–1854; born in Erlangen, Germany.

$s \to 0$ and obtain a solution. If one chooses $s = 0$ at the beginning—which means one uses Maxwell's equations in the form of Eqs. (1)–(3)—one runs into divergencies that prevent a solution.

The reason why this startling failure of Maxwell's equations has not been recognized earlier is that almost all published solutions assume a sinusoidal time variation. For a long time there were only two important exceptions: the radiation from a charged particle moving with arbitrary velocity studied by Schwarzschild (1903) and Abraham (1905, Vol. 2, Sec. 13), and the radiation of a Hertzian electric dipole with a current of arbitrary time variation discussed in the many editions of the book *Theorie der Elektrizität* by Abraham, Becker, and Sauter that apparently evolved from the solution of the charged particle.[4] Few other books on theoretical physics mention the radiation of waves with arbitrary time variation. Only very recently have these solutions for loss-free media been extended and applied to radio engineering (e.g., Spetner, 1974; Harmuth, 1981, 1984, 1985; Gómez *et al.*, 1984; Gómez and Morente, 1985). The new interest in electromagnetic waves with general time variation is due to the advances in radar and spread spectrum communication, the use of absorbing materials in the stealth technology, and the need to develop radar that neutralizes the stealth technology.

The generalization of Maxwell's equations has been a topic of discussion in theoretical physics for probably as long as the equations have existed. As an example of a printed comment on the matter we cite the textbook by Westphal (1953, p. 452, first paragraph).

[4] For English language editions of this book see Abraham and Becker (1932, 1950). A historical note on the matter may be found in a book by Harmuth (1984), p. 1.

2 Nonsinusoidal Waves in a Conducting Medium

2.1 Electric Step Function Excitation

The modified Maxwell equations of Section 1.8 without electric and magnetic charges ρ_e and ρ_m, having constant permittivity ϵ, permeability μ, electric conductivity σ, and magnetic conductivity s assume the following form if the electric and magnetic Ohm's laws

$$\mathbf{g}_e = \sigma \mathbf{E}, \qquad \mathbf{g}_m = s\mathbf{H} \tag{1}$$

are used:

$$\text{curl } \mathbf{H} = \epsilon\, \partial\mathbf{E}/\partial t + \sigma\mathbf{E} \tag{2}$$

$$-\text{curl } \mathbf{E} = \mu\, \partial\mathbf{H}/\partial t + s\mathbf{H} \tag{3}$$

$$\epsilon \text{ div } \mathbf{E} = \mu \text{ div } \mathbf{H} = 0 \tag{4}$$

Consider a planar, transverse electromagnetic (TEM) wave propagating in the direction y as shown in Fig. 2.1-1. A TEM wave requires

$$E_y = H_y = 0 \tag{5}$$

while a planar wave calls for the following relations:

$$\partial E_x/\partial x = \partial E_x/\partial z = \partial E_z/\partial x = \partial E_z/\partial z = 0 \tag{6}$$

$$\partial H_x/\partial x = \partial H_x/\partial z = \partial H_z/\partial x = \partial H_z/\partial z = 0 \tag{7}$$

Writing the operator curl in cartesian coordinates[1] and introducing the conditions of Eqs. (5)-(7) brings Eqs. (2) and (3) into the following form:

$$-\partial H_x/\partial y = \epsilon\, \partial E_z/\partial t + \sigma E_z \tag{8}$$

$$\partial H_z/\partial y = \epsilon\, \partial E_x/\partial t + \sigma E_x \tag{9}$$

$$\partial E_x/\partial y = \mu\, \partial H_z/\partial t + sH_z \tag{10}$$

$$-\partial E_z/\partial y = \mu\, \partial H_x/\partial t + sH_x \tag{11}$$

With the substitutions

$$E = E_x = E_z, \qquad H = H_x = -H_z \tag{12}$$

[1] See the formulas of vector analysis in Section 6.2.

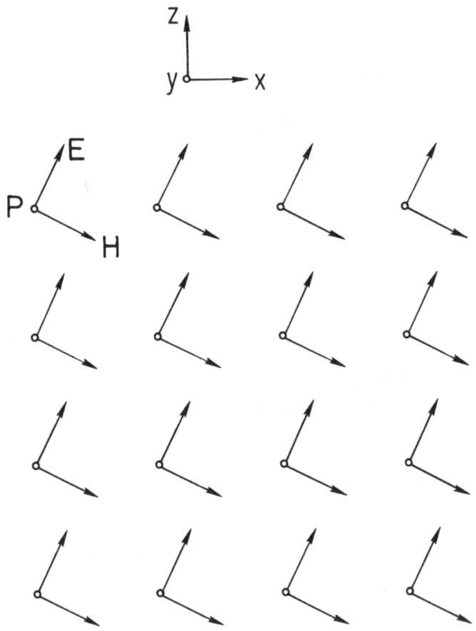

FIG. 2.1-1. General planar TEM wave propagating in the direction y, where y points into the paper plane.

one may rewrite the two pairs of Eqs. (8) and (11) as well as (9) and (10) as one pair:

$$\partial E/\partial y + \mu\, \partial H/\partial t + sH = 0 \tag{13}$$

$$\partial H/\partial y + \epsilon\, \partial E/\partial t + \sigma E = 0 \tag{14}$$

Instead of using the substitutions of Eq. (12) we may make the more general substitutions

$$E_x = E \cos \chi, \qquad E_z = E \sin \chi$$
$$H_x = H \sin \chi, \qquad H_z = -H \cos \chi$$

where χ is the *polarization angle* measured from the positive x axis to the vector **E** of the electric field strength or from the negative z axis to the vector **H** of the magnetic field strength, to obtain Eqs. (13) and (14). The physical meaning of E and H in Eqs. (13) and (14) is thus that of the magnitude of the field strengths **E** and **H**. Since the polarization angle χ is constant, the time variation of the field strengths **E** and **H** is the same as that of their magnitudes E and H. Hence, we can write our equations for E and H rather than for **E** and **H**.

Circularly polarized waves can be obtained by replacing the constant polarization angle χ by a time variable angle ωt:

$$E_x = E \cos \omega t, \qquad E_z = E \sin \omega t$$
$$H_x = H \sin \omega t, \qquad H_z = -H \cos \omega t$$

Substitution into Eqs. (8)–(11) yields again Eqs. (13) and (14). In this form one emphasizes functions with sinusoidal time variation. This distinction vanishes if one does not chose χ as a linear function of time, $\chi = \omega t$, but as a general function $\chi = f(t)$:

$$E_x = E\cos[f(t)], \qquad E_z = E\sin[f(t)]$$
$$H_x = H\sin[f(t)], \qquad H_z = -H\cos[f(t)]$$

Substitution into Eqs. (8)–(11) yields again Eqs. (13) and (14).

Every attempt to find a general solution for Eqs. (13) and (14) for $s = 0$ has led to divergencies. The problem does not occur if one wants particular solutions with periodic time variation of E and H. Furthermore, no problem is encountered if one wants general solutions of the specialized pair of equations for $s = \sigma = 0$. In radar and radio communication we are interested in waves that are zero for $t < t_0$ and have a certain time variation for $t \geqq t_0$. Waves with periodic time variation are not general enough to satisfy this condition, even though they have been used very successfully for the development of the theory of radar and radio communication as we find it in most textbooks. Waves with general time variation in a medium with conductivity $\sigma = 0$ have been studied to a much lesser extent.[2] We need waves with general time variation in a medium with conductivity $\sigma \neq 0$ if we want to study reflection and scattering of radar signals from the ground, water, or the absorbing materials used in the stealth technology. Sinusoidal waves and Fresnel's equations of reflection and refraction are of no help in this general case.

In order to overcome the problem of divergent solutions we retain the magnetic conductivity s. In this case one obtains convergent solutions for E and H from Eqs. (13) and (14), and *these solutions remain convergent if one makes in the end the transition* $s \to 0$.

Differentiation of Eq. (13) with respect to y and of Eq. (14) with respect to t yields:

$$\partial^2 E/\partial y^2 + \mu\, \partial^2 H/\partial y\, \partial t + s\, \partial H/\partial y = 0 \qquad (15)$$

$$\partial^2 H/\partial y\, \partial t + \epsilon\, \partial^2 E/\partial t^2 + \sigma\, \partial E/\partial t = 0 \qquad (16)$$

The terms containing H in Eq. (15) can be eliminated with the help of Eqs. (14) and (16):

$$\partial^2 E/\partial y^2 - \mu\epsilon\, \partial^2 E/\partial t^2 - (\mu\sigma + \epsilon s)\partial E/\partial t - s\sigma E = 0 \qquad (17)$$

If E is found from this equation one may obtain H from either Eq. (13) or

[2] The first general solution seems to be due to Schwarzschild (1903). Books by Abraham (1905), Abraham and Becker (1932, 1950), Becker (1957, 1964), and Landau and Lifshitz (1971) discuss such solutions. A historical review and some solutions of interest for antenna design are given by Harmuth (1984).

(14). Equation (14) is readily solved,

$$H(y, t) = -\int \left(\epsilon \frac{\partial E}{\partial t} + \sigma E\right) dy + H_y(t) \tag{18}$$

where $H_y(t)$ is an integration constant independent of y. One may also calculate $H(y, t)$ from Eq. (13). This is now an inhomogeneous ordinary differential equation of first order with constant coefficients, since $\partial E/\partial y$ is a known function:

$$dH/dt + s\mu^{-1}H = -\mu^{-1}\, \partial E/\partial y \tag{19}$$

The solution of the homogeneous equation yields:

$$H = Ce^{-st/\mu} \tag{20}$$

The inhomogeneous equation is solved by the method of variation of the constant. We assume that C is a function of t:

$$H(t) = C(t)e^{-st/\mu}, \qquad dH/dt = e^{-st/\mu}[dC(t)/dt - s\mu^{-1}C(t)] \tag{21}$$

Insertion of Eq. (21) into Eq. (19) yields

$$C(t) = -\mu^{-1} \int \frac{\partial E}{\partial y} e^{st/\mu} dt + H_t(y) \tag{22}$$

and H follows from Eq. (20):

$$H(y, t) = e^{-st/\mu}\left(-\mu^{-1} \int \frac{\partial E}{\partial y} e^{st/\mu}\, dt + H_t(y)\right) \tag{23}$$

Instead of deriving a differential equation for E from Eqs. (13) and (14) one may derive one for H. The interchange of E with H, ϵ with μ, and σ with s, reproduces these two equations. With the same interchanges one thus obtains from Eq. (17) a differential equation for H:

$$\partial^2 H/\partial y^2 - \mu\epsilon\, \partial^2 H/\partial t^2 - (\mu\sigma + \epsilon s)\partial H/\partial t - s\sigma H = 0 \tag{24}$$

This is the same equation as Eq. (17) with E replaced by H. Hence, a solution of Eq. (17) for E automatically gives us a solution for H.

If H is found from Eq. (24) one may obtain E from Eq. (13):

$$E(y, t) = -\int \left(\mu \frac{\partial H}{\partial t} + sH\right) dy + E_y(t) \tag{25}$$

Alternately, $E(y, t)$ may be determined from Eq. (14). This calls again for the solution of an inhomogeneous ordinary differential equation of first order with constant coefficients since $\partial H/\partial y$ is known:

$$dE/dt + \sigma\epsilon^{-1}E = -\epsilon^{-1}\, \partial H/\partial y \tag{26}$$

The solution by means of variation of the constant yields in analogy to Eq. (23):

$$E(y, t) = e^{-\sigma t/\epsilon}\left(-\epsilon^{-1}\int \frac{\partial H}{\partial y} e^{\sigma t/\epsilon}\, dt + E_t(y)\right) \tag{27}$$

At first glance it appears that the solution of Eq. (17) should yield for $E(y, t)$ the same function as Eqs. (25) and (27), but this is not so. Equations (17), (18), and (23) yield E and H, if boundary conditions are given for E, while Eqs. (24), (25), and (27) yield H and E if boundary conditions are given for H. Since boundary conditions can be defined independently for E and H, the general solution for E is represented by the sum of $E(y, t)$ obtained from Eq. (17) plus $E(y, t)$ of Eqs. (25) and (27), while the general solution for H is the sum of $H(y, t)$ of Eq. (24) plus the solution $H(y, t)$ of Eqs. (18) and (23). We will use the notation $E_E(y, t)$, $H_E(y, t)$ for the solutions due to boundary conditions for the electric field strength, and $E_H(y, t)$, $H_H(y, t)$ for the solutions due to boundary conditions for the magnetic field strength, whenever there is a risk of misunderstanding.

Let us turn to the boundary conditions. A partial differential equation by itself does not define a physical problem. One needs in addition boundary and initial conditions. Consider numerous electrodes in the plane $y = 0$. We may use them to apply a constant electric field strength with magnitude E_0 at the time $t = 0$ without applying a magnetic field strength:

$$\begin{aligned} E(0, t) = E_0 S(t) &= 0 \qquad \text{for} \quad t < 0 \\ &= E_0 \qquad \text{for} \quad t \geqq 0 \end{aligned} \tag{28}$$

At the plane $y \to \infty$ we have the further boundary condition

$$E(\infty, t) = \text{finite} \tag{29}$$

Let E and H initially be zero for $y > 0$ at the time $t = 0$. We have thus the initial conditions[3]

$$E(y, 0) = H(y, 0) = 0 \tag{30}$$

If $E(y, 0)$ and $H(y, 0)$ are zero for all values of $y > 0$, their derivatives with respect to y must be zero too:

$$\partial E(y, 0)/\partial y = \partial H(y, 0)/\partial y = 0 \tag{31}$$

[3] Even though we require $E(y, t) = H(y, t) = 0$ for $t = 0$ only, the implication is that $E(y, t)$ and $H(y, t)$ are zero for $t \leqq 0$, since $E(0, t)$ equals zero for $t < 0$ according to Eq. (28).

Equations (30) and (31) also imply the initial conditions

$$\partial E(y, t)/\partial t = \partial H(y, t)/\partial t = 0 \tag{32}$$

for $y > 0$ and $t = 0$ according to Eqs. (13) and (14).

We assume that the solution of Eq. (17) can be written in the form[4]

$$E(y, t) = E_E(y, t) = E_0[w(y, t) + F(y)] \tag{33}$$

Insertion of $F(y)$ into Eq. (17) yields the equation[5]

$$\partial^2 F/\partial y^2 - s\sigma F = 0 \tag{34}$$

with the general solution:

$$F(y) = A_{00}e^{-y/L} + A_{01}e^{y/L}, \qquad L = (s\sigma)^{-1/2} \tag{35}$$

The boundary conditions of Eqs. (28) and (29) require $A_{01} = 0$ and $A_{00} = 1$:

$$F(y) = e^{-y/L} \tag{36}$$

For the calculation of $w(y, t)$ of Eq. (33) we observe that the introduction of the function $F(y)$ transforms the boundary condition of Eq. (28) for $E = E_E$ into a homogeneous boundary condition for w,

$$E_E(0, t) = E_0 w(0, t) + E_0 = E_0 \qquad \text{for} \quad t \geqq 0 \tag{37}$$

$$w(0, t) = 0 \qquad \text{for} \quad t \geqq 0 \tag{38}$$

while Eq. (29) yields

$$w(\infty, t) = \text{finite} \tag{39}$$

The initial conditions of Eqs. (30) and (32) yield:

$$w(y, 0) + F(y) = 0, \qquad w(y, 0) = -e^{-y/L} \tag{40}$$

$$\partial w(y, t)/\partial t = 0 \qquad \text{for} \quad t = 0, y > 0 \tag{41}$$

Insertion of Eq. (33) into Eq. (17) yields for $w(y, t)$ the same equation as

---

[4] The assumption of a solution of the telegrapher's equation in the form $w(y, t) + F(y)$ for the voltages and currents along a transmission line of finite length is discussed by Smirnov (1964, Vol. 2, Ch. VII), who credits Krylov (1929) as the initiator of the method. The finite length of the transmission line leads to a Fourier series for $w(y, t)$ rather than a Fourier transform, which we will use. The telegrapher's equation uses the parameters inductance L, capacitance C, conductance G, and resistance R instead of μ, ϵ, σ, and s in Eq. (17). The need to introduce a parameter R equivalent to the magnetic conductance s never arose, since the resistance R was always part of the telegrapher's equation.

[5] Equation (34) as written follows from Eq. (17), but it is an ordinary differential equation and should be rewritten with d^2F/dy^2 replacing $\partial^2 F/\partial y^2$. We shall forgo such strictly cosmetic steps.

for $E(y, t)$:

$$\partial^2 w/\partial y^2 - \mu\epsilon\, \partial^2 w/\partial t^2 - (\mu\sigma + \epsilon s)\partial w/\partial t - s\sigma w = 0 \tag{42}$$

Particular solutions $w_\kappa(y, t)$ are obtained by the separation of variables using Bernoulli's product method,

$$w_\kappa(y, t) = \varphi(y)\psi(t) \tag{43}$$

$$\begin{aligned}\varphi^{-1}\, \partial^2\varphi/\partial y^2 &= \mu\epsilon\psi^{-1}\, \partial^2\psi/\partial t^2 + (\mu\sigma + \epsilon s)\psi^{-1}\, \partial\psi/\partial t + s\sigma \\ &= -(2\pi\kappa)^2\end{aligned} \tag{44}$$

which yields two ordinary differential equations

$$\partial^2\varphi/\partial y^2 + (2\pi\kappa)^2\varphi = 0 \tag{45}$$

and

$$\partial^2\psi/\partial t^2 + c^2(\mu\sigma + \epsilon s)\partial\psi/\partial t + [(2\pi\kappa c)^2 + s\sigma c^2]\psi = 0, \qquad c^2 = 1/\mu\epsilon \tag{46}$$

with the solutions:

$$\varphi(y) = A_{10} \sin 2\pi\kappa y + A_{11} \cos 2\pi\kappa y \tag{47}$$

$$\psi(t) = A_{20} \exp(\gamma_1 t) + A_{21} \exp(\gamma_2 t) \tag{48}$$

The coefficients γ_1 and γ_2 are the roots of the equation

$$\gamma^2 + c^2(\mu\sigma + \epsilon s)\gamma + [(2\pi\kappa)^2 + s\sigma]c^2 = 0 \tag{49}$$

which we write in the following form:

$$\begin{aligned}\gamma_1 &= -a + (a^2 - b^2c^2)^{1/2} \qquad \text{for} \quad a^2 > b^2c^2 \\ \gamma_2 &= -a - (a^2 - b^2c^2)^{1/2} \\ \gamma_1 &= -a + j(b^2c^2 - a^2)^{1/2} \qquad \text{for} \quad a^2 < b^2c^2 \\ \gamma_2 &= -a - j(b^2c^2 - a^2)^{1/2}\end{aligned} \tag{50}$$

$$a = (c^2/2)(\mu\sigma + \epsilon s) = (c/2)(Z\sigma + s/Z) = \sigma/2\epsilon + s/2\mu,$$

$$b^2 = (2\pi\kappa)^2 + s\sigma$$

$$c = (\mu\epsilon)^{-1/2}, \qquad Z = (\mu/\epsilon)^{1/2}, \qquad \epsilon = 1/Zc, \qquad \mu = Z/c$$

Note that Z is used as an abbreviation for $(\mu/\epsilon)^{1/2}$, which is *not* the impedance of a conducting medium as defined in the conventional theory of sinusoidal waves.

The boundary condition of Eq. (38) requires $A_{11} = 0$ in Eq. (47). The particular solution $w_\kappa(y, t)$ thus becomes:

$$w_\kappa(y, t) = [A_1 \exp(\gamma_1 t) + A_2 \exp(\gamma_2 t)]\sin 2\pi\kappa y \tag{51}$$

A general solution $w(y, t)$ is found by making A_1 and A_2 functions of the wavenumber κ, and then integrating over all possible values of κ:

$$w(y, t) = \int_0^\infty [A_1(\kappa)\exp(\gamma_1 t) + A_2(\kappa)\exp(\gamma_2 t)]\sin 2\pi\kappa y \, d\kappa \tag{52}$$

The time derivative $\partial w/\partial t$ equals:

$$\frac{\partial w}{\partial t} = \int_0^\infty [A_1(\kappa)\gamma_1 \exp(\gamma_1 t) + A_2(\kappa)\gamma_2 \exp(\gamma_2 t)]\sin 2\pi\kappa y \, d\kappa \tag{53}$$

The initial conditions of Eqs. (40) and (41) demand

$$\int_0^\infty [A_1(\kappa) + A_2(\kappa)]\sin 2\pi\kappa y \, d\kappa = -e^{-y/L} \tag{54}$$

$$\int_0^\infty [A_1(\kappa)\gamma_1 + A_2(\kappa)\gamma_2]\sin 2\pi\kappa y \, d\kappa = 0 \tag{55}$$

These two equations must be solved for the functions $A_1(\kappa)$ and $A_2(\kappa)$. To this end consider the Fourier sine transform in the following form:

$$\begin{aligned} g_s(\kappa) &= 2\int_0^\infty f_s(y)\sin 2\pi\kappa y \, dy \\ f_s(y) &= 2\int_0^\infty g_s(\kappa)\sin 2\pi\kappa y \, d\kappa \end{aligned} \tag{56}$$

If we identify $2g_s(\kappa)$ first with $A_1(\kappa) + A_2(\kappa)$ and then with $A_1(\kappa)\gamma_1 + A_2(\kappa)\gamma_2$ we obtain from Eqs. (54) and (55):

$$A_1(\kappa) + A_2(\kappa) = 2g_s(\kappa) = -4\int_0^\infty e^{-y/L} \sin 2\pi\kappa y \, dy \tag{57}$$

$$A_1(\kappa)\gamma_1 + A_2(\kappa)\gamma_2 = 0 \tag{58}$$

Using the tabulated integral

$$\int_0^\infty e^{uy} \sin 2\pi\kappa y \, dy = \frac{2\pi\kappa}{(2\pi\kappa)^2 + u^2} \tag{59}$$

one obtains from Eq. (57) with $L = (s\sigma)^{-1/2}$

$$A_1(\kappa) + A_2(\kappa) = -8\pi\kappa/[s\sigma + (2\pi\kappa)^2] = -8\pi\kappa/b^2 \tag{60}$$

and for the limit $s = 0$:

$$A_1(\kappa) + A_2(\kappa) = -2/\pi\kappa \tag{61}$$

The first reason for the need of a term $s\mathbf{H}$ in Eq. (3) and $\mathbf{g}_m$ in Eq. (1.8–5) becomes clear now. For $s = 0$, $\sigma \neq 0$, and $L = \infty$ the integral in Eq. (57) would not exist. We could have used the standard method of making $\sin 2\pi\kappa y$ integrable by multiplication with a term $e^{-y/L}$ and taking the limit $1/L = 0$. However, the physical meaning of this procedure would have remained unexplained, and the question would be raised why the factor $e^{-y/L}$ should be used and not some other factor or method to obtain integrability. Our approach brings out the physical significance of the integrability. No problem of integrability is encountered if Maxwell's equations are modified by the introduction of a magnetic current, and the integrability is maintained in the limit of a vanishing magnetic current.

Equations (58) and (60) are solved for $A_1(\kappa)$ and $A_2(\kappa)$:

$$\begin{aligned}
A_1(\kappa) &= -\frac{8\pi\kappa}{b^2}\frac{\gamma_2}{\gamma_2-\gamma_1} \\
&= -\frac{4\pi\kappa}{b^2}\left(1+\frac{a}{(a^2-b^2c^2)^{1/2}}\right) \qquad \text{for} \quad a^2 > b^2c^2 \\
&= -\frac{4\pi\kappa}{b^2}\left(1-\frac{ja}{(b^2c^2-a^2)^{1/2}}\right) \qquad \text{for} \quad a^2 < b^2c^2 \\
A_2(\kappa) &= -\frac{8\pi\kappa}{b^2}\frac{\gamma_1}{\gamma_1-\gamma_2} \\
&= -\frac{4\pi\kappa}{b^2}\left(1-\frac{a}{(a^2-b^2c^2)^{1/2}}\right) \qquad \text{for} \quad a^2 > b^2c^2 \\
&= -\frac{4\pi\kappa}{b^2}\left(1+\frac{ja}{(b^2c^2-a^2)^{1/2}}\right) \qquad \text{for} \quad a^2 < b^2c^2
\end{aligned} \tag{62}$$

Insertion of Eqs. (50) and (62) into Eq. (52) yields:

$$\begin{aligned}
w(y,t) = -e^{-at}\Bigg\{&\int_0^K\Bigg[\left(1+\frac{a}{(a^2-b^2c^2)^{1/2}}\right)\exp[(a^2-b^2c^2)^{1/2}t] \\
&+\left(1-\frac{a}{(a^2-b^2c^2)^{1/2}}\right)\exp[-(a^2-b^2c^2)^{1/2}t]\Bigg]\frac{\sin 2\pi\kappa y}{b^2/4\pi\kappa}\,d\kappa \\
&+\int_K^\infty\Bigg[\left(1+\frac{ja}{(b^2c^2-a^2)^{1/2}}\right)\exp[j(b^2c^2-a^2)^{1/2}t] \\
&+\left(1+\frac{ja}{(b^2c^2-a^2)^{1/2}}\right)\exp[-j(b^2c^2-a^2)^{1/2}t]\Bigg]\frac{\sin 2\pi\kappa y}{b^2/4\pi\kappa}\,d\kappa\Bigg\}
\end{aligned} \tag{63}$$

$$K = (2\pi)^{-1}(a^2/c^2 - s\sigma)^{1/2}, \quad b^2 = (2\pi\kappa)^2 + s\sigma, \quad a = (c/2)(Z\sigma + s/Z)$$

The imaginary terms in the second integral may be rewritten in real form by means of the formulas

$$e^{jq} + e^{-jq} = 2\cos q, \qquad -j(e^{jq} - e^{-jq}) = 2\sin q$$

while the first integral can be simplified with the help of hyperbolic functions:

$$e^{q} + e^{-q} = 2\,\text{ch}\,q, \qquad e^{q} - e^{-q} = 2\,\text{sh}\,q$$

One obtains:

$$\begin{aligned} w(y, t) = -\frac{2}{\pi} e^{-at}\Bigg[\int_0^{2\pi K} &\left(\text{ch}(a^2 - b^2c^2)^{1/2}t + \frac{a\,\text{sh}(a^2 - b^2c^2)^{1/2}t}{(a^2 - b^2c^2)^{1/2}}\right) \\ &\times \frac{\sin 2\pi\kappa y}{b^2}(2\pi\kappa)d(2\pi\kappa) \\ &+ \int_{2\pi K}^{\infty}\left(\cos(b^2c^2 - a^2)^{1/2}t + \frac{a\sin(b^2c^2 - a^2)^{1/2}t}{(b^2c^2 - a^2)^{1/2}}\right) \\ &\times \frac{\sin 2\pi\kappa y}{b^2}(2\pi\kappa)d(2\pi\kappa)\Bigg] \end{aligned} \tag{64}$$

To obtain $E_E(y, t)$ we still have to add $F(y)$ to $w(y, t)$ according to Eq. (33). With $F(y) = e^{-y/L}$ from Eq. (36) we get:

$$E_E(y, t) = E_0[e^{-y/L} + w(y, t)], \qquad L = (s\sigma)^{-1/2} \tag{65}$$

We now make the transition to $s = 0$. From Eqs. (50) and (63) we get in this limit:

$$b = \beta = 2\pi\kappa, \qquad a = \alpha = Zc\sigma/2, \qquad 2\pi K = \alpha/c = Z\sigma/2 \tag{66}$$

Equations (64) and (65) become:

$$\begin{aligned} w(y, t) = -\frac{2}{\pi} e^{-\alpha t}\Bigg[\int_0^{Z\sigma/2} &\left(\text{ch}(\alpha^2 - \beta^2c^2)^{1/2}t + \frac{\alpha\,\text{sh}(\alpha^2 - \beta^2c^2)^{1/2}t}{(\alpha^2 - \beta^2c^2)^{1/2}}\right) \\ &\times \frac{\sin \beta y}{\beta} d\beta \\ &+ \int_{Z\sigma/2}^{\infty}\left(\cos(\beta^2c^2 - \alpha^2)^{1/2}t + \frac{\alpha\sin(\beta^2c^2 - \alpha^2)^{1/2}t}{(\beta^2c^2 - \alpha^2)^{1/2}}\right) \\ &\times \frac{\sin \beta y}{\beta} d\beta\Bigg] \end{aligned} \tag{67}$$

$$E_E(y, t) = E_0[1 + w(y, t)] \tag{68}$$

In order to get some understanding of the physical content of these two equations, we observe that the first integral in Eq. (67) vanishes when the conductance σ approaches zero. We see that $\alpha = Zc\sigma/2$ also equals zero in this case:

$$E_E(y,t) = E_0\left(1 - \frac{2}{\pi}\int_0^\infty \frac{\cos\beta ct \sin\beta y}{\beta}\,d\beta\right)$$
$$= E_0\left[1 - \frac{1}{\pi}\left(\int_0^\infty \frac{\sin\beta(y+ct)}{\beta}\,d\beta + \int_0^\infty \frac{\sin\beta(y-ct)}{\beta}\,d\beta\right)\right] \quad (69)$$

From a table of integrals we find:

$$\begin{aligned} \int_0^\infty \frac{\sin pq}{q}\,dq &= \tfrac{1}{2}\pi &&\text{for } p>0 \\ &= 0 &&\text{for } p=0 \\ &= -\tfrac{1}{2}\pi &&\text{for } p<0 \end{aligned} \quad (70)$$

Equation (69) thus assumes the following form:

$$\begin{aligned} E_E(y,t) &= E_0 &&\text{for } y=0, && t=0 \\ &= 0 &&\text{for } y>0, && ct<y \\ &= E_0/2 &&\text{for } y>0, && ct=y \\ &= E_0 &&\text{for } y\geqq 0, && ct>y \end{aligned} \quad (71)$$

This represents a step function with amplitude E_0 propagating with velocity c toward increasing values of y. A look at Eq. (17) shows that for $s=0$ and $\sigma=0$ it becomes the one-dimensional wave equation which should yield the result of Eq. (71) for the boundary condition of Eq. (28), except for the case $ct=y$ when the wave equation yields the result E_0 rather than $E_0/2$. The discrepancy is due to the fact that the Fourier representation of a function converges to the median $(0+E_0)/2$, if the function has a discontinuity with two limits 0 and E_0 in a point.

Figure 2.1–2a shows $E_E(y,t)$ as function of t in the point y, while Fig. 2.1–2b shows it as function of y at the time t.

Let us determine the velocity of propagation of $E_E(y,t)$ in the general case $\sigma \neq 0$. In the second integral of Eq. (67) we can make the following transformations:

$$\begin{aligned} \cos(\beta^2c^2-\alpha^2)^{1/2}t \sin\beta y &= \tfrac{1}{2}\{\sin\beta[y - c(1-\alpha^2/\beta^2c^2)^{1/2}t] \\ &\quad + \sin\beta[y + c(1-\alpha^2/\beta^2c^2)^{1/2}t]\} \\ \sin(\beta^2c^2-\alpha^2)^{1/2}t \sin\beta y &= \tfrac{1}{2}\{\cos\beta[y - c(1-\alpha^2/\beta^2c^2)^{1/2}t] \\ &\quad - \cos\beta[y + c(1-\alpha^2/\beta^2c^2)^{1/2}t]\} \end{aligned} \quad (72)$$

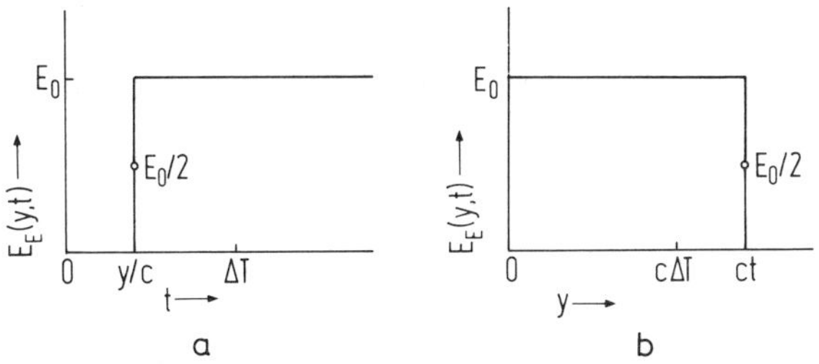

FIG. 2.1-2. The magnitude $E_E(y,t)$ of the field strength excited by an electric step function in a loss-free medium (a) at the location y as a function of the time variable and (b) at the time t as function of the space variable.

We recognize the velocities

$$c(1-\alpha^2/\beta^2c^2)^{1/2} = c(1 - Z^2\sigma^2/4\beta^2)^{1/2} \tag{73}$$

that vary between the limits

$$0 \leqq c(1 - Z^2\sigma^2/4\beta^2)^{1/2} \leqq c \tag{74}$$

when β varies between $Z\sigma/2$ and ∞. Hence, there is no velocity larger than c, and the maximum propagation velocity of $E_E(y, t)$ is c.

In the first integral of Eq. (67) we can also make two transformations:

$$\begin{aligned} \mathrm{ch}(\alpha^2 - \beta^2c^2)^{1/2}t \sin \beta y &= \tfrac{1}{2}\{\sin \beta[y - jc(\alpha^2/\beta^2c^2 - 1)^{1/2}t] \\ &\quad + \sin \beta[y + jc(\alpha^2/\beta^2c^2 - 1)^{1/2}t]\} \\ \mathrm{sh}(\alpha^2 - \beta^2c^2)^{1/2}t \sin \beta y &= \tfrac{1}{2}j\{\cos \beta[y - jc(\alpha^2/\beta^2c^2 - 1)^{1/2}t] \\ &\quad - \cos \beta[y + jc(\alpha^2/\beta^2c^2 - 1)^{1/2}t]\} \end{aligned} \tag{75}$$

We have now imaginary velocities

$$jc(\alpha^2/\beta^2c^2 - 1) = jc(Z^2\sigma^2/4\beta^2 - 1)^{1/2} \tag{76}$$

whose magnitude varies between the limits

$$\infty > c(Z^2\sigma^2/4\pi^2 - 1)^{1/2} \geqq 0 \tag{77}$$

when β varies between 0 and $Z\sigma/2$. There is no objection to an imaginary velocity having values larger than c, the limit c applies only to the velocity with which information and energy can be transmitted.

For $\sigma \neq 0$ one may rewrite Eq. (67) into a more compact normalized form. We make the following substitutions

$$\begin{aligned} \eta &= \beta c/\alpha = 4\pi\kappa/Z\sigma = (4\pi\kappa/\sigma)(\epsilon/\mu)^{1/2} \\ \theta &= \alpha t = Z\sigma ct/2 = \sigma t/2\epsilon \\ \xi &= \alpha y/c = Z\sigma y/2 = (\mu/\epsilon)^{1/2}\sigma y/2 \end{aligned} \tag{78}$$

and obtain with

$$\beta y = \eta\xi, \qquad d\beta = \alpha c^{-1}\,d\eta, \qquad d\beta/\beta = d\eta/\eta, \qquad \eta = 1 \quad \text{for} \quad \beta = Z\sigma/2 \tag{79}$$

the result:

$$E_E(\xi, \theta) = E_0[1 + w(\xi, \theta)]$$
$$w(\xi, \theta) = -\frac{2}{\pi}e^{-\theta}\left[\int_0^1 \left(\text{ch}(1-\eta^2)^{1/2}\theta + \frac{\text{sh}(1-\eta^2)^{1/2}\theta}{(1-\eta^2)^{1/2}}\right)\frac{\sin \xi\eta}{\eta}\,d\eta \right. \tag{80}$$
$$\left. + \int_1^\infty \left(\cos(\eta^2-1)^{1/2}\theta + \frac{\sin(\eta^2-1)^{1/2}\theta}{(\eta^2-1)^{1/2}}\right)\frac{\sin \xi\eta}{\eta}\,d\eta\right]$$

This equation contains only the normalized space and time variables ξ and θ, since the variable η is eliminated by the integration. All three parameters μ, ϵ, and σ of Eqs. (2) and (3) have been made part of the normalized variables ξ and θ. A major drawback of Eq. (80) is that η becomes infinite for $\sigma = 0$. Hence, the equation is not well suited for the study of the transition to the loss-free medium.

Figure 2.1–3 gives a typical computer plot of $E_E(\xi, \theta)$ for $\xi = 2$ and θ in the range $0 \leqq \theta \leqq 60$. The function is zero for $\theta < \xi$, or $t < y/c$ in nonnormalized notation. At $\theta = \xi$ is a jump that decreases monotonously from 1 at $\xi = 0$ to 0 for $\xi \to \infty$. From this jump the function increases to 1 for $\theta \to \infty$. The increase is very slow for large values of ξ. The derivative at $\theta = \xi$ can be shown analytically to be infinite, while the derivative at $\theta \to \infty$ can be shown

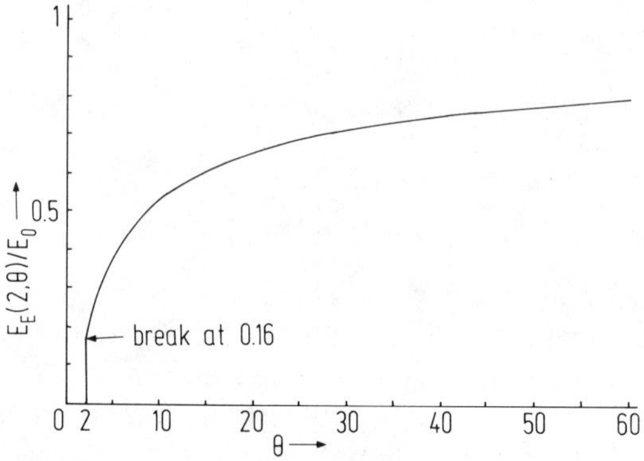

FIG. 2.1-3. The magnitude $E_E(\xi,\theta)$ of the normalized electric field strength excited by an electric step function in a lossy medium at the normalized location $\xi = \alpha y/c = 2$ as a function of the normalized time $\theta = \alpha t$. This illustration is based on computer plots by R. N. Boules, University of Alexandria, Egypt. More plots are shown in Fig. 5.2-1.

to be zero (Boules, 1987). We will discuss this and similar plots in more detail in Sections 5.2 and 5.3.

Let us consider the magnetic field strength. In the general case $\sigma \neq 0$ the calculation is quite involved and must be postponed to a later section. For $\sigma = 0$ we may put $s = 0$ in Eqs. (18) and (23) without causing divergencies. We thus obtain from Eq. (69):

$$\begin{aligned} H_E(y, t) &= -\epsilon \int \left(\frac{\partial E_E}{\partial t}\right) dy + H_y(t) \\ &= 2E_0 \epsilon c \pi^{-1} \int_0^\infty \left(\sin \beta ct \int \sin \beta y \, dy\right) d\beta + H_y(t) \\ &= \frac{2E_0}{Z\pi} \int_0^\infty \frac{\sin \beta ct \cos \beta y}{\beta} d\beta + H_y(t) \end{aligned} \tag{81}$$

$$\begin{aligned} H_E(y, t) &= -\mu^{-1} \int \left(\frac{\partial E_E}{\partial y}\right) dt + H_t(y) \\ &= \frac{2E_0}{\pi\mu} \int_0^\infty \left(\cos \beta y \int \cos \beta ct \, dt\right) d\beta + H_t(y) \\ &= \frac{2E_0}{Z\pi} \int_0^\infty \frac{\sin \beta ct \cos \beta y}{\beta} d\beta + H_t(y) \end{aligned} \tag{82}$$

Since Eqs. (81) and (82) must yield the same value for $H(y, t)$, we have

$$H_y(t) = H_t(y) = H_c = 0 \tag{83}$$

where $H_c = 0$ follows from Eq. (30), and

$$H_E(y, t) = \frac{E_0}{Z\pi} \left(\int_0^\infty \frac{\sin \beta(y + ct)}{\beta} d\beta - \int_0^\infty \frac{\sin \beta(y - ct)}{\beta} d\beta \right) \tag{84}$$

With the help of Eq. (70) we obtain:

$$\begin{aligned} H_E(y, t) &= 0 && \text{for} \quad y = 0, \quad t = 0 \\ &= 0 && \text{for} \quad y > 0, \quad ct < y \\ &= E_0/2Z && \text{for} \quad y > 0, \quad ct = y \\ &= E_0/Z && \text{for} \quad y \geqq 0, \quad ct > y \end{aligned} \tag{85}$$

We have the same space–time variation for $H_E(y, t)$ as for $E_E(y, t)$ in Eq. (71), except for the point $y = 0, t = 0$. The electric field strength E_0 is applied at $t = 0$ to the plane $y = 0$, but the magnetic field strength E_0/Z is produced by the electric field strength. Hence, we have $H_E(y, t) = E_0/Z$ for $y = 0$, $ct > 0$—or $t > 0$—but not for $y = 0, t = 0$.

2.2 Initial Conditions and Fourier Transform

Since it is widely believed that general solutions of Maxwell's equations can be obtained from the usual solutions with sinusoidal time variation by means of a Fourier transform, we will show by an example why this simple approach does not work.

Consider the solution of Eq. (2.1 – 17) for a wave with sinusoidal time variation. We use the exponential notation $e^{j\omega t}$ in order to cover both a sinusoidal and a cosinusoidal time variation,

$$E(y, t) = u(y)e^{j\omega t} \tag{1}$$

and obtain for $s = 0$:

$$d^2u/dy^2 - j\omega\mu(j\omega\epsilon + \sigma)u = 0 \tag{2}$$

The complex constant[1]

$$\gamma'' = [j\omega\mu(j\omega\epsilon + \sigma)]^{1/2} = \alpha'' + j\beta \tag{3}$$

is called the propagation constant of the sinusoidal wave. Instead of γ'' one may also use the (complex circular) wave number

$$k = j\gamma'' = -\beta + j\alpha'' \tag{4}$$

For $\sigma = 0$ one obtains:

$$-k = \beta = \omega\sqrt{\mu\epsilon} = \omega/c = 2\pi\kappa, \qquad \kappa = f/c = 1/\lambda \tag{5}$$

The attenuation and phase constant α'' and β can be obtained from Eq. (3) by squaring and solving the resulting two equations for the real and imaginary components:

$$\begin{aligned} \alpha'' &= \omega\{(\mu\epsilon/2)[(1 + \sigma^2/\omega^2\epsilon^2)^{1/2} - 1]\}^{1/2} \\ \beta &= \omega\{(\mu\epsilon/2)[(1 + \sigma^2/\omega^2\epsilon^2)^{1/2} + 1]\}^{1/2} \end{aligned} \tag{6}$$

The solution of Eq. (2) is given by

$$u(y) = E_1 e^{-(\alpha''+j\beta)y} + E_2 e^{(\alpha''+j\beta)y} \tag{7}$$

In a medium with $\sigma > 0$ the constant E_2 must be zero to avoid arbitrarily large values of u for large values of y. The solution $E(y, t)$ thus assumes the form

$$E(y, t) = E_1 e^{-\gamma'' y + j\omega t} = E_1 e^{j(ky+\omega t)} = E_1 e^{-\alpha'' y} e^{-j(\beta y - \omega t)} \tag{8}$$

or

$$E(y, t) = [E_{01}\cos(\beta y - \omega t) + E_{02}\sin(\beta y - \omega t)]e^{-\alpha'' y} \tag{9}$$

---

[1] The notation γ'' and α'' is used since γ and α already denote other quantities.

Let us note that we could derive this result by making the magnetic conductivity s zero at the beginning and not after carrying the calculation through for a general value of s as in Section 2.1. This is typical for sinusoidal solutions of Maxwell's equations, and for general *periodic* solutions obtained from sums of sinusoidal solutions.

The magnetic field strength $H(y, t)$ follows either from Eq. (2.1–18),

$$H(y, t) = -\int \left(\epsilon \frac{\partial E}{\partial t} + \sigma E\right) dy + H_y(t) \tag{10}$$

or from Eq. (2.1–23) for $s = 0$:

$$H(y, t) = -\mu^{-1} \int \left(\frac{\partial E}{\partial y}\right) dt + H_t(y) \tag{11}$$

We obtain from Eqs. (8) and (11):

$$\begin{aligned} H(y, t) &= -\frac{jk}{\mu} E_1 \int e^{j(ky+\omega t)}\, dt + H_t(y) \\ &= -(k/\omega\mu) E_1 e^{j(ky+\omega t)} + H_t(y) \end{aligned} \tag{12}$$

Alternately we get from Eqs. (8) and (10):

$$\begin{aligned} H(y, t) &= -(j\omega\epsilon + \sigma) E_1 \int e^{j(ky+\omega t)}\, dy + H_y(t) \\ &= -[(j\omega\epsilon + \sigma)/jk] E_1 e^{j(ky+\omega t)} + H_y(t) \\ &= -(k/\omega\mu) E_1 e^{j(ky+\omega t)} + H_y(t) \end{aligned} \tag{13}$$

The comparison of the last lines of Eqs. (12) and (13) yields

$$H_y(t) = H_t(y) \tag{14}$$

which means these functions can neither vary with t nor with y but must be constant. We assume that this constant magnetic field strength is zero. In this case the ratio $E(y, t)/H(y, t)$ becomes the usual wave impedance Z_s for sinusoidal waves:

$$Z_s = -\omega\mu/k = j\omega\mu/\gamma'' = [j\omega\mu/(j\omega\epsilon + \sigma)]^{1/2} \tag{15}$$

Note that we did not encounter any problem by setting $s = 0$ in Eq. (2.1–23) to calculate $H(y, t)$.

Let us now see whether our solutions $E(y, t)$ and $H(y, t)$ can be used to represent the propagation of a step function as defined by the boundary and initial conditions of Eqs. (2.1–28) and (2.1–30), or any other function that is zero for $t < 0$ and has the time variation $E_g(0, t)$ for $t \geqq 0$. We use the Fourier transform to produce a general solution $E_g(y, t)$ from the particular solution of Eq. (9):

$$E_g(y, t) = e^{-\alpha'' y} \int_0^\infty [E_{01}(\beta)\cos(\beta y - \omega t) + E_{02}(\beta)\sin(\beta y - \omega t)]\, d\beta \quad (16)$$

The function $E_g(0, t)$ represents the boundary condition[2] for $y = 0$:

$$E_g(0, t) = \int_0^\infty [E_{01}(\beta)\cos \omega t - E_{02}(\beta)\sin \omega t]\, d\beta \quad (17)$$

Since β is a function of ω, we have here a Fourier cosine and a Fourier sine integral to represent the function $E_g(0, t)$. The choice of $E_{01}(\beta)$ and $E_{02}(\beta)$ according to Eq. (17) will thus satisfy the boundary condition.

The initial condition $E_g(y, 0)$, required for any wave that is zero for $t \leqq 0$, yields:

$$E_g(y, 0) = e^{-\alpha'' y} \int_0^\infty [E_{01}(\beta)\cos \beta y + E_{02}(\beta)\sin \beta y]\, d\beta = 0 \quad (18)$$

This condition can be satisfied for arbitrary values of y only if $E_{01}(\beta)$ and $E_{02}(\beta)$ are zero for any value[3] of β.

We see thus that a superposition of the usual harmonic solutions of Maxwell's equations cannot be used to construct transient solutions. Even though the boundary condition can be satisfied by means of the Fourier transform according to Eq. (17), the initial condition $E_g(y, 0) = 0$ cannot be satisfied. The much more elaborate calculations of Section 2.1 are needed for the study of the transmission of signals with beginning and end.

2.3 Electric Exponential Ramp Function Excitation

The solution of Maxwell's equations for a step function excitation discussed in Section 2.1 may be used for a more general time variation of an excitation function by means of the series expansion developed in connection with Fig. 1.3-3. The question arises whether there are other basic solutions besides the step function from which one could derive further solutions by means of series expansions. It is known that the solution of linear, partial differential equations with constant coefficients by means of separation of variables favors the exponential function. If we want periodic solutions we use automatically the functions $e^{j\omega t}$ and $e^{-j\omega t}$. For transient

[2] If $E_g(0, t)$ is a step function, we may use a factor $e^{-t/\tau}$ and let τ approach infinity after integration.

[3] This property of a Fourier transform is usually referred to as the uniqueness of a Fourier transform.

solutions one will thus try the real exponential functions $e^{t/\tau}$ and $e^{-t/\tau}$ multiplied with the unit step function $S(t)$. Since we have found that the step function $S(t)$ provides us with a basic solution, we use the functions $S(t)(1-e^{-t/\tau})$ and $S(t)(e^{t/\tau}-1)$ shown in Figs. 1.2–4b and c instead of $S(t)e^{-t/\tau}$ and $S(t)e^{t/\tau}$.

Consider the function $QS(t)(1-e^{-t/\tau})$. This is the function of Fig. 1.2–4b for $Q>0$, $\tau>0$, or the function of Fig. 1.2–4c for $Q<0$, $\tau<0$. Let Q have the absolute values

$$|Q|=(1-e^{-\Delta T/\tau})^{-1} \tag{1}$$

so that the function $QS(t)(1-e^{-t/\tau})$ equals 1 for positive values of τ and $t=\Delta T$. The exponential ramp function $r(t)$ may then be defined:

$$\begin{aligned} r(t)=QS(t)(1-e^{-t/\tau}) &= 0 && \text{for} \quad t<0 \\ &= Q(1-e^{-t/\tau}) && \text{for} \quad t \geqq 0 \end{aligned} \tag{2}$$

We repeat the calculation of Section 2.1 for this transient function. The boundary condition of Eq. (2.1–28) is replaced:

$$\begin{aligned} E(0,t)=E_0 r(t) &= 0 && \text{for} \quad t<0 \\ &= E_0 Q(1-e^{-t/\tau})=E_1(1-e^{-t/\tau}) && \text{for} \quad t \geqq 0 \end{aligned} \tag{3}$$

At the plane $y\to\infty$ again the boundary condition holds

$$E(\infty,t)=\text{finite} \tag{4}$$

If E and H are zero for $y \geqq 0$ at the time $t=0$ we have the initial conditions

$$E(y,0)=H(y,0)=0, \qquad y \geqq 0 \tag{5}$$

This equation implies that the derivatives with respect to y must be zero too:

$$\partial E(y,0)/\partial y=\partial H(y,0)/\partial y=0 \tag{6}$$

According to Eqs. (2.1–13) and (2.1–14) the derivatives of the field strengths with respect to t must vanish, if the field strengths and their derivatives with respect to y are zero:

$$\partial E(y,t)/\partial t=\partial H(y,t)/\partial t=0 \qquad \text{for} \quad t=0, \quad y \geqq 0 \tag{7}$$

We assume that the solution of the previously derived differential equation Eq. (2.1–17)

$$\partial^2 E/\partial y^2-\mu\epsilon\, \partial^2 E/\partial t^2-(\mu\sigma+\epsilon s)\partial E/\partial t-s\sigma E=0 \tag{8}$$

can be written in the form:

$$E(y,t)=E_E(y,t)=E_0 Q[u(y,t)+(1-e^{-t/\tau})F(y)], \qquad t \geqq 0 \tag{9}$$

Insertion of the term $E_0Q(1-e^{-t/\tau})F(y)$ into Eq. (8) yields:

$$(1-e^{-t/\tau})\partial^2F/\partial y^2+\tau^{-2}\mu\epsilon e^{-t/\tau}F-\tau^{-1}(\mu\sigma+\epsilon s)e^{-t/\tau}F - s\sigma(1-e^{-t/\tau})F=0 \quad (10)$$

Since $F(y)$ is assumed to be a function of y but not of t, the terms with different functions of t must vanish separately. We get thus an equation for the first and the last term of Eq. (10)

$$\partial^2F/\partial y^2-s\sigma F=0 \quad (11)$$

and a second equation with the two remaining terms:

$$\tau^{-1}(\tau^{-1}\mu\epsilon-\mu\sigma-\epsilon s)e^{-t/\tau}=0 \quad (12)$$

For Eq. (11) we again get the solution

$$F(y)=A_{00}e^{-y/L}, \qquad L=(s\sigma)^{-1/2} \quad (13)$$

that satisfies the boundary condition of Eq. (4). Equation (12) has the trivial solution $\tau=\infty$, a nontrivial solution

$$\tau=0 \quad (14)$$

from the exponential factor $e^{-t/\tau}$, and the solution of primary interest

$$\tau^{-1}=c^2(\mu\sigma+\epsilon s)=2a, \qquad c^2=1/\mu\epsilon \quad (15)$$

Since τ can have nonnegative values only, the boundary condition of Eq. (3) can have the form $E_1(1-e^{-t/|\tau|})$ only, while the form $E_1(e^{-t/|\tau|}-1)$ is excluded. Only the function of Fig. 1.2–4b can satisfy Eq. (15); the function of Fig. 1.2–4c must be discarded. The important improvement over the solution of Section 2.1 is that the function of Fig. 1.2–4b rises continuously from its value 0 at $t\leqq 0$.

The special solution for $\tau=0$ yields

$$\lim_{\tau\to 0} S(t)(1-e^{-t/\tau})/(1-e^{-\Delta T/\tau})=S(t) \quad (16)$$

which is the step function used in Section 2.1. Hence, the step function is a special case of the more general theory developed here.

We substitute Eq. (13) into Eq. (9)

$$E_E(y,t)=E_0Q[u(y,t)+A_{00}(1-e^{-t/\tau})e^{-y/L}] \quad (17)$$

and equate $E_E(0,t)$ with the boundary condition of Eq. (3):

$$E_0Q[u(0,t)+A_{00}(1-e^{-t/\tau})]=E_0Q(1-e^{-t/\tau}), \qquad t\geqq 0 \quad (18)$$

With

$$A_{00} = 1 \tag{19}$$

we obtain a homogeneous boundary condition for $u(y, t)$:

$$u(0, t) = 0 \tag{20}$$

At the plane $y \to \infty$ we obtain from Eq. (4) with $F(\infty) = 0$ the boundary condition

$$u(\infty, t) = \text{finite} \tag{21}$$

The initial condition of Eq. (5) yields

$$E_E(y, 0) = E_0 Q u(y, 0) = 0 \tag{22}$$

while the second initial condition of Eq. (7) requires:

$$\partial u/\partial t + 2ae^{-y/L} = 0 \qquad \text{for} \quad t = 0, \quad y \geqq 0 \tag{23}$$

The calculation of $u(y, t)$ proceeds as in Section 2.1 for $w(y, t)$ until Eqs. (2.1-52) and (2.1-53) are reached:

$$u(y, t) = \int_0^\infty [A_1(\kappa) \exp(\gamma_1 t) + A_2(\kappa) \exp(\gamma_2 t)] \sin 2\pi\kappa y \, d\kappa \tag{24}$$

$$\frac{\partial u}{\partial t} = \int_0^\infty [A_1(\kappa)\gamma_1 \exp(\gamma_1 t) + A_2(\kappa)\gamma_2 \exp(\gamma_2 t)] \sin 2\pi\kappa y \, d\kappa \tag{25}$$

The values of γ_1 and γ_2 are given by Eq. (2.1-50).

We substitute Eqs. (24) and (25) into Eqs. (22) and (23):

$$\int_0^\infty [A_1(\kappa) + A_2(\kappa)] \sin 2\pi\kappa y \, d\kappa = 0 \tag{26}$$

$$\int_0^\infty [A_1(\kappa)\gamma_1 + A_2(\kappa)\gamma_2] \sin 2\pi\kappa y \, d\kappa = -2ae^{-y/L} \tag{27}$$

These two equations should be compared with Eqs. (2.1-54) and (2.1-55).

Using the Fourier sine transform pair of Eq. (2.1-56) we obtain from Eqs. (26) and (27):

$$A_1(\kappa) + A_2(\kappa) = 0 \tag{28}$$

$$A_1(\kappa)\gamma_1 + A_2(\kappa)\gamma_2 = 2g_s(\kappa) = -8a \int_0^\infty e^{-y/L} \sin 2\pi\kappa y \, dy \tag{29}$$

With the help of Eq. (2.1-59) we get:

$$A_1(\kappa)\gamma_1 + A_2(\kappa)\gamma_2 = -16a\pi\kappa/[s\sigma + (2\pi\kappa)^2] = -16a\pi\kappa/b^2 \tag{30}$$

The solution of Eqs. (28) and (30) yields with the help of Eq. (2.1–50):

$$A_1(\kappa) = -A_2(\kappa) = \frac{8\beta a}{\gamma_2 - \gamma_1} = -\frac{4\beta a}{(a^2 - b^2c^2)^{1/2}}, \qquad a^2 > b^2c^2$$
$$= j\frac{4\beta a}{(b^2c^2 - a^2)^{1/2}}, \qquad a^2 < b^2c^2 \tag{31}$$

We substitute Eqs. (31) and (2.1–50) into Eq. (24):

$$u(y, t) = -4ae^{-at}\Bigg(\int_0^K \frac{\exp[(a^2 - b^2c^2)^{1/2}t] - \exp[-(a^2 - b^2c^2)^{1/2}t]}{(a^2 - b^2c^2)^{1/2}} \times \frac{\beta \sin 2\pi\kappa y}{b^2}\, d\kappa$$
$$- j\int_K^\infty \frac{\exp[j(b^2c^2 - a^2)^{1/2}t] - \exp[-j(b^2c^2 - a^2)^{1/2}t]}{(b^2c^2 - a^2)^{1/2}} \times \frac{\beta \sin 2\pi\kappa y}{b^2}\, d\kappa\Bigg)$$
$$K = (2\pi)^{-1}(a^2/c^2 - s\sigma)^{1/2} \tag{32}$$

This equation may be rewritten with the help of hyperbolic and trigonometric functions:

$$u(y, t) = -\frac{4ae^{-at}}{\pi}\Bigg(\int_0^{2\pi K} \frac{\operatorname{sh}(a^2 - b^2c^2)^{1/2}t}{(a^2 - b^2c^2)^{1/2}}\, \frac{\beta \sin \beta y}{b^2}\, d\beta + \int_{2\pi K}^\infty \frac{\sin(b^2c^2 - a^2)^{1/2}t}{(b^2c^2 - a^2)^{1/2}}\, \frac{\beta \sin \beta y}{b^2}\, d\beta\Bigg) \tag{33}$$

The solution $E_E(y, t)$ follows from Eq. (17) with the help of Eqs. (1), (19), and (33):

$$E_E(y, t) = E_1[(1 - e^{-2at})e^{-y/L} + u(y, t)], \qquad E_1 = E_0 Q \tag{34}$$

We may now make the transition $s \to 0$. If we had made this transition before evaluating the integral of Eq. (29) we would have gotten no solution. We must carry the general value s as far as Eq. (34) or the calculation of the magnetic field strength $H_E(y, t)$ in Section 2.9 would fail. Equations (33) and (34) yield:

$$u(y, t) = -\frac{4\alpha e^{-\alpha t}}{\pi}\Bigg(\int_0^{Z\sigma/2} \frac{\operatorname{sh}(\alpha^2 - \beta^2c^2)^{1/2}t}{(\alpha^2 - \beta^2c^2)^{1/2}}\, \frac{\sin \beta y}{\beta}\, d\beta + \int_{Z\sigma/2}^\infty \frac{\sin(\beta^2c^2 - \alpha^2)^{1/2}t}{(\beta^2c^2 - \alpha^2)^{1/2}}\, \frac{\sin \beta y}{\beta}\, d\beta\Bigg) \tag{35}$$

$$E_E(y, t) = E_1[1 - e^{-2\alpha t} + u(t)], \qquad E_1 = E_0 Q \tag{36}$$

One may rewrite Eqs. (35) and (36) for $\sigma \neq 0$ in a more compact normalized form. Using the substitutions of Eqs. (2.1–78) and (2.1–79) we obtain:

$$E_E(\xi, \theta) = E_1[1 - e^{-2\theta} + u(\xi, \theta)], \qquad E_1 = E_0 Q \tag{37}$$

$$\begin{aligned} u(\xi, \theta) = -\frac{4}{\pi} e^{-\theta} \bigg(& \int_0^1 \frac{\operatorname{sh}(1-\eta^2)^{1/2}\theta}{(1-\eta^2)^{1/2}} \frac{\sin \xi\eta}{\eta} \, d\eta \\ & + \int_1^\infty \frac{\sin(\eta^2 - 1)^{1/2}\theta}{(\eta^2 -)^{1/2}} \frac{\sin \xi\eta}{\eta} \, d\eta \bigg) \end{aligned} \tag{38}$$

These two equations should be compared with Eq. (2.1–80). Again, all three parameters μ, ϵ, and σ have been made part of the normalized variables ξ and θ. As before, the integration variable η becomes infinite for $\sigma \to 0$, and the equations are thus not well suited for the study of the transition to the loss-free medium.

For a check of Eq. (36) consider the limit of a vanishing conductance σ which implies $\alpha = \sigma/2\epsilon \to 0$. With

$$\lim_{\alpha \to 0} Q = (1 - e^{-\Delta T/\tau})^{-1} = \tau/\Delta T = 1/2\alpha\, \Delta T \tag{39}$$

we obtain

$$\lim_{\alpha \to 0} u(y, t) = -\frac{2}{\pi c\, \Delta T} \int_0^\infty \frac{\sin \beta ct \sin \beta y}{\beta^2} \, d\beta \tag{40}$$

Furthermore we have

$$\lim_{\alpha \to 0} Q(1 - e^{-2\alpha t}) = t/\Delta T \tag{41}$$

The exciting exponential ramp function of Eq. (3) becomes a linear ramp function

$$\begin{aligned} E(0, t) &= 0 && \text{for} \quad t < 0 \\ &= E_0 t/\Delta T && \text{for} \quad t \geqq 0 \end{aligned} \tag{42}$$

due to the coupling of τ with a and α by Eq. (15). Equation (36) becomes:

$$E_E(y, t) = E_0 \left(\frac{t}{\Delta T} - \frac{2}{\pi c\, \Delta T} \int_0^\infty \frac{\sin \beta ct \sin \beta y}{\beta^2} \, d\beta \right) \tag{43}$$

In a table of integrals we find

$$\begin{aligned} \int_0^\infty \frac{\sin \beta ct \sin \beta y}{\beta^2} \, d\beta &= \frac{\pi ct}{2} && \text{for} \quad 0 \leqq ct \leqq y \\ &= \frac{\pi y}{2} && \text{for} \quad 0 \leqq y \leqq ct \end{aligned} \tag{44}$$

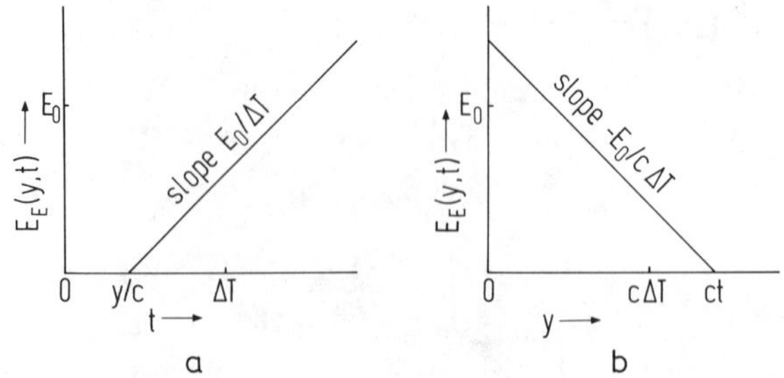

FIG. 2.3-1. Magnitude of the electric field strength according to Eq. (45) (a) at the point y as a function of time, and (b) at the time t as a function of the space coordinate. Note that ΔT has here a certain finite value as in Fig. 1.2-4a.

Substitution into Eq. (43) yields the expected result:

$$\begin{aligned} E_E(y, t) &= 0 && \text{for} \quad t \leqq y/c \\ &= E_0(t/\Delta T - y/c\,\Delta T) && \text{for} \quad t \geqq y/c \end{aligned} \tag{45}$$

The function $E_E(y, t)$ is shown at the point y in space in Fig. 2.3-1a, and for the point t in time in Fig. 2.3-1b.

For the limit $\sigma = 0$ we may calculate the magnetic field strength $H_E(y, t)$ from Eqs. (2.1-18) and (2.1-23), since one may put $s = 0$ in this case without causing any problem of convergence:

$$H_E(y, t) = -\epsilon \int \frac{\partial E_E}{\partial t} dy + H_y(t) \tag{46}$$

$$H_E(y, t) = -\mu^{-1} \int \frac{\partial E_E}{\partial y} dt + H_t(y) \tag{47}$$

From Eq. (45) follows:

$$\begin{aligned} \epsilon \int \frac{\partial E_E}{\partial t} dy &= 0 && \text{for} \quad t \leqq y/c \\ &= \epsilon E_0 y/\Delta T && \text{for} \quad t \geqq y/c \end{aligned} \tag{48}$$

$$\begin{aligned} \mu^{-1} \int \frac{\partial E_E}{\partial y} dt &= 0 && \text{for} \quad t \leqq y/c \\ &= -E_0 t/c\mu\,\Delta T && \text{for} \quad t \geqq y/c \end{aligned} \tag{49}$$

With $\epsilon = 1/Zc$ and $c\mu = Z$ we obtain from Eqs. (46) and (47):

$$\begin{aligned} H_E(y, t) &= H_y(t) && \text{for} \quad t \leqq y/c \\ &= -E_0 Z^{-1} y/c\,\Delta T + H_y(t) && \text{for} \quad t \geqq y/c \end{aligned} \tag{50}$$

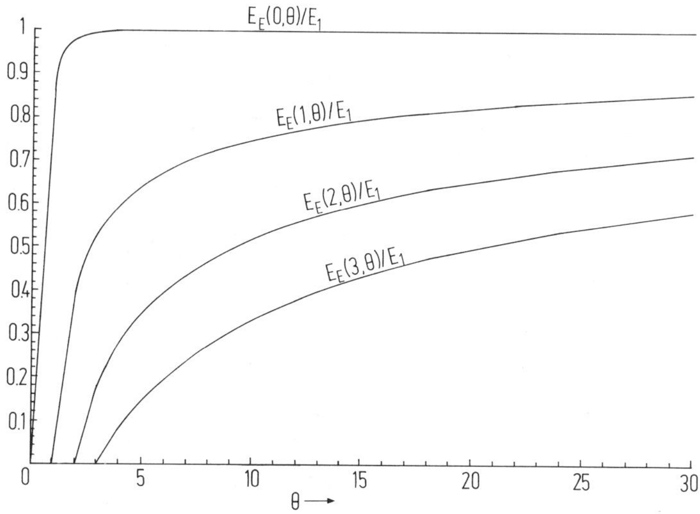

FIG. 2.3-2. Magnitude of the electric field strength due to excitation by an electric exponential ramp function for the locations $\xi = 0, 1, 2, 3$ in the time range $0 \leqq \theta \leqq 30$. This illustration is based on computer plots by M. Hussain of Kuwait University.

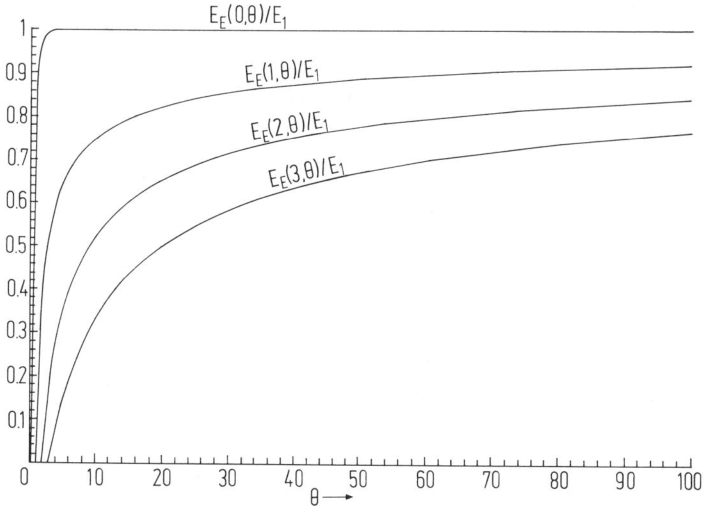

FIG. 2.3-3. Magnitude of the field strength due to excitation by an electric exponential ramp function for the locations $\xi = 0, 1, 2, 3$ in the time range $0 \leqq \theta \leqq 100$. This illustration is based on computer plots by M. Hussain of Kuwait University.

$$\begin{aligned} H_E(y,t) &= H_t(y) && \text{for} \quad t \leqq y/c \\ &= E_0 Z^{-1} t/\Delta T + H_t(y) && \text{for} \quad t \geqq y/c \end{aligned} \tag{51}$$

Since both equations must yield the same result for $H_E(y, t)$ we obtain the integration constants $H_y(t)$ and $H_t(y)$:

$$\begin{aligned} &H_y(t) = H_t(y) = H_{E0} && \text{for} \quad t \leqq y/c \\ &H_y(t) = E_0 Z^{-1} t/\Delta T, \qquad H_t(y) = -E_0 Z^{-1} y/c\,\Delta T && \text{for} \quad t \geqq y/c \end{aligned} \tag{52}$$

Since the constant H_{E0} must be zero to satisfy the initial condition of Eq. (5) we get for the magnetic field strength:

$$H_E(y,t) = S(t - y/c) E_0 Z^{-1}(t/\Delta T - y/c\,\Delta T) \tag{53}$$

Computer plots of the electric field strength $E_E(\xi, \theta)/E_1$ of Eq. (37) are shown for the locations $\xi = 0, 1, 2, 3$ in Fig. 2.3–2 in the time range $0 \leqq \theta \leqq 30$ and in Fig. 2.3–3 in the time range $0 \leqq \theta \leqq 100$.

2.4 Magnetic Field Due to Electric Step Function

In Section 2.1 we had derived the electric field strength caused by excitation with an electric step function at the plane $y = 0$. For $s \neq 0$ the electric field strength is given by Eq. (2.1–65):

$$E_E(y,t) = E_0[e^{-y/L} + w(y,t)], \qquad L = (s\sigma)^{-1/2} \tag{1}$$

The magnetic field strength is obtained by means of Eqs. (2.1–18) and (2.1–23):

$$H_E(y,t) = -\int \left(\epsilon \frac{\partial E_E}{\partial t} + \sigma E_E \right) dy + H_y(t) \tag{2}$$

$$H_E(y,t) = -\mu^{-1} e^{-st/\mu} \int \frac{\partial E_E}{\partial y} e^{st/\mu}\, dt + H_t(y) e^{-st/\mu} \tag{3}$$

Introduction of Eq. (1) into Eq. (3) yields with the help of Eq. (2.1–52):

$$\begin{aligned} H_E(y,t) &= -E_0 \mu^{-1} e^{-st/\mu} \int \left(-(s\sigma)^{1/2} e^{-y/L} + \frac{\partial w}{\partial y} \right) e^{st/\mu}\, dt + H_t(y) e^{-st/\mu} \\ &= -E_0 \mu^{-1} e^{-st/\mu} \Bigg(-\mu(\sigma/s)^{1/2} e^{-y/L + st/\mu} \\ &\quad + 2\pi \int_0^\infty \{A_1(\kappa)\kappa(\gamma_1 + s/\mu)^{-1} \exp[(\gamma_1 + s/\mu)t] \\ &\quad + A_2(\kappa)\kappa(\gamma_2 + s/\mu)^{-1} \exp[(\gamma_2 + s/\mu)t]\} \cos 2\pi\kappa y\, d\kappa \Bigg) + H_t(y) e^{-st/\mu} \end{aligned}$$

$$= Z^{-1}E_0\left[Z\left(\frac{\sigma}{s}\right)^{1/2} e^{-y/L} - \frac{1}{2\pi}\int_0^\infty [A_1(\kappa)(\gamma_1 + s/\mu)^{-1}\exp(\gamma_1 t)\right.$$

$$\left. + A_2(\kappa)(\gamma_2 + s/\mu)^{-1}\exp(\gamma_2 t)]\beta c\cos\beta y\, d\beta\right] + H_t(y)e^{-st/\mu} \qquad (4)$$

With the help of the identities

$$2a - s/\mu = 2\alpha, \qquad b^2c^2 - 2as/\mu + s^2/\mu^2 = c^2\beta^2,$$
$$2a^2 - b^2c^2 - as/\mu = 2a\alpha - b^2c^2 \qquad (5)$$

one obtains from Eqs. (2.1–62) and (2.1–50):

$$\frac{A_1(\kappa)\beta c}{\gamma_1 + s/\mu} = -\frac{4\beta^2\gamma_2 c}{b^2(\gamma_2 - \gamma_1)(\gamma_1 + s/\mu)} = \frac{4\alpha}{b^2c}(1 + q_s), \qquad a^2 > b^2c^2$$
$$= \frac{4\alpha}{b^2c}(1 - jq_s'), \qquad a^2 < b^2c^2$$

$$\frac{A_2(\kappa)\beta c}{\gamma_2 + s/\mu} = -\frac{4\beta^2\gamma_1 c}{b^2(\gamma_1 - \gamma_2)(\gamma_2 + s/\mu)} = \frac{4\alpha}{b^2c}(1 - q_s), \qquad a^2 > b^2c^2 \qquad (6)$$
$$= \frac{4\alpha}{b^2c}(1 + jq_s'), \qquad a^2 < b^2c^2$$

$$q_s = \frac{2a\alpha - b^2c^2}{2\alpha(a^2 - b^2c^2)^{1/2}}, \qquad q_s' = \frac{2a\alpha - b^2c^2}{2\alpha(b^2c^2 - a^2)^{1/2}}$$

Insertion into Eq. (4) yields:

$$H_E(y, t) = Z^{-1}E_0\left[Z\left(\frac{\sigma}{s}\right)^{1/2}\exp[-(s\sigma)^{1/2}y]\right.$$
$$- 2Z\sigma e^{-at}\left(\int_0^K \{(1 + q_s)\exp[(a^2 - b^2c^2)^{1/2}t]\right.$$
$$+ (1 - q_s)\exp[-(a^2 - b^2c^2)^{1/2}t]\}\frac{\cos\beta y}{b^2}\, d\kappa$$
$$+ \int_K^\infty \{(1 - jq_s')\exp[j(b^2c^2 - a^2)^{1/2}t]$$
$$\left.\left. + (1 + jq_s')\exp[-j(b^2c^2 - a^2)^{1/2}t]\}\frac{\cos\beta y}{b^2}\, d\kappa\right)\right] + H_t(y)e^{-st/\mu}$$

$$K = (2\pi)^{-1}(a^2/c^2 - s\sigma)^{1/2} \qquad (7)$$

Using hyperbolic and trigonometric functions we obtain:

$$H_E(y,t) = Z^{-1}E_0\Big\{Z\left(\frac{\sigma}{s}\right)^{1/2} \exp[-(s\sigma)^{1/2}y]$$
$$-\frac{2Z\sigma}{\pi}e^{-at}\Big[\int_0^{2\pi K}\Big(\mathrm{ch}(a^2-b^2c^2)^{1/2}t$$
$$+\frac{(2a\alpha-b^2c^2)\mathrm{sh}(a^2-b^2c^2)^{1/2}t}{2\alpha(a^2-b^2c^2)^{1/2}}\Big)\frac{\cos\beta y}{b^2}d\beta$$
$$+\int_{2\pi K}^{\infty}\Big(\cos(b^2c^2-a^2)^{1/2}t$$
$$+\frac{(2a\alpha-b^2c^2)\sin(b^2c^2-a^2)^{1/2}t}{2\alpha(b^2c^2-a^2)^{1/2}}\Big)\frac{\cos\beta y}{b^2}d\beta\Big]\Big\}$$
$$+H_t(y)e^{-st/\mu} \tag{8}$$

We make now the transition $s \to 0$ for $\sigma > 0$. The first term yields:

$$\lim_{s\to 0} Z(\sigma/s)^{1/2}\exp[-(s\sigma)^{1/2}y] = Z[(\sigma/s)^{1/2} - \sigma y] \tag{9}$$

In order to calculate the first integral of Eq. (8) we observe that the denominator $b^2 = \beta^2 + s\sigma$ causes a problem for $s = 0$, but the denominator $2\alpha(a^2 - b^2c^2)^{1/2}$ does not since we get

$$\lim_{s\to 0}\frac{2a\alpha - b^2c^2}{2\alpha(a^2-b^2c^2)^{1/2}}\mathrm{sh}(a^2-b^2c^2)^{1/2}t$$
$$= \frac{2\alpha^2-\beta^2c^2}{2\alpha(\alpha^2-\beta^2c^2)^{1/2}}\mathrm{sh}(\alpha^2-\beta^2c^2)^{1/2}t \tag{10}$$

which yields $\alpha t/2$ for $\beta^2c^2 = \alpha^2$. Hence, the first integral in Eq. (8) becomes for small values of s:

$$I_{E1}(y,t) = \frac{2Z\sigma}{\pi}e^{-\alpha t}\int_0^{Z\sigma/2}\Big(\mathrm{ch}(\alpha^2-\beta^2c^2)^{1/2}t$$
$$+\frac{(2\alpha^2-\beta^2c^2)\mathrm{sh}(\alpha^2-\beta^2c^2)^{1/2}t}{2\alpha(\alpha^2-\beta^2c^2)^{1/2}}\Big)\frac{\cos\beta y}{\beta^2+s\sigma}d\beta \tag{11}$$

This integral may be split into two parts by means of the identity

$$\cos\beta y = 1 - 2\sin^2(\beta y/2) \tag{12}$$

to yield:

$$I_{E1}(y,t)=\frac{2Z\sigma}{\pi}e^{-\alpha t}\Bigg[\int_0^{Z\sigma/2}\Bigg(\mathrm{ch}(\alpha^2-\beta^2c^2)^{1/2}t + \frac{(2\alpha^2-\beta^2c^2)\mathrm{sh}(\alpha^2-\beta^2c^2)^{1/2}t}{2\alpha(\alpha^2-\beta^2c^2)^{1/2}}\Bigg)\frac{d\beta}{\beta^2+s\sigma} - 2\int_0^{Z\sigma/2}\Bigg(\mathrm{ch}(\alpha^2-\beta^2c^2)^{1/2}t + \frac{(2\alpha^2-\beta^2c^2)\mathrm{sh}(\alpha^2-\beta^2c^2)^{1/2}t}{2\alpha(\alpha^2-\beta^2c^2)^{1/2}}\Bigg)\frac{\sin^2(\beta y/2)}{\beta^2+s\sigma}d\beta\Bigg] \tag{13}$$

We split this integral into four components

$$I_{E1}(y,t)=-I_{11}(y,t)+I_{12}(t)+I_{13}(t)-I_{14}(t) \tag{14}$$

where $I_{11}(y,t)$ is the second integral in Eq. (13), which remains finite for $s=0$,

$$I_{11}(y,t)=\frac{Z\sigma}{\pi}e^{-\alpha t}\int_0^{Z\sigma/2}\Bigg(\mathrm{ch}(\alpha^2-\beta^2c^2)^{1/2}t + \frac{(2\alpha^2-\beta^2c^2)\mathrm{sh}(\alpha^2-\beta^2c^2)^{1/2}t}{2\alpha(\alpha^2-\beta^2c^2)^{1/2}}\Bigg) \times\left(\frac{\sin(\beta y/2)}{\beta/2}\right)^2 d\beta \tag{15}$$

and can be evaluated by computer. The other three components of Eq. (14) are:

$$I_{12}(t)=\frac{2Z\sigma}{\pi}e^{-\alpha t}\int_0^{Z\sigma/2}\frac{\mathrm{ch}(\alpha^2-\beta^2c^2)^{1/2}t}{\beta^2+s\sigma}d\beta \tag{16}$$

$$I_{13}(t)=\frac{2Z\sigma\alpha}{\pi}e^{-\alpha t}\int_0^{Z\sigma/2}\frac{\mathrm{sh}(\alpha^2-\beta^2c^2)^{1/2}t}{(\alpha^2-\beta^2c^2)^{1/2}(\beta^2+s\sigma)}d\beta \tag{17}$$

$$I_{14}(t)=\frac{2c}{\pi}e^{-\alpha t}\int_0^{Z\sigma/2}\frac{\beta^2\,\mathrm{sh}(\alpha^2-\beta^2c^2)^{1/2}t}{(\alpha^2-\beta^2c^2)^{1/2}(\beta^2+s\sigma)}d\beta \to \frac{2c}{\pi}e^{-\alpha t}\int_0^{Z\sigma/2}\frac{\mathrm{sh}(\alpha^2-\beta^2c^2)^{1/2}}{(\alpha^2-\beta^2c^2)^{1/2}}d\beta \tag{18}$$

The integral $I_{14}(t)$ remains finite for $s\to 0$, but $I_{12}(t)$ and $I_{13}(t)$ pose problems when β approaches zero. Hence, we divide the integration interval

$0 \leqq \beta \leqq Z\sigma/2$ into two subintervals $0 \leqq \beta \leqq d$ and $d \leqq \beta \leqq Z\sigma/2$, where d is a finite constant small compared with $Z\sigma/2 = \alpha/c$:

$$I_{12}(t) = \frac{2Z\sigma}{\pi} e^{-\alpha t} \left(\int_0^d \frac{\mathrm{ch}(\alpha^2 - \beta^2 c^2)^{1/2} t}{\beta^2 + s\sigma} d\beta + \int_d^{Z\sigma/2} \frac{\mathrm{ch}(\alpha^2 - \beta^2 c^2)^{1/2} t}{\beta^2} d\beta \right) \tag{19}$$

$$I_{13}(t) = \frac{2Z\alpha\sigma}{\pi} e^{-\alpha t} \left(\int_0^d \frac{\mathrm{sh}(\alpha^2 - \beta^2 c^2)^{1/2} t}{(\alpha^2 - \beta^2 c^2)^{1/2} (\beta^2 + s\sigma)} d\beta + \int_d^{Z\sigma/2} \frac{\mathrm{sh}(\alpha^2 - \beta^2 c^2)^{1/2} t}{(\alpha^2 - \beta^2 c^2)^{1/2} \beta^2} d\beta \right) \tag{20}$$

In the interval $d \leqq \beta \leqq Z\sigma/2$ the integrals are finite and they can be evaluated numerically by computer. In the interval $0 \leqq \beta \leqq d$ the variable β is small compared with α/c, and we can resort to series expansions:

$$\begin{aligned}
(\alpha^2 - \beta^2 c^2)^{1/2} &\doteqdot \alpha(1 - \beta^2 c^2/2\alpha^2), \\
(\alpha^2 - \beta^2 c^2)^{-1/2} &\doteqdot \alpha^{-1}(1 + \beta^2 c^2/2\alpha^2) \\
\mathrm{ch}(\alpha^2 - \beta^2 c^2)^{1/2} t &\doteqdot \mathrm{ch}[(1 - \beta^2 c^2/2\alpha^2)\alpha t] \\
&= \mathrm{ch}\,\alpha t\, \mathrm{ch}(\beta^2 c^2 t/2\alpha) - \mathrm{sh}\,\alpha t\, \mathrm{sh}(\beta^2 c^2 t/2\alpha) \qquad (21) \\
\mathrm{sh}(\alpha^2 - \beta^2 c^2)^{1/2} t &\doteqdot \mathrm{sh}[(1 - \beta^2 c^2/2\alpha^2)\alpha t] \\
&= \mathrm{sh}\,\alpha t\, \mathrm{ch}(\beta^2 c^2 t/2\alpha) - \mathrm{ch}\,\alpha t\, \mathrm{sh}(\beta^2 c^2 t/2\alpha)
\end{aligned}$$

The two problem integrals in Eqs. (19) and (20) become:

$$\begin{aligned}
I_{12p}(t) &= \frac{2Z\sigma}{\pi} e^{-\alpha t} \int_0^d \frac{\mathrm{ch}(\alpha^2 - \beta^2 c^2)^{1/2} t}{\beta^2 + s\sigma} d\beta \\
&= \frac{2Z\sigma}{\pi} e^{-\alpha t} \left(\mathrm{ch}\,\alpha t \int_0^d \frac{\mathrm{ch}(\beta^2 c^2 t/2\alpha)}{\beta^2 + s\sigma} d\beta - \mathrm{sh}\,\alpha t \int_0^d \frac{\mathrm{sh}(\beta^2 c^2 t/2\alpha) d\beta}{\beta^2 + s\sigma} \right) \qquad (22)
\end{aligned}$$

$$\begin{aligned}
I_{13p}(t) &= \frac{2Z\alpha\sigma}{\pi} e^{-\alpha t} \int_0^d \frac{\mathrm{sh}(\alpha^2 - \beta^2 c^2)^{1/2} t}{(\alpha^2 - \beta^2 c^2)^{1/2} (\beta^2 + s\sigma)} d\beta \\
&= \frac{2Z\sigma}{\pi} e^{-\alpha t} \int_0^d \left(1 + \frac{\beta^2 c^2}{2\alpha^2}\right) \frac{\mathrm{sh}(\alpha^2 - \beta^2 c^2)^{1/2} t}{\beta^2 + s\sigma} d\beta
\end{aligned}$$

$$= \frac{2Z\sigma}{\pi} e^{-\alpha t} \left(\operatorname{sh} \alpha t \int_0^d \frac{\operatorname{ch}(\beta^2 c^2 t/2\alpha) d\beta}{\beta^2 + s\sigma} \right.$$
$$- \operatorname{ch} \alpha t \int_0^d \frac{\operatorname{sh}(\beta^2 c^2 t/2\alpha) d\beta}{\beta^2 + s\sigma}$$
$$\left. + \frac{c^2}{2\alpha^2} \int_0^d \frac{\beta^2 \operatorname{sh}(\alpha^2 - \beta^2 c^2)^{1/2} t}{\beta^2 + s\sigma} d\beta \right) \tag{23}$$

For two of these integrals we can make the transition $s \to 0$ without creating a pole:

$$\lim_{s \to 0} \int_0^d \frac{\beta^2 \operatorname{sh}(\alpha^2 - \beta^2 c^2)^{1/2} t}{\beta^2 + s\sigma} d\beta = \int_0^d \operatorname{sh}(\alpha^2 - \beta^2 c^2)^{1/2} t \, d\beta \tag{24}$$

$$\lim_{s \to 0} \int_0^d \frac{\operatorname{sh}(\beta^2 c^2 t/2\alpha) d\beta}{\beta^2 + s\sigma} = \int_0^d \frac{\operatorname{sh}(\beta^2 c^2 t/2\alpha) d\beta}{\beta^2} \tag{25}$$

The problem has been reduced to just one integral. Using the relation

$$\operatorname{ch} x = 2 \operatorname{ch}^2(x/2) - 1 = 1 + 2 \operatorname{sh}^2(x/2) \tag{26}$$

we obtain

$$\int_0^d \frac{\operatorname{ch}(\beta^2 c^2 t/2\alpha) d\beta}{\beta^2 + s\sigma} = \int_0^d \frac{d\beta}{\beta^2 + s\sigma} + 2 \int_0^d \frac{\operatorname{sh}^2(\beta^2 c^2 t/4\alpha) d\beta}{\beta^2 + s\sigma} \tag{27}$$

The second integral has again no pole for $s = 0$,

$$\lim_{s \to 0} \int_0^d \frac{\operatorname{sh}^2(\beta^2 c^2 t/4\alpha) d\beta}{\beta^2 + s\sigma} = \int_0^d \frac{\operatorname{sh}^2(\beta^2 c^2 t/4\alpha) d\beta}{\beta^2} \tag{28}$$

while the first is tabulated:

$$\int_0^d \frac{d\beta}{\beta^2 + s\sigma} = \frac{1}{(s\sigma)^{1/2}} \operatorname{arctg} \frac{\beta}{(s\sigma)^{1/2}} \Bigg|_0^d \tag{29}$$

With the series expansion

$$\operatorname{arctg} x = \pi/2 - 1/x + 1/3x^2 - \cdots, \qquad x^2 \geqq 1 \tag{30}$$

we get:

$$\lim_{s \to 0} \int_0^d \frac{d\beta}{\beta^2 + s\sigma} = \frac{\pi}{2(s\,\sigma)^{1/2}} - \frac{1}{d} \tag{31}$$

Converting ch αt and sh αt to exponential functions we get for $I_{12p}(t)$ and

$I_{13p}(t)$:

$$I_{12p}(t) = \frac{Z\sigma}{\pi}\left[(1 + e^{-2\alpha t})\left(\frac{\pi}{2(s\sigma)^{1/2}} - \frac{1}{d} + 2\int_0^d \frac{\mathrm{sh}^2(\beta^2 c^2 t/4\alpha)d\beta}{\beta^2}\right)\right.$$
$$\left. - (1 - e^{-2\alpha t})\int_0^d \frac{\mathrm{sh}(\beta^2 c^2 t/2\alpha)d\beta}{\beta^2}\right] \tag{32}$$

$$I_{13p}(t) = \frac{Z\sigma}{\pi}\left[(1 - e^{-2\alpha t})\left(\frac{\pi}{2(s\sigma)^{1/2}} - \frac{1}{d} + 2\int_0^d \frac{\mathrm{sh}^2(\beta^2 c^2 t/4\alpha)d\beta}{\beta^2}\right)\right.$$
$$- (1 + e^{-2\alpha t})\int_0^d \frac{\mathrm{sh}(\beta^2 c^2 t/2\alpha)d\beta}{\beta^2}$$
$$\left. + \frac{c^2}{2\alpha^2} e^{-\alpha t}\int_0^d \mathrm{sh}[(1 - \beta^2 c^2/\alpha^2)^{1/2}\alpha t]d\beta\right] \tag{33}$$

The last integral of $I_{13p}(t)$ becomes insignificant for small values of d, and we may simplify $I_{13p}(t)$:

$$I_{13p}(t) = \frac{Z\sigma}{\pi}\left[(1 - e^{-2\alpha t})\left(\frac{\pi}{2(s\sigma)^{1/2}} - \frac{1}{d} + 2\int_0^d \frac{\mathrm{sh}^2(\beta^2 c^2 t/4\alpha)d\beta}{\beta^2}\right)\right.$$
$$\left. - (1 + e^{-2\alpha t})\int_0^d \frac{\mathrm{sh}(\beta^2 c^2 t/2\alpha)d\beta}{\beta^2}\right] \tag{34}$$

The integrals I_{12p} and I_{13p} become equal when t approaches infinity.

Integrals $I_{12}(t)$ and $I_{13}(t)$ of Eqs. (19) and (20) assume the form:

$$I_{12}(t) = \frac{Z\sigma}{\pi}\left[(1 + e^{-2\alpha t})\left(\frac{\pi}{2(s\sigma)^{1/2}} - \frac{1}{d} + 2\int_0^d \frac{\mathrm{sh}^2(\beta^2 c^2 t/4\alpha)d\beta}{\beta^2}\right)\right.$$
$$- (1 - e^{-2\alpha t})\int_0^d \frac{\mathrm{sh}(\beta^2 c^2 t/2\alpha)d\beta}{\beta^2}$$
$$\left. + 2e^{-\alpha t}\int_d^{Z\sigma/2} \frac{\mathrm{ch}(\alpha^2 - \beta^2 c^2)^{1/2} t}{\beta^2} d\beta\right] \tag{35}$$

$$I_{13}(t) = \frac{Z\sigma}{\pi}\left[(1 - e^{-2\alpha t})\left(\frac{\pi}{2(s\sigma)^{1/2}} - \frac{1}{d} + 2\int_0^d \frac{\mathrm{sh}^2(\beta^2 c^2 t/4\alpha)d\beta}{\beta^2}\right)\right.$$
$$- (1 + e^{-2\alpha t})\int_0^d \frac{\mathrm{sh}(\beta^2 c^2 t/2\alpha)d\beta}{\beta^2}$$
$$\left. + 2\alpha e^{-\alpha t}\int_d^{Z\sigma/2} \frac{\mathrm{sh}(\alpha^2 - \beta^2 c^2)^{1/2} t}{(\alpha^2 - \beta^2 c^2)^{1/2}\beta^2} d\beta\right] \tag{36}$$

The sum of these two integrals minus integrals $I_{14}(t)$ of Eq. (18) and $I_{11}(y, t)$ of Eq. (15) yields integral $I_{E1}(y, t)$ of Eq. (14). We sum first $I_{12}(t)$ and $I_{13}(t)$:

$$
\begin{aligned}
I_{12}(t) + I_{13}(t) = Z\left(\frac{\sigma}{s}\right)^{1/2} + \frac{2Z\sigma}{\pi}\Bigg[e^{-\alpha t}\int_d^{Z\sigma/2}\Bigg(&\mathrm{ch}(\alpha^2-\beta^2c^2)^{1/2}t \\
&+ \frac{\alpha\,\mathrm{sh}(\alpha^2-\beta^2c^2)^{1/2}t}{(\alpha^2-\beta^2c^2)^{1/2}}\Bigg)\frac{d\beta}{\beta^2} \\
&- \frac{1}{d} + \int_0^d [2\,\mathrm{sh}^2(\beta^2c^2t/4\alpha) - \mathrm{sh}(\beta^2c^2t/2\alpha)]\frac{d\beta}{\beta^2}\Bigg] \qquad (37)
\end{aligned}
$$

With the relation $2\,\mathrm{sh}^2(x/2) = \mathrm{ch}\,x - 1$ we may rewrite the last term

$$
\begin{aligned}
2\,\mathrm{sh}^2(\beta^2c^2t/4\alpha) - \mathrm{sh}(\beta^2c^2t/2\alpha) &= \mathrm{ch}(\beta^2c^2t/2\alpha) - \mathrm{sh}(\beta^2c^2t/2\alpha) - 1 \\
&= \exp(-\beta^2c^2t/2\alpha) - 1
\end{aligned}
$$

and obtain for $I_{E1}(y, t)$:

$$
\begin{aligned}
I_{E1}(y, t) = Z\left(\frac{\sigma}{s}\right)^{1/2} &+ \frac{2Z\sigma}{\pi}\Bigg\{e^{-\alpha t}\int_d^{Z\sigma/2}\Bigg[\mathrm{ch}(\alpha^2-\beta^2c^2)^{1/2}t \\
&+ \frac{\alpha\,\mathrm{sh}(\alpha^2-\beta^2c^2)^{1/2}t}{(\alpha^2-\beta^2c^2)^{1/2}}\Bigg]\frac{d\beta}{\beta^2} - \frac{1}{d} + \int_0^d [\exp(-\beta^2c^2t/2\alpha) - 1]\frac{d\beta}{\beta^2}\Bigg\} \\
&- \frac{2c}{\pi}e^{-\alpha t}\int_0^{Z\sigma/2}\frac{\mathrm{sh}(\alpha^2-\beta^2c^2)^{1/2}t}{(\alpha^2-\beta^2c^2)^{1/2}}\,d\beta \\
&- \frac{Z\sigma}{\pi}e^{-\alpha t}\int_0^{Z\sigma/2}\Bigg[\mathrm{ch}(\alpha^2-\beta^2c^2)^{1/2}t \\
&+ \frac{(2\alpha^2-\beta^2c^2)\mathrm{sh}(\alpha^2-\beta^2c^2)^{1/2}t}{2\alpha(\alpha^2-\beta^2c^2)^{1/2}}\Bigg]\left(\frac{\sin(\beta y/2)}{\beta/2}\right)^2 d\beta, \\
d \ll Z\sigma/2 = \alpha/c \qquad\qquad & \qquad (38)
\end{aligned}
$$

For finite values of t and sufficiently small values of d one may write:

$$
\int_0^d [\exp(-\beta^2c^2t/2\alpha) - 1]\frac{d\beta}{\beta^2} = -\frac{c^2t}{2\alpha}\int_0^d d\beta = -\frac{c^2t}{2\alpha}d
$$

This integral is thus negligible for finite values of t and small values of d.

The function $I_{E1}(y, t)$ without the term $Z(\sigma/s)^{1/2}$ is denoted $-I'_{E1}(y, t)$:

$$
-I'_{E1}(y, t) = I_{E1}(y, t) - Z(\sigma/s)^{1/2} \qquad (39)
$$

We turn to the second integral in Eq. (8). It remains finite for $s \to 0$ if the electric conductivity σ is larger than zero:

$$I_{E2}(y,t)=\frac{2Z\sigma}{\pi}e^{-\alpha t}\int_{Z\sigma/2}^{\infty}\left(\cos(\beta^2c^2-\alpha^2)^{1/2}t\right.$$
$$\left.+\frac{(2\alpha^2-\beta^2c^2)\sin(\beta^2c^2-\alpha^2)^{1/2}t}{2\alpha(\beta^2c^2-\alpha^2)^{1/2}}\right)\frac{\cos\beta y}{\beta^2}\,d\beta \qquad (40)$$

We may now rewrite $H_E(y, t)$ of Eq. (8):

$$H_E(y,t)=Z^{-1}E_0[Z(\sigma/s)^{1/2}-Z\sigma y+I'_{E1}(y,t)-Z(\sigma/s)^{1/2}$$
$$-I_{E2}(y,t)]+H_t(y)$$
$$=Z^{-1}E_0[-Z\sigma y+I'_{E1}(y,t)-I_{E2}(y,t)]+H_t(y) \qquad (41)$$

We see that the two terms $Z(\sigma/s)^{1/2}$ and $-Z(\sigma/s)^{1/2}$ in Eq. (41) cancel, and that $H_E(y, t)$ thus remains finite for $s\to 0$.

We turn to the integral of Eq. (2). Insertion of Eq. (1) yields:

$$H_E(y,t)=-E_0\left(\epsilon\int\frac{\partial w}{\partial t}dy+\sigma\int(e^{-y/L}+w)dy\right)+H_y(t)$$
$$=-E_0\left(\epsilon\int\frac{\partial w}{\partial t}dy-(\sigma/s)^{1/2}e^{-y/L}+\sigma\int w\,dy\right)+H_y(t) \qquad (42)$$

Using Eqs. (2.1-52) and (2.1-53) we get:

$$H_E(y,t)=E_0\left[\left(\frac{\sigma}{s}\right)^{1/2}e^{-y/L}+\frac{\epsilon}{2\pi}\int_0^\infty[A_1(\kappa)\gamma_1\exp(\gamma_1t)\right.$$
$$+A_2(\kappa)\gamma_2\exp(\gamma_2t)]\frac{\cos\beta y}{\beta}d\beta$$
$$\left.+\frac{\sigma}{2\pi}\int_0^\infty[A_1(\kappa)\exp(\gamma_1t)+A_2(\kappa)\exp(\gamma_2t)]\frac{\cos\beta y}{\beta}d\beta\right]+H_y(t)$$
$$=Z^{-1}E_0\left[Z\left(\frac{\sigma}{S}\right)^{1/2}e^{-y/L}-\frac{1}{2\pi}\int_0^\infty[-A_1(\kappa)(\gamma_1+2\alpha)\exp(\gamma_1t)\right.$$
$$\left.-A_2(\kappa)(\gamma_2+2\alpha)\exp(\gamma_2t)]\frac{\cos\beta y}{c\beta}d\beta\right]+H_y(t) \qquad (43)$$

From Eqs. (2.1-62) and (2.1-50) follows, with q_s and q_s' defined in Eq. (6):

$$-\frac{A_1(\kappa)\gamma_1+2\alpha)}{\beta c}=\frac{4\gamma_2(\gamma_1+2\alpha)}{b^2(\gamma_2-\gamma_1)c}=\frac{4\alpha}{b^2c}(1+q_s),\qquad a^2>b^2c^2$$

$$= \frac{4\alpha}{b^2 c}(1 - jq_s'), \qquad a^2 < b^2c^2$$

$$-\frac{A_2(\kappa)(\gamma_2 + 2\alpha)}{\beta c} = \frac{4\gamma_1(\gamma_2 + 2\alpha)}{b^2(\gamma_1 - \gamma_2)c} = \frac{4\alpha}{b^2 c}(1 - q_s), \qquad a^2 > b^2c^2 \tag{44}$$

$$= \frac{4\alpha}{b^2 c}(1 + jq_s'), \qquad a^2 < b^2c^2$$

The comparison of Eqs. (4) and (6) with Eqs. (43) and (44) shows that they are equal for

$$H_t(y)e^{-st/\mu} = H_y(t) = H_{E0}e^{-st/\mu} \tag{45}$$

The initial condition of Eq. (2.1-30) requires $H_{E0} = 0$.

We conclude from this result that Maxwell's equations with a magnetic conductance term $-s\mathbf{H}$ added have a solution for any value of s, including $s = 0$, if one uses excitation with an electric step function. Since the step function permits the representation of any transient of physical interest according to Section 1.3, we can generalize this result to any electric transient that is zero for $t < 0$ and has a physically realizable time variation for $t \geqq 0$. One of our tasks will be to show that such a result can also be obtained for excitation by a magnetic step function, since an electromagnetic wave is excited by either an electric field strength, or a magnetic field strength, or a combination of both.

We rewrite $H_E(y, t)$ of Eq. (41) in terms of the normalized variables ξ and θ, using the definitions of Eqs. (2.1-78) and (2.1-79), and dropping the constant $H_t(y) = H_{E0} = 0$:

$$H_E(\xi, \theta) = Z^{-1}E_0[-2\xi + I_{E1}'(\xi, \theta) - I_{E2}(\xi, \theta)] \tag{46}$$

$$\begin{aligned} I_{E1}'(\xi, \theta) = \frac{2}{\pi}\Bigg\{ 2\Bigg[\frac{1}{\delta} - e^{-\theta}\int_\delta^1 \Bigg(\mathrm{ch}(1-\eta^2)^{1/2}\theta \\ + \frac{\mathrm{sh}(1-\eta^2)^{1/2}\theta}{(1-\eta^2)^{1/2}} \Bigg) \frac{d\eta}{\eta^2} \\ - \int_0^\delta [\exp(-\eta^2\theta/2) - 1]\frac{d\eta}{\eta^2} \Bigg] + e^{-\theta}\Bigg[\int_0^1 \frac{\mathrm{sh}(1-\eta^2)^{1/2}\theta}{(1-\eta^2)^{1/2}}\, d\eta \\ + \int_0^1 \Bigg(\mathrm{ch}(1-\eta^2)^{1/2}\theta + \frac{(2-\eta^2)\mathrm{sh}(1-\eta^2)^{1/2}\theta}{2(1-\eta^2)^{1/2}} \Bigg) \\ \times \left(\frac{\sin(\xi\eta/2)}{\eta/2} \right)^2 d\eta \Bigg] \Bigg\} \end{aligned} \tag{47}$$

$$I_{E2}(\xi, \theta) = \frac{4}{\pi} e^{-\theta} \int_1^\infty \left(\cos(\eta^2 - 1)^{1/2}\theta + \frac{(2 - \eta^2)\sin(\eta^2 - 1)^{1/2}\theta}{2(\eta^2 - 1)^{1/2}} \right) \frac{\cos \xi\eta}{\eta^2} \, d\eta \tag{48}$$

$$\delta = 2d/Z\sigma \ll 1$$

Plots of $H_E(\xi, \theta)$ are shown in Fig. 2.4-1 for $\xi = \alpha y/c = 0, 1, 2, 3$ and $\theta = \alpha t$ in the range $0 \leqq \theta \leqq 26$. The functions are zero for $\theta < \xi$ or $t < y/c$, and they have a jump at $\theta = \xi$ that decreases rapidly with increasing values of ξ. For large values of θ the functions increase linearly with θ. The reason for this increase is the term $\sigma\mathbf{E}$ in Maxwell's equation

$$\text{curl } \mathbf{H} = \epsilon \, \partial\mathbf{E}/\partial t + \sigma\mathbf{E} \tag{49}$$

For explanation consider the circuit in Fig. 2.4-2. The voltage step function $VS(t)$ is applied to a chain of resistors connected by switches that are closed at the times $t = \Delta T, 2\,\Delta T, \cdots$. The current $i = V/R$ flows at the time $t = 0$, while the current $i = V(1 + k)/R$ flows at the time $t = k\,\Delta T$. Hence, if the step voltage $VS(t)$ is applied at the time $t = 0$, we get first a jump of the current from 0 to V/R, and then a current linearly increasing with time for integer multiples of ΔT. In the case of the electromagnetic wave, the wave

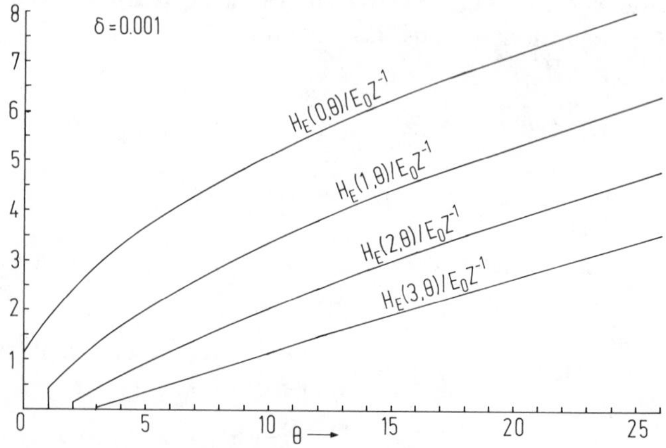

FIG. 2.4-1. Magnitude H_E of the magnetic field strength according to Eq. (46) caused by an electric step function excitation at the plane $\xi = 0$ of a lossy medium. The value $\delta = 0.001$ was used for the computation of these plots. The illustration is based on computer plots by R. Boules, University of Alexandria, Egypt.

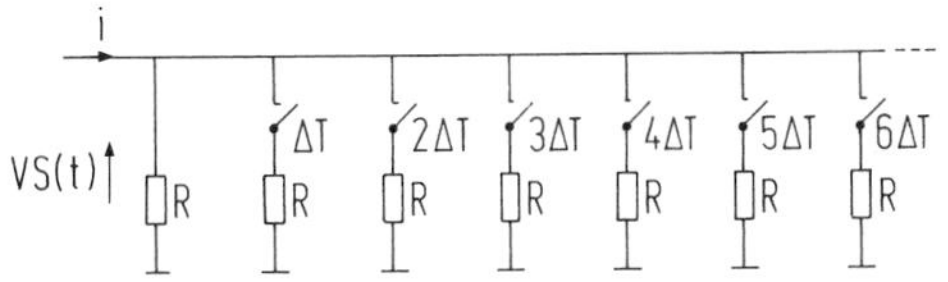

FIG. 2.4-2. Electric circuit for the elucidation of the plots for the magnetic field strength in Fig. 2.4-1.

exists in a volume of the lossy medium that increases with time like ct. The increasing losses call for an increasing power supplied to the medium. Since the electric field strength is kept constant, the magnetic field strength increases to supply this power.

One may increase the similarity between the circuit of Fig. 2.4-2 and wave propagation in a lossy medium still more by making the distance between the resistors equal to $c\,\Delta T$ and the spatial dimensions of the resistors small compared with $c\,\Delta T$. The switches in Fig. 2.4-2 may then be left out.

2.5 Magnetic Field Without Magnetic Conductivity

When we calculated the electric field strengths due to either an electric step function or an electric exponential ramp function excitation in Sections 2.1 and 2.3, it was evident that the choice $s = 0$ at the beginning of the calculations would cause the integrals in Eqs. (2.1-57) and (2.3-29) to diverge. The effect of the choice $s = 0$ at the beginning of the calculation for the magnetic field strength in Section 2.4 is not evident, and we want to elaborate this effect.

We make the transition $s \to 0$ in Eqs. (2.4-1) and (2.4-3):

$$E_E(y, t) = E_0[1 + w(y, t)] \tag{1}$$

$$H_E(y, t) = -\mu^{-1} \int \frac{\partial E_E}{\partial y}\, dt + H_t(y) \tag{2}$$

Introduction of Eq. (1) into Eq. (2) yields with the help of Eq. (2.1-52):

$$\begin{aligned} H_E(y, t) &= -E_0\mu^{-1} \int \frac{\partial w}{\partial y}\, dt + H_t(y) \\ &= -\frac{E_0}{2\pi Z} \int_0^\infty [A_1(\kappa)\gamma_1^{-1} \exp(\gamma_1 t) + A_2(\kappa)\gamma_2^{-1} \exp(\gamma_2 t)]\beta c \\ &\quad \times \cos \beta y\, d\beta + H_y(t) \end{aligned} \tag{3}$$

In analogy to Eq. (2.4-6) we obtain from Eqs. (2.1-62) and (2.1-50):

$$
\begin{aligned}
\frac{A_1(\kappa)\beta c}{\gamma_1} &= -\frac{4\gamma_2 c}{(\gamma_2-\gamma_1)\gamma_1} = \frac{4\alpha}{\beta^2 c}(1+q_0), &\quad \alpha^2 > \beta^2 c^2 \\
&= \frac{4\alpha}{\beta^2 c}(1-jq_0'), &\quad \alpha^2 < \beta^2 c^2 \\
\frac{A_2(\kappa)\beta c}{\gamma_2} &= -\frac{4\gamma_1 c}{(\gamma_1-\gamma_2)\gamma_2} = \frac{4\alpha}{\beta^2 c}(1-q_0), &\quad \alpha^2 > \beta^2 c^2 \qquad (4) \\
&= \frac{4\alpha}{\beta^2 c}(1+jq_0'), &\quad \alpha^2 < \beta^2 c^2
\end{aligned}
$$

$$
q_0 = \frac{2\alpha^2-\beta^2c^2}{2\alpha(\alpha^2-\beta^2c^2)^{1/2}}, \qquad q_0' = \frac{2\alpha^2-\beta^2c^2}{2\alpha(\beta^2c^2-\alpha^2)^{1/2}}
$$

Insertion into Eq. (3) yields:

$$
\begin{aligned}
H_E(y,t) = -\frac{2E_0\sigma}{\pi}e^{-\alpha t}\Bigg[\int_0^{Z\sigma/2}\Bigg(&\mathrm{ch}(\alpha^2-\beta^2c^2)^{1/2}t \\
&+\frac{(2\alpha^2-\beta^2c^2)\mathrm{sh}(\alpha^2-\beta^2c^2)^{1/2}t}{2\alpha(\alpha^2-\beta^2c^2)^{1/2}}\Bigg)\frac{\cos\beta y}{\beta^2}\,d\beta \\
+\int_{Z\sigma/2}^{\infty}\Bigg(&\cos(\beta^2c^2-\alpha^2)^{1/2}t \\
&+\frac{(2\alpha^2-\beta^2c^2)\sin(\beta^2c^2-\alpha^2)^{1/2}t}{2\alpha(\beta^2c^2-\alpha^2)^{1/2}}\Bigg)\frac{\cos\beta y}{\beta^2}\,d\beta\Bigg] + H_t(y) \qquad (5)
\end{aligned}
$$

The comparison with Eq. (2.4-8) shows that the term

$$Z(\sigma/s)^{1/2}\exp[-(s\sigma)^{1/2}y]$$

is missing. This term contributed the expression

$$Z[(\sigma/s)^{1/2}-\sigma y]$$

to the magnetic field strength according to Eq. (2.4-9). Its first term $Z(\sigma/s)^{1/2}$ compensated a term of equal magnitude but opposite sign in Eq. (2.4-41), that came from the infinity of the first integral in Eq. (2.4-8) for $\beta = 0$. The second term $-Z\sigma y$ shows up in the final result of Eq. (2.4-41). These are the two effects caused by the choice $s = 0$ at the beginning of the calculation.

In Section 1.8 we had claimed that Maxwell's equations must be modified by the addition of a magnetic current density $\mathbf{g}_\mathrm{m} = s\mathbf{H}$, and that this current density or the magnetic conductivity s could be made zero only at the end of the calculation. In Section 2.1 we showed that this modification of Maxwell's theory was sufficient to derive the electric field strength of Eq. (2.1-68). However, it might be possible that someone could derive the very same result from the original Maxwell equations with $\mathbf{g}_m = 0$ and $s = 0$. If this indeed were so, one could not derive the magnetic field strength from Eq. (2.1-68) by means of Eq. (2). This proves that our modification of Maxwell's equations is not only sufficient but also necessary.[1]

2.6 Magnetic Step Function Excitation

In Sections 2.1, 2.3, and 2.4 we considered the excitation of a TEM wave by the application of an electric field strength to the plane $y = 0$ at the time $t = 0$. Consider now the case where a magnetic but no electric field strength is applied. This condition can be created, at least in theory, by placing many current loops in the plane $y = 0$. Later on we will be interested in the case when both an electric and a magnetic field strength are applied. This general condition is encountered when a planar electromagnetic wave hits the surface of a conducting medium.

Consider the following step function for the magnetic field strength as a boundary condition:

$$\begin{aligned} H(0, t) = H_0 S(t) &= 0 \qquad \text{for} \quad t < 0 \\ &= H_0 \qquad \text{for} \quad t \geqq 0 \end{aligned} \tag{1}$$

At the plane $y \to \infty$ we have again the boundary condition corresponding to Eq. (2.1-29):

$$H(\infty, t) = \text{finite} \tag{2}$$

Let $H(y, t)$ be zero for $y > 0$ at the time $t = 0$. We have thus exactly the same problem that we solved in Section 2.1, except that we must write H, H_0, and H_H for E, E_0, and E_E. According to Eqs. (2.1-17) and (2.1-24) we may rewrite Eq. (2.1-65)

$$H_H(y, t) = H_0[e^{-y/L} + w(y, t)], \qquad L = (s\sigma)^{-1/2} \tag{3}$$

where $w(y, t)$ is defined by Eq. (2.1-64).

The electric field strength $E_H(y, t)$ caused by the magnetic field strength

[1] It is perhaps more lucid to interpret the proof of sufficiency and necessity by reading it in the reverse direction: It is necessary to modify Maxwell's equations, and the addition of a magnetic current density term is a sufficient modification. It is quite possible that there are other methods of modification for Maxwell's equations that will yield physically acceptable solutions. Without introducing some assumption one cannot prove that the addition of a magnetic current density term is the only mathematically possible and physically acceptable modification.

$H(0, t)$ at the boundary plane $y = 0$ follows from Eqs. (2.1-25) and (2.1-27):

$$E_H(y, t) = -\int \left(\mu \frac{\partial H_H}{\partial t} + sH_H\right) dy + E_y(t) \tag{4}$$

$$E_H(y, t) = -\epsilon^{-1} e^{-\sigma t/\epsilon} \int \frac{\partial H_H}{\partial y} e^{\sigma t/\epsilon}\, dt + E_t(y) e^{-\sigma t/\epsilon} \tag{5}$$

We do not get simply a repetition of the calculations of Section 2.4 with $E_E(y, t)$ replaced by $H_H(y, t)$, since we want a solution for $\sigma > 0$, $s = 0$ rather than for $s > 0$, $\sigma = 0$.

Let us first resolve the simple limit case $s = 0$, $\sigma = 0$. We get from Eqs. (3) and (2.1-67) with $\alpha = Z\sigma c/2 = 0$:

$$H_H(y, t) = H_0 \left(1 - \frac{2}{\pi} \int_0^\infty \frac{\cos \beta ct \sin \beta y}{\beta} d\beta\right) \tag{6}$$

From Eq. (2.1-70) follows:

$$\begin{aligned} H_H(y, t) &= H_0 && \text{for} \quad y = 0, \quad t = 0 \\ &= 0 && \text{for} \quad y > 0, \quad ct < y \\ &= H_0/2 && \text{for} \quad y > 0, \quad ct = y \\ &= H_0 && \text{for} \quad y \geqq 0, \quad ct > y \end{aligned} \tag{7}$$

These relations are equivalent to Eq. (2.1-71).

From Eqs. (4) and (6) we get for $s = 0$ and $\sigma = 0$

$$E_H(y, t) = \frac{2H_0 Z}{\pi} \int_0^\infty \frac{\sin \beta ct \cos \beta y}{\beta} d\beta + E_y(t) \tag{8}$$

while Eqs. (5) and (6) yield:

$$E_H(y, t) = \frac{2H_0 Z}{\pi} \int_0^\infty \frac{\sin \beta ct \cos \beta y}{\beta} d\beta + E_t(y) \tag{9}$$

The equality of Eqs. (8) and (9) demands that $E_y(t)$ and $E_t(y)$ equal a constant E_c, which must be zero if $E_H(y, t)$ is zero for $t < 0$. In analogy to Eq. (2.1-84) we get:

$$\begin{aligned} E_H(y, t) &= \frac{H_0 Z}{\pi} \left(\int_0^\infty \frac{\sin \beta(y + ct) d\beta}{\beta} - \int_0^\infty \frac{\sin \beta(y - ct) d\beta}{\beta} \right) \\ &= 0 && \text{for} \quad y = 0, \quad t = 0 \\ &= 0 && \text{for} \quad y > 0, \quad ct < y \\ &= H_0 Z/2 && \text{for} \quad y > 0, \quad ct = 0 \\ &= H_0 Z && \text{for} \quad y \geqq 0, \quad ct > y \end{aligned} \tag{10}$$

We turn now to the more general case $\sigma > 0$. Equations (3) and (5) yield:

$$E_H(y, t) = H_0 Z\left(Z^{-1}(s/\sigma)^{1/2} e^{-y/L} - c e^{-\sigma t/\epsilon} \int \frac{\partial w}{\partial y} e^{\sigma t/\epsilon}\, dt \right) + E_t(y) e^{-\sigma t/\epsilon} \tag{11}$$

From Eq. (2.1-52) follows:

$$\begin{aligned} H_0 Z c e^{-\sigma t/\epsilon} \int \frac{\partial w}{\partial y} e^{\sigma t/\epsilon}\, dt &= 2\pi H_0 Z c e^{-\sigma t/\epsilon} \int_0^\infty \left(\int \{A_1(\kappa)\exp[(\gamma_1 + \sigma/\epsilon)t] \right. \\ &\quad \left. + A_2(\kappa)\exp[(\gamma_2 + \sigma/\epsilon)t]\}dt \right) \kappa \cos 2\pi\kappa y\, d\kappa \\ &= \frac{H_0 Z}{2\pi} \int_0^\infty [A_1(\kappa)(\gamma_1 + \sigma/\epsilon)^{-1} \exp(\gamma_1 t) \\ &\quad + A_2(\kappa)(\gamma_2 + \sigma/\epsilon)^{-1} \exp(\gamma_2 t)]\beta c \cos \beta y\, d\beta \end{aligned} \tag{12}$$

With the help of Eqs. (2.1-62) and (2.1-50) one obtains:

$$\begin{aligned} \frac{A_1(\kappa)\beta c}{\gamma_1 + \sigma/\epsilon} &= \frac{-4\beta^2\gamma_2 c}{b^2(\gamma_2 - \gamma_1)(\gamma_1 + \sigma/\epsilon)} = \frac{2s}{b^2 c\mu}(1 - p_s), \qquad a^2 > b^2c^2 \\ &= \frac{2s}{b^2 c\mu}(1 + jp_s'), \qquad a^2 < b^2c^2 \\ \frac{A_2(\kappa)\beta c}{\gamma_2 + \sigma/\epsilon} &= \frac{4\beta^2\gamma_1 c}{b^2(\gamma_1 - \gamma_2)(\gamma_2 + \sigma/\epsilon)} = \frac{2s}{b^2 c\mu}(1 + p_s), \qquad a^2 > b^2c^2 \\ &= \frac{2s}{b^2 c\mu}(1 - jp_s'), \qquad a^2 < b^2c^2 \end{aligned} \tag{13}$$

$$p_s = \frac{\mu\beta^2c^2 + (\sigma/\epsilon - s/\mu)s/2}{(a^2 - b^2c^2)^{1/2}s}, \qquad p_s' = \frac{\mu\beta^2c^2 + (\sigma/\epsilon - s/\mu)s/2}{(b^2c^2 - a^2)^{1/2}s}$$

Equation (11) assumes the form:

$$E_H(y,t) = H_0 Z\bigg[Z^{-1}\left(\frac{s}{\sigma}\right)^{1/2}\exp[-(s\sigma)^{1/2}y]$$
$$-\frac{s}{\pi c\mu}e^{-at}\bigg(\int_0^{2\pi K}\{(1-p_s)\exp[(a^2-b^2c^2)^{1/2}t]$$
$$+(1+p_s)\exp[-(a^2-b^2c^2)^{1/2}t]\}\frac{\cos\beta y}{b^2}\,d\beta$$
$$+\int_{2\pi K}^{\infty}\{(1+jp_s')\exp[j(b^2c^2-a^2)^{1/2}t]$$
$$+(1-jp_s')\exp[-j(b^2c^2-a^2)^{1/2}t]\}\frac{\cos\beta y}{b^2}\,d\beta\bigg)\bigg] + E_t(y)e^{-\sigma t/\epsilon} \quad (14)$$

Using hyperbolic and trigonometric functions we obtain:

$$E_H(y,t) = H_0 Z\bigg\{Z^{-1}\left(\frac{s}{\sigma}\right)^{1/2}\exp[-(s\sigma)^{1/2}y]$$
$$-\frac{2s}{\pi c\mu}e^{-at}\bigg[\int_0^{2\pi K}\bigg(\mathrm{ch}(a^2-b^2c^2)^{1/2}t - \frac{\beta^2c^2+(\sigma/\epsilon - s/\mu)s/2\mu}{(a^2-b^2c^2)^{1/2}s/\mu}$$
$$\times\,\mathrm{sh}(a^2-b^2c^2)^{1/2}t\bigg)\frac{\cos\beta y}{b^2}\,d\beta$$
$$+\int_{2\pi K}^{\infty}\bigg(\cos(b^2c^2-a^2)^{1/2}t - \frac{\beta^2c^2+(\sigma/\epsilon - s/\mu)s/2\mu}{(b^2c^2-a^2)^{1/2}s/\mu}$$
$$\times\sin(b^2c^2-a^2)^{1/2}t\bigg)\frac{\cos\beta y}{b^2}\,d\beta\bigg]\bigg\} + E_t(t)e^{-\sigma t/\epsilon} \quad (15)$$

We make now the transition $s \to 0$ for $\sigma > 0$. The first term in Eq. (15) vanishes. This result should be compared with Eq. (2.4-9). The first integral assumes for $s \to 0$ the form:

$$I_{H1}(y,t) = \frac{2}{\pi}e^{-\alpha t}\int_0^{Z\sigma/2}\frac{\mathrm{sh}(\alpha^2-\beta^2c^2)^{1/2}t}{(\alpha^2/c^2-\beta^2)^{1/2}}\cos\beta y\,d\beta \quad (16)$$

This integral can be evaluated by computer, if one observes the relation

$$\lim_{\beta\to Z\sigma/2}\frac{\mathrm{sh}(\alpha^2-\beta^2c^2)^{1/2}t}{(\alpha^2-\beta^2c^2)^{1/2}} = t \quad (17)$$

We turn to the second integral in Eq. (15):

$$I_{H2}(y,t) = -\frac{2}{\pi} e^{-\alpha t} \int_{Z\sigma/2}^{\infty} \frac{\sin(\beta^2c^2 - \alpha^2)^{1/2}t}{(\beta^2 - \alpha^2/c^2)^{1/2}} \cos \beta y \, d\beta \tag{18}$$

This integral can again be evaluated numerically. With Eqs. (16) and (18) we obtain from Eq. (15):

$$E_H(y,t) = H_0 Z[I_{H1}(y,t) + I_{H2}(y,t)] + E_t(y)e^{-\sigma t/\epsilon} \tag{19}$$

We turn to the integral of Eq. (4). Introduction of Eq. (3) yields:

$$\begin{aligned} E_H(y,t) &= -H_0 \int \left(\mu \frac{\partial w}{\partial t} + se^{-y/L} + sw \right) dy + E_y(t) \\ &= -H_0 \left(\mu \int \frac{\partial w}{\partial t} dy - (s/\sigma)^{1/2} e^{-y/L} + s \int w \, dy \right) + E_y(t) \end{aligned} \tag{20}$$

Using Eqs. (2.1-52) and (2.1-53) we get:

$$\begin{aligned} E_H(y,t) = H_0 Z \Bigg[& Z^{-1} \left(\frac{s}{\sigma} \right)^{1/2} e^{-y/L} + \frac{1}{2\pi c} \int_0^{\infty} [A_1(\kappa)\gamma_1 \exp(\gamma_1 t) \\ & + A_2(\kappa)\gamma_2 \exp(\gamma_2 t)] \frac{\cos \beta y}{\beta} d\beta \\ & + \frac{s}{2\pi Z} \int_0^{\infty} [A_1(\kappa)\exp(\gamma_1 t) + A_2(\kappa)\exp(\gamma_2 t)] \frac{\cos \beta y}{\beta} d\beta \Bigg] + E_y(t) \\ = H_0 Z \Bigg[& Z^{-1} \left(\frac{s}{\sigma} \right)^{1/2} e^{-y/L} + \frac{1}{2\pi c} \int_0^{\infty} [A_1(\kappa)(\gamma_1 + s/\mu)\exp(\gamma_1 t) \\ & + A_2(\kappa)(\gamma_2 + s/\mu)\exp(\gamma_2 t)] \frac{\cos \beta y}{\beta} d\beta \Bigg] + E_y(t) \end{aligned} \tag{21}$$

From Eqs. (2.1-62) and (2.1-50) follows with p_s and p_s' defined in Eq. (13):

$$\begin{aligned} \frac{A_1(\kappa)(\gamma_1 + s/\mu)}{c\beta} &= -\frac{4\gamma_2(\gamma_1 + s/\mu)}{b^2(\gamma_2 - \gamma_1)} = -\frac{2s}{b^2 c\mu}(1 - p_s), \qquad a^2 > b^2c^2 \\ &= -\frac{2s}{b^2 c\mu}(1 + jp_s'), \qquad a^2 < b^2c^2 \\ \frac{A_2(\kappa)(\gamma_2 + s/\mu)}{c\beta} &= -\frac{4\gamma_1(\gamma_2 + s/\mu)}{b^2(\gamma_1 - \gamma_2)} = -\frac{2s}{b^2 c\mu}(1 + p_s), \qquad a^2 > b^2c^2 \\ &= -\frac{2s}{b^2 c\mu}(1 - jp_s'), \qquad a^2 < b^2c^2 \end{aligned} \tag{22}$$

Substitution of Eq. (22) into Eq. (21) yields:

$$E_H(y,t) = H_0 Z\Bigg[Z^{-1}\left(\frac{s}{\sigma}\right)^{1/2} e^{-y/L}$$

$$-\frac{s}{\pi c\mu} e^{-at}\Bigg(\int_0^{2\pi K}\{(1-p_s)\exp[(a^2-b^2c^2)^{1/2}t]$$

$$+(1+p_s)\exp[-(a^2-b^2c^2)^{1/2}t]\}\frac{\cos\beta y}{b^2}\,d\beta$$

$$+\int_{2\pi K}^{\infty}\{(1+jp'_s)\exp[j(b^2c^2-a^2)^{1/2}t]$$

$$+(1-jp'_s)\exp[-j(b^2c^2-a^2)^{1/2}t]\}\frac{\cos\beta y}{b^2}\,d\beta\Bigg)\Bigg] + E_y(t) \quad (23)$$

The comparison of Eqs. (23) and (14) shows that they are equal for any value of s or σ if we choose

$$E_t(y)e^{-\sigma t/\epsilon} = E_y(t) = E_{H0}e^{-\sigma t/\epsilon} \quad (24)$$

The initial condition of Eq. (2.1-30) requires $E_{H0} = 0$.

We rewrite Eq. (19) in terms of the normalized variables ξ and θ using the definitions of Eqs. (2.1-78) and (2.1-79) and dropping the function $E_t(y)$:

$$E_H(\xi,\theta) = H_0 Z\frac{2}{\pi}e^{-\theta}\Bigg(\int_0^1 \frac{\operatorname{sh}(1-\eta^2)^{1/2}\theta}{(1-\eta^2)^{1/2}}\cos\eta\xi\,d\eta$$

$$+\int_0^1 \frac{\sin(\eta^2-1)^{1/2}\theta}{(\eta^2-1)^{1/2}}\cos\eta\xi\,d\eta\Bigg) \quad (25)$$

The magnetic field strength of Eq. (3) assumes for $s = 0$ in normalized notation the form:

$$H_H(\xi,\theta) = H_0[1 + w(\xi,\theta)] \quad (26)$$

where $w(\xi, \theta)$ is defined in Eq. (2.1-80).

Plots of $E_H(\xi, \theta)/H_0Z$ are shown in Fig. 2.6-1 for $\xi = \alpha y/c = 0, 1, 2, 3$ and $\theta = \alpha t$ in the range $0 \leqq \theta \leqq 26$. The functions are zero for $\theta < \xi$ or $t < y/c$. There is a jump at $\theta = \xi$. For $\theta \to \infty$ all plots approach zero.

The delay and the jump correspond to the result for the magnetic field strength caused by electric excitation in Fig. 2.4-1. But the curves now drop to zero for $\theta \to \infty$ rather than increase linearly with θ. The reason is that the magnetic field strength is not attenuated directly by losses. It must first be transformed into an electric field strength, which is then attenuated by ohmic losses. As the electric field strength approaches zero, the ohmic losses

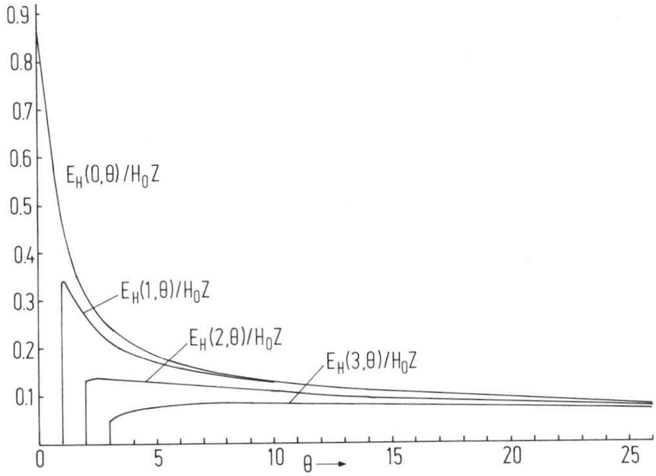

FIG. 2.6-1. Magnitude E_H of the electric field strength according to Eq. (25) caused by a magnetic step function excitation at the plane $\xi = 0$ of a lossy medium. Note that the maximum of E_H is at the jump for $E_H(0,\theta)$ but shifts to the right for $\xi > 0$, as clearly shown by $E_H(3,\theta)$. The illustration is based on computer plots by R. Boules, University of Alexandria, Egypt.

also approach zero, and a constant magnetic field without electric field remains. This is, of course, what we expect to happen when an electromagnet is switched on.

For $H_H(\xi, \theta)/H_0$ of Eq. (26) we get the same plots as for $E_E(\xi, \theta)/E_0$ in Figs. 2.1-3, 5.2-1, and 5.5-2.

The plots of Fig. 2.6-1 may be further elucidated by a modification of the circuit diagram of Fig. 2.4-2 as shown in Fig. 2.6-2. A current step function $IS(t)$ applied to the input will make the voltage jump from zero to $v = IR$. If the switches are then closed at the times ΔT, $2\,\Delta T$, . . . , the voltage will drop to $v = IR/(1 + k)$ at the time $t = k\,\Delta T$. One may increase the similarity between this circuit and wave propagation in a medium with ohmic losses still more by making the distance between the resistors equal to $c\,\Delta T$ and the spatial dimensions of the resistors small compared with $c\,\Delta T$. The switches may then be left out.

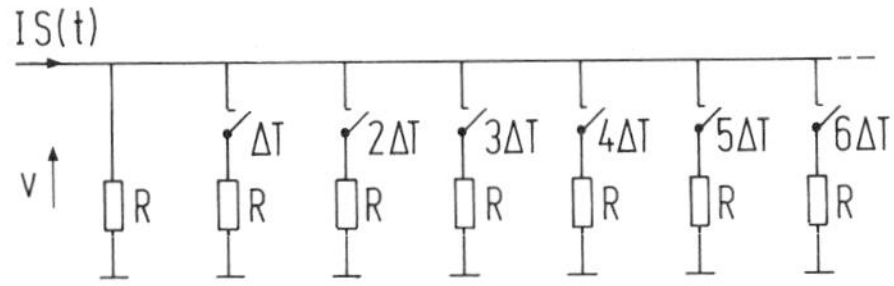

FIG. 2.6-2. Electric circuit for the elucidation of the plots for the electric field strength in Fig. 2.6-1.

2.7 Electric Field Without Magnetic Conductivity

We repeat the calculations of Section 2.6 but choose $s = 0$ from the beginning. Instead of $H_H(y, t)$ according to Eq. (2.6-3), we have thus:

$$H_H(y, t) = H_0[1 + w(y, t)] \tag{1}$$

Substitution into Eq. (2.6-5) yields:

$$\begin{aligned} E_H(y, t) &= -H_0 Zce^{-\sigma t/\epsilon} \int \frac{\partial w}{\partial y} e^{\sigma t/\epsilon}\, dt + E_t(y)e^{-\sigma t/\epsilon} \\ &= -\frac{ZH_0 c}{2\pi} \int_0^\infty [A_1(\kappa)(\gamma_1 + \sigma/\epsilon)^{-1} \exp(\gamma_1 t) \\ &\quad + A_2(\kappa)(\gamma_2 + \sigma/\epsilon)^{-1} \exp(\gamma_2 t)]\beta \cos \beta y\, d\beta + E_t(y)e^{-\sigma t/\epsilon} \end{aligned} \tag{2}$$

From Eq. (2.6-13) we obtain for $s = 0$:

$$\begin{aligned} \frac{A_1(\kappa)\beta c}{\gamma_1 + \sigma/\epsilon} &= -\frac{2c}{(\alpha^2 - \beta^2 c^2)^{1/2}}, \qquad \alpha^2 > \beta^2 c^2 \\ &= +j\frac{2c}{(\beta^2 c^2 - \alpha^2)^{1/2}}, \qquad \alpha^2 < \beta^2 c^2 \\ \frac{A_2(\kappa)\beta c}{\gamma_2 + \sigma/\epsilon} &= +\frac{2c}{(\alpha^2 - \beta^2 c^2)^{1/2}}, \qquad \alpha^2 > \beta^2 c^2 \\ &= -j\frac{2c}{(\beta^2 c^2 - \alpha^2)^{1/2}}, \qquad \alpha^2 < \beta^2 c^2 \end{aligned} \tag{3}$$

Insertion into Eq. (2) brings:

$$\begin{aligned} E_H(y, t) &= H_0 Z\pi^{-1} e^{-\alpha t}\Bigg(\int_0^{Z\sigma/2} \{\exp[(\alpha^2 - \beta^2 c^2)^{1/2} t] \\ &\quad - \exp[-(\alpha^2 - \beta^2 c^2)^{1/2} t]\} \frac{\cos \beta y}{(\alpha^2/c^2 - \beta^2)^{1/2}}\, d\beta \\ &\quad - j\int_{Z\sigma/2}^\infty \{\exp[j(\beta^2 c^2 - \alpha^2)^{1/2} t] - \exp[-j(\beta^2 c^2 - \alpha^2)^{1/2} t]\} \\ &\quad \times \frac{\cos \beta y}{(\beta^2 - \alpha^2/c^2)^{1/2}}\, d\beta\Bigg) + E_t(y)e^{-\sigma t/\epsilon} \\ &= H_0 Z\frac{2}{\pi} e^{-\alpha t}\Bigg(\int_0^{Z\sigma/2} \frac{\operatorname{sh}(\alpha^2 - \beta^2 c^2)^{1/2} t}{(\alpha^2/c^2 - \beta^2)^{1/2}} \cos \beta y\, d\beta \\ &\quad + \int_{Z\sigma/2}^\infty \frac{\sin(\beta^2 c^2 - \alpha^2)^{1/2} t}{(\beta^2 - \alpha^2/c^2)^{1/2}} \cos \beta y\, d\beta\Bigg) + E_t(y)e^{-\sigma t/\epsilon} \end{aligned} \tag{4}$$

This is the same result as Eq. (2.6-19), but the calculation was much simpler. Let us see whether the transition $s \to 0$ before rather than after the calculation can also be made in the case of Eq. (2.6-4). We get from Eqs. (1) and (2.6-4) for $s = 0$:

$$
\begin{aligned}
E_H(y, t) &= -\mu \int \frac{\partial H_H}{\partial t} dy + E_y(t) \\
&= -\mu H_0 \int \frac{\partial w}{\partial t} dy + E_y(t) \\
&= \frac{H_0 Z}{2\pi c} \int_0^\infty [A_1(\kappa)\gamma_1 \exp(\gamma_1 t) + A_2(\kappa)\gamma_2 \exp(\gamma_2 t)] \\
&\quad \times \frac{\cos \beta y}{\beta} d\beta + E_y(t)
\end{aligned} \tag{5}
$$

Equation (2.6-22) yields for $s = 0$:

$$
\begin{aligned}
\frac{A_1(\kappa)\gamma_1}{\beta c} &= \frac{2c}{(\alpha^2 - \beta^2 c^2)^{1/2}}, &\quad \alpha^2 > \beta^2 c^2 \\
&= -j\frac{2c}{(\beta^2 c^2 - \alpha^2)^{1/2}}, &\quad \alpha^2 < \beta^2 c^2 \\
\frac{A_2(\kappa)\gamma_2}{\beta c} &= -\frac{2c}{(\alpha^2 - \beta^2 c^2)^{1/2}}, &\quad \alpha^2 > \beta^2 c^2 \\
&= +j\frac{2c}{(\beta^2 c^2 - \alpha^2)^{1/2}}, &\quad \alpha^2 < \beta^2 c^2
\end{aligned} \tag{6}
$$

Insertion into Eq. (5) yields:

$$
\begin{aligned}
E_H(y, t) = H_0 Z \frac{2}{\pi} e^{-\alpha t} \Bigg(\int_0^{Z\sigma/2} \frac{\operatorname{sh}(\alpha^2 - \beta^2 c^2)^{1/2} t}{(\alpha^2 - \beta^2 c^2)^{1/2}} \cos \beta y \, d\beta \\
+ \int_{Z\sigma/2}^{\infty} \frac{\sin(\beta^2 c^2 - \alpha^2)^{1/2} t}{(\beta^2 c^2 - \alpha^2)^{1/2}} \cos \beta y \, d\beta \Bigg) + E_y(t)
\end{aligned} \tag{7}
$$

Again we obtain in the short way the result of Eq. (2.6-19) with $E_t(y)e^{-\sigma t/\epsilon}$ replaced by $E_y(t)$.

2.8 Boundary Conditions with Arbitrary Time Variation

In Sections 2.1 and 2.4 we determined the magnitudes $E_E(y, t)$ and $H_E(y, t)$ of the electric and magnetic field strength in a conducting medium if an electric field strength of magnitude $E(0, t)$ with the time variation of a step function was applied at the plane $y = 0$:

$$E(0, t) = E_0 S(t) \tag{1}$$

Then we calculated in Section 2.6 the field strengths $H_H(y, t)$ and $E_H(y, t)$ due to a magnetic step function

$$H(0, t) = H_0 S(t) \tag{2}$$

at the plane $y = 0$. We consider now the general case where $E(0, t)$ and $H(0, t)$ are zero for $t < 0$, but may have any time variation for $t \geqq 0$ that permits a representation by a superposition of step functions according to Fig. 1.3-3 and Eqs. (1.3-1) or (1.3-7):

$$E(0, t) = \int_0^t \frac{dE(0, t'')}{dt''} S(t - t'')dt'' \tag{3}$$

$$H(0, t) = \int_0^t \frac{dH(0, t'')}{dt''} S(t - t'')dt'' \tag{4}$$

It is not necessary to assume an arbitrary magnetic conductivity s, and we consider thus the simpler case $s = 0$. Electric and magnetic field strengths due to excitation by an electric step function are then given by Eqs. (2.1-68) and (2.4-41). Dropping the constant $H_t(y)$, we have:

$$E_E(y, t) = E_0[1 + w(y, t)] \tag{5}$$

$$H_E(y, t) = E_0 Z^{-1}[-Z\sigma y + I'_{E1}(y, t) - I_{E2}(y, t)] \tag{6}$$

where $w(y, t)$, $I'_{E1}(y, t)$, and $I_{E2}(y, t)$ are defined by Eqs. (2.1-67), (2.4-39), and (2.4-40). The terms in brackets in Eqs. (5) and (6) are the changed step functions $S(t)g(t)$ used in Eqs. (1.3-8) and (1.3-9). The electric field strength $E(0, t)S(t)$ with general time variation applied at the plane $y = 0$ will thus produce the following electric and magnetic field strengths at the plane y:

$$E_E(y, t) = \int_0^t \frac{dE(0, t'')}{dt''} S(t - t'')[1 + w(y, t - t'')]dt'' \tag{7}$$

$$H_E(y, t) = Z^{-1} \int_0^t \frac{dE(0, t'')}{dt''} S(t - t'')[-Z\sigma y + I'_{E1}(y, t - t'') - I_{E2}(y, t - t'')]dt'' \tag{8}$$

A step function $H_0 S(t)$ applied at the plane $y = 0$ produces the following electric and magnetic field strengths in a conducting medium at the plane y according to Eqs. (2.6-3) for $s = 0$ and (2.6-19) for $E_t(y) = 0$:

$$H_H(y, t) = H_0[1 + w(y, t)] \tag{9}$$

$$E_H(y, t) = H_0 Z[I_{H1}(y, t) + I_{H2}(y, t)] \tag{10}$$

A magnetic field strength $H(0, t)S(t)$ with general time variation applied at the plane $y = 0$ will produce the following field strengths at the plane y in analogy to Eqs. (7) and (8):

$$H_H(y, t) = \int_0^t \frac{dH(0, t'')}{dt''} S(t - t'')[1 + w(y, t - t'')]dt'' \tag{11}$$

$$E_H(y, t) = Z \int_0^t \frac{dH(0, t'')}{dt''} S(t - t'')[I_{H1}(y, t - t'') + I_{H2}(y, t - t'')]dt'' \tag{12}$$

Finally, if both an electric and a magnetic field strength with general time variation $E(0, t)$ and $H(0, t)$ are applied at the plane $y = 0$ at the time $t = 0$, we obtain the sum of the field strengths produced by them at the plane y:

$$E(y, t) = E_E(y, t) + E_H(y, t) = \int_0^t \frac{dE(0, t'')}{dt''} S(t - t'')[1 + w(y, t - t'')]dt'' + Z \int_0^t \frac{dH(0, t'')}{dt''} S(t - t'')[I_{H1}(y, t - t'') + I_{H2}(y, t - t'')]dt'' \tag{13}$$

$$H(y, t) = H_H(y, t) + H_E(y, t) = \int_0^t \frac{dH(0, t'')}{dt''} S(t - t'') \times [1 + w(y, t - t'')]dt'' + Z^{-1} \int_0^t \frac{dE(0, t'')}{dt''} S(t - t'') \times [-Z\sigma y + I'_{E1}(y, t - t'') - I_{E2}(y, t - t'')]dt'' \tag{14}$$

Let us emphasize that the value of Eqs. (13) and (14) lies primarily in the easy evaluation by computer once $w(y, t)$, $I_{H1}(y, t)$, $I_{H2}(y, t)$, $I'_{E1}(y, t)$, and $I_{E2}(y, t)$ have been found analytically, as we did in Sections 2.1, 2.4, and 2.6. The transformation of the integrals and differentials to sums and differences needed for the computer evaluation follows readily from Eqs. (1.3-1) and (1.3-9).

The ratio $Z(y, t) = E(y, t)/H(y, t)$ is the *impedance* as a function of y and t for a plane wave excited by the electric field strength $E(0, t)$ and the magnetic field strength $H(0, t)$ in the plane $y = 0$. We are particularly interested in the impedance $Z(0, t)$. For its calculation we obtain from Eqs. (2.1-67), (2.6-16), (2.6-18), (2.4-39), and (2.4-40):

$$w(0, t) = 0 \tag{15}$$

$$I_{H1}(0, t) = \frac{2}{\pi} e^{-\alpha t} \int_0^{Z\sigma/2} \frac{\operatorname{sh}(\alpha^2 - \beta^2 c^2)^{1/2} t}{(\alpha^2/c^2 - \beta^2)^{1/2}} d\beta \tag{16}$$

$$I_{H2}(0, t) = \frac{2}{\pi} e^{-\alpha t} \int_{Z\sigma/2}^{\infty} \frac{\sin(\beta^2 c^2 - \alpha^2)^{1/2} t}{(\beta^2 - \alpha^2/c^2)^{1/2}} d\beta \tag{17}$$

$$I'_{E1}(0, t) = \frac{2Z\sigma}{\pi}\left[\frac{1}{d} - e^{-\alpha t}\int_d^{Z\sigma/2}\left(\operatorname{ch}(\alpha^2 - \beta^2c^2)^{1/2}t\right.\right.$$
$$\left. + \frac{\alpha\operatorname{sh}(\alpha^2 - \beta^2c^2)^{1/2}t}{(\alpha^2 - \beta^2c^2)^{1/2}}\right)\frac{d\beta}{\beta}$$
$$\left. - \int_0^d [\exp(-\beta^2c^2t/2\alpha) - 1]\frac{d\beta}{\beta^2}\right]$$
$$+ \frac{2}{\pi}e^{-\alpha t}\int_0^{Z\sigma/2}\frac{\operatorname{sh}(\alpha^2 - \beta^2c^2)^{1/2}t}{(\alpha^2/c^2 - \beta^2)^{1/2}}d\beta \tag{18}$$

$$I_{E2}(0, t) = \frac{2Z\sigma}{\pi}e^{-\alpha t}\int_{Z\sigma/2}^{\infty}\left(\cos(\beta^2c^2 - \alpha^2)^{1/2}t\right.$$
$$\left. + \frac{(2\alpha^2 - \beta^2c^2)\sin(\beta^2c^2 - \alpha^2)^{1/2}t}{2\alpha(\beta^2c^2 - \alpha^2)^{1/2}}\right)\frac{d\beta}{\beta^2} \tag{19}$$

The impedance $Z(0, t)$ at the plane $y = 0$ becomes

$$Z(0, t) = \left(E(0, t) + Z\int_0^t \frac{dH(0, t'')}{dt''}S(t - t'')[I_{H1}(0, t - t'')\right.$$
$$\left. + I_{H2}(0, t - t'')]dt''\right) \Big/ \left(H(0, t) + Z^{-1}\int_0^t \frac{dE(0, t'')}{dt''}\right.$$
$$\left. \times S(t - t'')[I'_{E1}(0, t - t'') - I_{E2}(0, t - t'')]dt''\right) \tag{20}$$

The impedance $Z(y, t)$ defined here must be related to the conventional impedance of field strengths having the time variation of a sinusoidal function and its generalization by an operator used for field strengths with general time variation (Harmuth, 1984, Sec. 3.1).

Let a wave be excited by an electric step function $E_0S(t)$ and a magnetic step function $H_0S(t)$ in the plane $y = 0$. The wave has the field strengths

$$E(y, t) = E_E(y, t) + E_H(y, t)$$
$$= E_0[1 + w(y, t)] + ZH_0[I_{H1}(y, t) + I_{H2}(y, t)] \tag{21}$$

$$H(y, t) = H_H(y, t) + H_E(y, t)$$
$$= H_0[1 + w(y, t)] + Z^{-1}E_0[-Z\sigma y + I'_{E1}(y, t) - I_{E2}(y, t)] \tag{22}$$

The ratio of E_E and H_H is a constant

$$E_E(y, t)/H_H(y, t) = E_0/H_0 \tag{23}$$

but the ratio of E_H and H_E is a function of y and t:

$$E_H(y, t)/H_E(y, t) = Z^2(H_0/E_0)[I_{H1}(y, t) + I_{H2}(y, t)] \div [-Z\sigma y + I'_{E1}(y, t) - I_{E2}(y, t)] \qquad (24)$$

As a result, the ratio $E(y, t)/H(y, t)$ is in general a function of y and t. Consider first the limit of a vanishing conductivity $\sigma = 0$. We obtain from Eqs. (2.1-71) and (2.6-10)

$$\begin{aligned} E(y, t) &= E_0 && \text{for } y = 0, \; t = 0 \\ &= 0 && \text{for } y > 0, \; ct < y \\ &= (E_0 + ZH_0)/2 && \text{for } y > 0, \; ct = y \\ &= E_0 + ZH_0 && \text{for } y \geqq 0, \; ct > y \end{aligned} \qquad (25)$$

while Eqs. (2.1-85) and (2.6-7) yield:

$$\begin{aligned} H(y, t) &= H_0 && \text{for } y = 0, \; t = 0 \\ &= 0 && \text{for } y > 0, \; ct < y \\ &= (E_0 + ZH_0)/2Z && \text{for } y > 0, \; ct = y \\ &= (E_0 + ZH_0)/Z && \text{for } y \geqq 0, \; ct > y \end{aligned} \qquad (26)$$

For $t > 0$ and $ct > y$ the electric field strength equals $E_0 + ZH_0$, while the magnetic field strength is proportionate to $E_0 + ZH_0$ regardless of the independently choosable boundary values E_0 and H_0. Hence, one has a constant impedance $E(y, t)/H(y, t) = Z$. One cannot obtain E_0 and H_0 separately from a measurement of $E(y, t)$ and $H(y, t)$ for $y > 0$. In the limit $\sigma = 0$, we thus again obtain the result of Section 1.5, that planar waves in a loss-free medium require the same time variation for electric and magnetic field strength.

Consider now the general case $\sigma \neq 0$. The ratio $E(y, t)/H(y, t) = Z(y, t)$ is a function of y and t according to Eqs. (21) to (24), even if one chooses $E_0 = ZH_0$:

$$Z(y, t) = \frac{E_0[1 + w(y, t)] + ZH_0[I_{H1}(y, t) + I_{H2}(y, t)]}{H_0[1 + w(y, t)] + (E_0/Z)[-Z\sigma y + I'_{E1}(y, t) - I_{E2}(y, t)]} \qquad (27)$$

The field strengths E_0 and H_0 applied to the boundary $y = 0$ can be determined *independently* from an observation of $E(y, t)$ and $H(y, t)$ for $y > 0$, since Eqs. (21) and (22) can be solved for E_0 and H_0:

$$E_0 = \frac{E(y, t)[1 + w(y, t)] - ZH(y, t)[I_{H1}(y, t) + I_{H2}(y, t)]}{[1 + w(y, t)]^2 + [I_{H1}(y, t) + I_{H2}(y, t)][Z\sigma y - I'_{E1}(y, t) + I_{E2}(y, t)]} \qquad (28)$$

$$H_0 = \frac{H(y,t)[1+w(y,t)] - Z^{-1}E(y,t)[-Z\sigma y + I'_{E1}(y,t) - I_{E2}(y,t)]}{[1+w(y,t)]^2 + [I_{H1}(y,t) + I_{H2}(y,t)][Z\sigma y - I'_{E1}(y,t) + I_{E2}(y,t)]} \tag{29}$$

A time variation $f(t)$ in the plane $y=0$ becomes generally a time variation $f_E(y, t)$ for the electric field strength and a time variation $f_H(y, t)$ for the magnetic field strength. If the transmission medium is linear, which means the proportionality and the superposition law are satisfied, one may use an orthogonal system of functions $f(i, t)$ to represent the general function $f(t)$:

$$f(t) = \sum_{i=0}^{\infty} a(i) f(i, t) \tag{30}$$

If the function $f(i, t)$ at the plane $y=0$ is transformed into the function $f_E(i, y, t)$ or $f_H(i, y, t)$ at the plane y, one obtains:

$$f_E(y, t) = \sum_{i=0}^{\infty} a(i) f_E(i, y, t) \tag{31}$$

$$f_H(y, t) = \sum_{i=0}^{\infty} a(i) f_H(i, y, t) \tag{32}$$

The ratio $E_0 f_E(i, y, t)/H_0 f_H(i, y, t)$ is the impedance at the plane y for the function i of the orthogonal system $f(i, t)$. If we use the functions of the Fourier series for $f(i, t)$,

$$\begin{aligned} f(0, t) &= 1 \\ f(2i, t) &= \sqrt{2}\cos 2\pi i t/T \\ f(2i-1, t) &= \sqrt{2}\sin 2\pi i t/T \\ i &= 1, 2, \ldots \end{aligned} \tag{33}$$

the functions $f_E(i, y, t)$ and $f_H(i, y, t)$ will differ from $f(i, t)$ only by a different amplitude and a time or phase shift. As a result, the ratio $E_0 f_E(i, y, t)/H_0 f_H(i, y, t)$ can be represented by a complex quantity Z as shown, e.g., by Eq. (2.2-15) for $\omega = 2\pi i/T$. In general, however, this ratio will be a function of i, y, and t:

$$Z(i, y, t) = E_0 f_E(i, y, t)/H_0 f_H(i, y, t) \tag{34}$$

One may call $Z(i, y, t)$ an operator $\mathbf{Z}$ that changes the function $f_H(i, y, t)$ into the function $f_E(i, y, t)$:

$$\mathbf{Z} f_H(i, y, t) = (E_0/H_0) f_E(i, y, t) \tag{35}$$

The sinusoidal functions have the obvious advantage of requiring only a

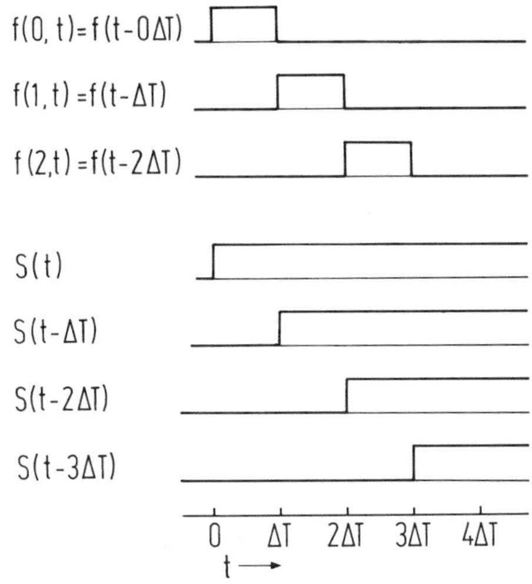

FIG. 2.8-1. Systems of orthogonal rectangular pulses and of linearly independent step functions for the representation of signals.

complex quantity rather than an operator **Z**, but a superposition of sinusoidal functions *alone* cannot be used to study the propagation of signals. Let us repeat here that a signal has a beginning and end, which means it is represented by a function—usually a time function—that is zero outside a finite interval. We have seen in Section 2.1 that the initial conditions $E(y, t) = H(y, t) = 0$ for $t \leqq 0$ have to be satisfied. A superposition of sinusoidal waves is generally not able to satisfy these initial conditions, and we had to introduce the additional function $F(y)$ in Eq. (2.1-33) to succeed. For this reason, the orthogonal system of sinusoidal functions is not well suited for the investigation of electromagnetic waves that represent signals. More suitable functions are those that are zero outside a finite interval, and which in addition have all the same time variation. The system of rectangular pulses shown on top of Fig. 2.8-1 is an example.[1] Any superposition of these functions will be zero outside a finite interval. In addition, the impedance $Z(i, y, t)$ of Eq. (34) becomes $Z(y, t - i\,\Delta T)$, which means that the variable i is essentially eliminated. The step functions $S(t - i\,\Delta T)$ in Fig. 2.8-1 are not quite as good for the representation of signals as the rectangular pulses, since

[1] From the standpoint of abstract mathematics this system of functions has the drawback of not being complete, but this is of no interest in an experimental science. We cannot make continuous observations at the times $t, t + dt, t + 2dt, \ldots$, but only discrete observations at the times $t, t + \Delta T, t + 2\Delta T \ldots$. An observable electric field strength can thus always be represented by a superposition of rectangular pulses of duration ΔT.

they are not zero for large values of t, and they are linearly independent rather than orthogonal. However, they retain the simplification of the impedance from $Z(i, y, t)$ to $Z(y, t - i\,\Delta T)$, and they yield formulas for $w(y, t)$ in Eqs. (2.1-33) and (2.1-67) that are half as long as those for the rectangular pulses, since each rectangular pulse $f(t - i\,\Delta T)$ can be represented by a superposition $S(t - i\,\Delta T) - S[t - (i + 1)\Delta T]$ of step functions. For rectangular pulses one gets thus the function $w(y, t) - w(y, t - \Delta T)$ instead of $w(y, t)$.

2.9 Magnetic Field Due to Electric Ramp Function

From the electric field strength of Eq. (2.3-34) caused by an electric exponential ramp function follows the magnetic field strength by means of Eqs. (2.1-18) and (2.1-23):

$$H_E(y, t) = \int \left(\epsilon \frac{\partial E_E}{\partial t} + \sigma E_E\right) dy + H_y(t) \tag{1}$$

$$H_E(y, t) = e^{-st/\mu}\left(-\mu^{-1} \int \frac{\partial E_E}{\partial y} e^{st/\mu} dt + H_t(y)\right) \tag{2}$$

For the calculation of the various derivatives and integrals we start from Eq. (2.3-34). First, we determine $H_E(y, t)$ from Eq. (2), using the notation $Q = (1 - e^{-2a\,\Delta T})^{-1}$ and $E_0 Q = E_1$:

$$\begin{aligned} &-E_1 \mu^{-1} e^{-st/\mu} \int (1 - e^{-2at}) \frac{d}{dy} e^{-y/L} e^{st/\mu}\, dt \\ &= \frac{E_1}{\mu L} e^{-y/L} e^{-st/\mu} \int (e^{st/\mu} - e^{-(2a - s/\mu)t}) dt \\ &= E_1 (\sigma/s)^{1/2} [1 + (s/Z^2\sigma) e^{-2at}] e^{-y/L} \end{aligned} \tag{3}$$

With the help of Eqs. (2.3-24) we get further from Eq. (2):

$$\begin{aligned} -E_1 \mu^{-1} e^{-st/\mu} \int \frac{\partial u}{\partial y} e^{st/\mu}\, dt = &-\frac{2\pi E_1}{\mu} \int_0^\infty [A_1(\kappa) \kappa (\gamma_1 + s/\mu)^{-1} \exp(\gamma_1 t) \\ &+ A_2(\kappa) \kappa (\gamma_2 + s/\mu)^{-1} \exp(\gamma_2 t)] \cos 2\pi\kappa y\, d\kappa \end{aligned} \tag{4}$$

Substitution of Eqs. (3) and (4) into Eq. (2) brings:

$$\begin{aligned} H_E(y, t) = E_1 \Bigg[&\left(\frac{\sigma}{s}\right)^{1/2} \left(1 + \frac{s}{Z^2\sigma} e^{-2at}\right) e^{-y/L} \\ &- 2\pi \int_0^\infty [A_1(\kappa)(\gamma_1 \mu + s)^{-1} \exp(\gamma_1 t) + A_2(\kappa)(\gamma_2 \mu + s)^{-1} \exp(\gamma_2 t)] \\ &\times \kappa \cos 2\pi\kappa y\, d\kappa \Bigg] + H_t(y) e^{-st/\mu} \end{aligned} \tag{5}$$

From Eqs. (2.3-31) and (2.1-50) follows:

$$
\begin{aligned}
\frac{2\pi A_1(\kappa)\kappa}{\gamma_1\mu + s} &= \frac{4\epsilon a}{b^2}(1 + q_e), \qquad a^2 > b^2c^2 \\
&= \frac{4\epsilon a}{b^2}(1 - jq_e'), \qquad a^2 < b^2c^2 \\
\frac{2\pi A_2(\kappa)\kappa}{\gamma_2\mu + s} &= \frac{4\epsilon a}{b^2}(1 - q_e), \qquad a^2 > b^2c^2 \\
&= \frac{4\epsilon a}{b^2}(1 + jq_e'), \qquad a^2 < b^2c^2 \\
q_e = \frac{\alpha - s/2\mu}{(a^2 - b^2c^2)^{1/2}}&, \qquad q_e' = \frac{\alpha - s/2\mu}{(b^2c^2 - a^2)^{1/2}}
\end{aligned}
\tag{6}
$$

Equations (5) and (6) yield:

$$
\begin{aligned}
H_E(y, t) = Z^{-1}E_1\Bigg[& Z\left(\frac{\sigma}{s}\right)^{1/2}\left(1 + \frac{s}{Z^2\sigma}e^{-2at}\right)e^{-y/L} \\
& - \frac{2a}{\pi c}e^{-at}\Bigg(\int_0^{2\pi K}\{(1 + q_e)\exp[(a^2 - b^2c^2)^{1/2}t] \\
& + (1 - q_e)\exp[-(a^2 - b^2c^2)^{1/2}t]\}\frac{\cos\beta y}{b^2}\,d\beta \\
& + \int_0^{2\pi K}\{(1 - jq_e')\exp[j(b^2c^2 - a^2)^{1/2}t] \\
& + (1 + jq_e')\exp[-j(b^2c^2 - a^2)^{1/2}t]\}\frac{\cos\beta y}{b^2}\,d\beta\Bigg)\Bigg] + H_t(y)e^{-st/\mu} \\
2\pi K = (a^2/c^2 - s\sigma)^{1/2} &
\end{aligned}
\tag{7}
$$

Using hyperbolic and trigonometric functions we obtain:

$$
\begin{aligned}
H_E(y, t) = Z^{-1}E_1\Bigg\{ & Z\left(\frac{\sigma}{s}\right)^{1/2}\left(1 + \frac{s}{Z^2\sigma}e^{-2at}\right)\exp[-(s\sigma)^{1/2}y] \\
& - \frac{4a}{\pi c}e^{-at}\Bigg[\int_0^{2\pi K}\Bigg(\operatorname{ch}(a^2 - b^2c^2)^{1/2}t
\end{aligned}
$$

$$
+\frac{\alpha - s/2\mu}{(a^2 - b^2c^2)^{1/2}} \operatorname{sh}(a^2 - b^2c^2)^{1/2}t\Bigg) \frac{\cos \beta y}{b^2} d\beta
$$

$$
+\int_{2\pi K}^{\infty} \left(\cos(b^2c^2 - a^2)^{1/2}t + \frac{\alpha - s/2\mu}{(b^2c^2 - a^2)^{1/2}} \sin(b^2c^2 - a^2)^{1/2}t \right)
$$

$$
\times \frac{\cos \beta y}{b^2} d\beta \Bigg]\Bigg\} + H_t(y)e^{-st/\mu} \tag{8}
$$

We now make the transition $s \to 0$ for $\sigma > 0$. The first term yields:

$$
\lim_{s\to 0} Z\left(\frac{\sigma}{s}\right)^{1/2} \left(1 + \frac{s}{Z^2\sigma} e^{-2at}\right) \exp[-(s\sigma)^{1/2}y] = Z\left[\left(\frac{\sigma}{s}\right)^{1/2} - \sigma y\right] \tag{9}
$$

The calculation of the two integrals closely follows the calculation in Section 2.4. The first integral is denoted I_{E3} for $s \to 0$:

$$
I_{E3}(y, t) = \frac{2Z\sigma}{\pi} e^{-\alpha t} \int_0^{Z\sigma/2} \left(\operatorname{ch}(\alpha^2 - \beta^2c^2)^{1/2}t + \frac{\alpha \operatorname{sh}(\alpha^2 - \beta^2c^2)^{1/2}t}{(\alpha^2 - \beta^2c^2)^{1/2}} \right)
$$

$$
\times \frac{\cos \beta y}{\beta^2 + s\sigma} d\beta \tag{10}
$$

Using the identity

$$
\cos \beta y = 1 - 2\sin^2(\beta y/2) \tag{11}
$$

we rewrite the integral:

$$
I_{E3}(y, t) = \frac{2Z\sigma}{\pi} e^{-\alpha t} \Bigg[\int_0^{Z\sigma/2} \Bigg(\operatorname{ch}(\alpha^2 - \beta^2c^2)^{1/2}t
$$

$$
+ \frac{\alpha \operatorname{sh}(\alpha^2 - \beta^2c^2)^{1/2}t}{(\alpha^2 - \beta^2c^2)^{1/2}} \Bigg) \frac{d\beta}{\beta^2 + s\sigma}
$$

$$
- 2\int_0^{Z\sigma/2} \Bigg(\operatorname{ch}(\alpha^2 - \beta^2c^2)^{1/2}
$$

$$
+ \frac{\alpha \operatorname{sh}(\alpha^2 - \beta^2c^2)^{1/2}t}{(\alpha^2 - \beta^2c^2)^{1/2}} \Bigg) \frac{\sin^2(\beta y/2)}{\beta^2 + s\sigma} d\beta \Bigg] \tag{12}
$$

We split I_{E3} into three components

$$
I_{E3}(y, t) = -I_{31}(y, t) + I_{32}(t) + I_{33}(t) \tag{13}
$$

where $I_{31}(y, t)$ is the second integral in Eq. (12), which remains finite for $s = 0$

$$I_{31}(y, t) = \frac{Z\sigma}{\pi} e^{-\alpha t} \int_0^{Z\sigma/2} \left(\operatorname{ch}(\alpha^2 - \beta^2 c^2)^{1/2} t + \frac{\alpha \operatorname{sh}(\alpha^2 - \beta^2 c^2)^{1/2} t}{(\alpha^2 - \beta^2 c^2)^{1/2}} \right) \times \left(\frac{\sin(\beta y/2)}{\beta/2} \right)^2 d\beta \tag{14}$$

and can be evaluated by computer. The other two components of Eq. (13) are:

$$I_{32}(t) = I_{12}(t) = \frac{2Z\sigma}{\pi} e^{-\alpha t} \int_0^{Z\sigma/2} \frac{\operatorname{ch}(\alpha^2 - \beta^2 c^2)^{1/2} t}{\beta^2 + s\sigma} d\beta \tag{15}$$

$$I_{33}(t) = I_{13}(t) = \frac{2Z\sigma\alpha}{\pi} e^{-\alpha t} \int_0^{Z\sigma/2} \frac{\operatorname{sh}(\alpha^2 - \beta^2 c^2)^{1/2} t}{(\alpha^2 - \beta^2 c^2)^{1/2} (\beta^2 + s\sigma)} d\beta \tag{16}$$

The integrals $I_{32}(t)$ and $I_{33}(t)$ are the same as $I_{12}(t)$ and $I_{13}(t)$ in Eqs. (2.4-16) and (2.4-17). Using their sum of Eq. (2.4-37) we may write I_{E3} in analogy to Eq. (2.4-38) by leaving out I_{14} of Eq. (2.4-18) and replacing I_{11} of Eq. (2.4-15) by I_{31} of Eq. (14):

$$\begin{aligned} I_{E3}(y, t) = {} & Z\left(\frac{\sigma}{s}\right)^{1/2} + \frac{2Z\sigma}{\pi} \left[e^{-\alpha t} \int_d^{Z\sigma/2} \left(\operatorname{ch}(\alpha^2 - \beta^2 c^2)^{1/2} t \right.\right. \\ & \left.\left. + \frac{\alpha \operatorname{sh}(\alpha^2 - \beta^2 c^2)^{1/2} t}{(\alpha^2 - \beta^2 c^2)^{1/2}} \right) \frac{d\beta}{\beta^2} - \frac{1}{d} + \int_0^d [\exp(-\beta^2 c^2 t/2\alpha) - 1] \frac{d\beta}{\beta^2} \right] \\ & - \frac{Z\sigma}{\pi} e^{-\alpha t} \int_0^{Z\sigma/2} \left(\operatorname{ch}(\alpha^2 - \beta^2 c^2)^{1/2} t \right. \\ & \left. + \frac{\alpha \operatorname{sh}(\alpha^2 - \beta^2 c^2)^{1/2} t}{(\alpha^2 - \beta^2 c^2)^{1/2}} \right) \left(\frac{\sin(\beta y/2)}{\beta/2} \right)^2 d\beta \end{aligned} \tag{17}$$

$$d \ll Z\sigma/2 = \alpha/c$$

The function I_{E3} without the term $Z(\sigma/s)^{1/2}$ is denoted $-I'_{E3}(y, t)$:

$$-I'_{E3}(y, t) = I_{E3}(y, t) - Z(\sigma/s)^{1/2} \tag{18}$$

We turn to the second integral in Eq. (8). It remains finite for $s \to 0$ if the electric conductivity σ is larger than zero:

$$I_{E4}(y, t) = \frac{2Z\sigma}{\pi} e^{-\alpha t} \int_{Z\sigma/2}^{\infty} \left(\cos(\beta^2 c^2 - \alpha^2)^{1/2} t + \frac{\alpha \sin(\beta^2 c^2 - \alpha^2)^{1/2} t}{(\beta^2 c^2 - \alpha^2)^{1/2}} \right) \frac{\cos \beta y}{\beta^2} d\beta \tag{19}$$

The magnetic field strength $H_E(y, t)$ of Eq. (8) becomes:

$$\begin{aligned} H_E(y, t) &= Z^{-1}E_1[Z(\sigma/s)^{1/2} - Z\sigma y - Z(\sigma/s)^{1/2} \\ &\quad + I'_{E3}(y, t) - I_{E4}(y, t)] + H_t(y) \\ &= Z^{-1}E_1[-Z\sigma y + I'_{E3}(y, t) - I_{E4}(y, t)] + H_t(y) \end{aligned} \tag{20}$$

The two terms $Z(\sigma/s)^{1/2}$ and $-Z(\sigma/s)^{1/2}$ cancel and $H_E(y, t)$ remains finite for $s = 0$.

We turn to Eq. (1). With Eq. (2.3-34) follows:

$$\begin{aligned} H_E(y, t) = -E_1 \int \bigg(& 2a\epsilon e^{-2at} e^{-y/L} + \epsilon \frac{\partial u}{\partial t} \\ & + \sigma(1 - e^{-2at})e^{-y/L} + \sigma u \bigg) dy + H_y(t) \end{aligned} \tag{21}$$

One obtains from Eqs. (2.3-25) and (2.3-24):

$$\begin{aligned} -\epsilon \int \frac{\partial u}{\partial t} dy = \frac{\epsilon}{2\pi} \int_0^\infty & [A_1(\kappa)\gamma_1 \exp(\gamma_1 t) \\ & + A_2(\kappa)\gamma_2 \exp(\gamma_2 t)]\kappa^{-1} \cos 2\pi\kappa y \, d\kappa \end{aligned} \tag{22}$$

$$-\sigma \int u \, dy = \frac{\sigma}{2\pi} \int_0^\infty [A_1(\kappa)\exp(\gamma_1 t) + A_2(\kappa)\exp(\gamma_2 t)]\kappa^{-1} \cos 2\pi\kappa y \, d\kappa \tag{23}$$

These two integrals are summed:

$$\begin{aligned} -\left(\epsilon \int \frac{\partial u}{\partial t} dy + \sigma \int u \, dy\right) = \frac{1}{2\pi} \int_0^\infty & [A_1(\kappa)(\epsilon\gamma_1 + \sigma)\exp(\gamma_1 t) \\ & + A_2(\kappa)(\epsilon\gamma_2 + \sigma)\exp(\gamma_2 t)] \frac{\cos \beta y}{\beta} d\beta \end{aligned} \tag{24}$$

One obtains with the help of Eqs. (2.3-31) and (2.1-50):

$$\begin{aligned} \frac{A_1(\kappa)(\epsilon\gamma_1 + \sigma)}{\beta} &= \frac{8a(\epsilon\gamma_1 + \sigma)}{b^2(\gamma_2 - \gamma_1)} = -\frac{4\epsilon a}{b^2}(1 + q_e), && a^2 > b^2c^2 \\ &= -\frac{4\epsilon a}{b^2}(1 - jq'_e), && a^2 < b^2c^2 \\ \frac{A_2(\kappa)(\epsilon\gamma_2 + \sigma)}{\beta} &= \frac{8a(\epsilon\gamma_2 + \sigma)}{b^2(\gamma_1 - \gamma_2)} = -\frac{4\epsilon a}{b^2}(1 - q_e), && a^2 > b^2c^2 \\ &= -\frac{4\epsilon a}{b^2}(1 + jq'_e), && a^2 < b_2c^2 \end{aligned} \tag{25}$$

The quantities q_e and q'_e are defined in Eq. (6).

Substitution of Eqs. (24) and (25) into Eq. (21) and integration of $e^{-y/L}$ brings:

$$
\begin{aligned}
H_E(y, t) = Z^{-1}E_1\Bigg[& Z\left(\frac{\sigma}{s}\right)^{1/2}\left(1+\frac{s}{Z^2\sigma}\right)e^{-2at}e^{-y/L} \\
& -\frac{2a}{\pi c}e^{-at}\Bigg(\int_0^{2\pi K}\{(1+q_e)\exp[(a^2-b^2c^2)^{1/2}t] \\
& +(1-q_e)\exp[-(a^2-b^2c^2)^{1/2}t]\}\frac{\cos\beta y}{b^2}\,d\beta \\
& +\int_{2\pi K}^{\infty}\{(1-jq_e')\exp[j(b^2c^2-a^2)^{1/2}t] \\
& +(1+jq_e')\exp[-j(b^2c^2-a^2)^{1/2}t]\}\frac{\cos\beta y}{b^2}\,d\beta\Bigg)\Bigg] + H_y(t)
\end{aligned}
\tag{26}
$$

A comparison with Eq. (7) shows that the condition

$$H_t(y)e^{-st/\mu} = H_y(t) = H_{E0}e^{-st/\mu} \tag{27}$$

must be satisfied. The initial condition of Eq. (2.3-5) requires $H_{E0} = 0$.

We rewrite $H_E(y, t)$ of Eq. (20) in normalized form using Eqs. (2.1-78) and (2.1-79):

$$H_E(\xi, \theta) = Z^{-1}E_1[-2\xi + I'_{E3}(\xi, \theta) - I_{E4}(\xi, \theta)] \tag{28}$$

$$
\begin{aligned}
I'_{E3}(\xi, \theta) = \frac{2}{\pi}\Bigg\{ & 2\Bigg[\frac{1}{\delta} - e^{-\theta}\int_\delta^1\Bigg(\mathrm{ch}(1-\eta^2)^{1/2}\theta \\
& +\frac{\mathrm{sh}(1-\eta^2)^{1/2}\theta}{(1-\eta^2)^{1/2}}\Bigg)\frac{d\eta}{\eta^2} \\
& -\int_0^\delta[\exp(-\eta^2\theta/2)-1]\frac{d\eta}{\eta^2}\Bigg] \\
& +e^{-\theta}\int_0^1\Bigg(\mathrm{ch}(1-\eta^2)^{1/2}\theta + \frac{\mathrm{sh}(1-\eta^2)^{1/2}\theta}{(1-\eta^2)^{1/2}}\Bigg) \\
& \times\left(\frac{\sin(\xi\,\eta/2)}{\eta/2}\right)^2 d\eta\Bigg\}
\end{aligned}
\tag{29}
$$

$$
\begin{aligned}
I_{E4}(\xi, \theta) = \frac{4}{\pi}e^{-\theta}\int_1^\infty & \Bigg(\cos(\eta^2-1)^{1/2}\theta + \frac{\sin(\eta^2-1)^{1/2}\theta}{(\eta^2-1)^{1/2}}\Bigg) \\
& \times\frac{\cos\xi\eta}{\eta^2}\,d\eta
\end{aligned}
\tag{30}
$$

$$\delta = 2d/Z\sigma \ll 1$$

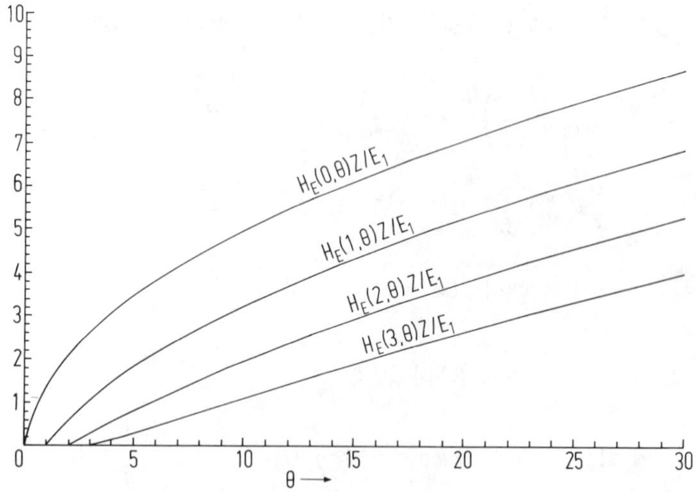

FIG. 2.9-1. Magnitude of the magnetic field strength due to excitation by an electric exponential ramp function for the locations $\xi = 0, 1, 2, 3$ in the time range $0 \leqq \theta \leqq 30$. This illustration is based on computer plots by M. Hussain of Kuwait University.

Plots of $H_E(\xi, \theta)Z/E_1$ for the locations $\xi = 0, 1, 2, 3$ are shown in Fig. 2.9-1 for the time range $0 \leqq \theta \leqq 30$ and in Fig. 2.9-2 for the time range $0 \leqq \theta \leqq 100$. The functions are zero for $\theta \leqq \xi$, and they vary proportionately to θ for large values of θ.

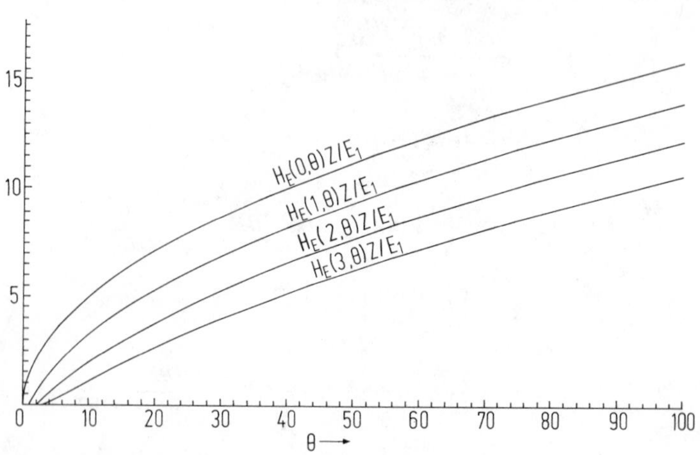

FIG. 2.9-2. Magnitude of the magnetic field strength due to excitation by an electric exponential ramp function for the locations $\xi = 0, 1, 2, 3$ in the time range $0 \leqq \theta \leqq 100$. This illustration is based on computer plots by M. Hussain of Kuwait University.

2.10 Magnetic Exponential Ramp Function Excitation

Let a magnetic exponential ramp function be applied at the time $t=0$ at the plane $y=0$,

$$\begin{aligned} H(0,t) &= H_0 r(t) = 0 && \text{for} \quad t<0 \\ &= H_0 Q(1-e^{-t/\tau}) = H_1(1-e^{-t/\tau}) && \text{for} \quad t \geqq 0 \end{aligned} \tag{1}$$

where $Q=(1-e^{-2a\Delta T})^{-1}$ is a constant and $r(t)$ is defined by Eq. (2.3-2). The calculation is the same as in Section 2.3, except that the electric field strength is replaced everywhere by the magnetic field strength. We obtain from Eq. (2.3-34)

$$H_H(y,t) = H_1[(1-e^{-2at})e^{-y/L} + u(y,t)], \qquad H_1 = H_0 Q \tag{2}$$

where $u(y,t)$ is defined by either Eq. (2.3-24) or (2.3-33). We first resolve the simple limit case $s=0$, $\sigma=0$. Equations (2), (2.3-1), and (2.3-39)–(2.3-42) yield:

$$\begin{aligned} H_H(y,t) &= H_0[t/\Delta T + u(y,t)] \\ &= H_0\left(\frac{t}{\Delta T} - \frac{2}{\pi c\,\Delta T}\int_0^\infty \frac{\sin\beta ct \sin\beta y}{\beta^2}\,d\beta\right) \end{aligned} \tag{3}$$

With the help of Eq. (2.3-44) we obtain:

$$\begin{aligned} H_H(y,t) &= 0 && \text{for} \quad t<y/c \\ &= \mathrm{H}_0(t/\Delta T - y/c\,\Delta T) && \text{for} \quad t \geqq y/c \end{aligned} \tag{4}$$

The electric field strength $E_H(y,t)$ follows from Eqs. (2.1-25), (2.1-27), and (4):

$$\begin{aligned} E_H(y,t) &= -\mu\int \frac{\partial H_H}{\partial t}\,dy + E_y(t) \\ &= E_y(t) && \text{for} \quad t<y/c \\ &= -ZH_0 y/c\,\Delta T + E_y(t) && \text{for} \quad t \geqq y/c \end{aligned} \tag{5}$$

$$\begin{aligned} E_H(y,t) &= -\epsilon^{-1}\int \frac{\partial H_H}{\partial y}\,dt + E_t(y) \\ &= E_t(y) && \text{for} \quad t<y/c \\ &= ZH_0 t/\Delta T + E_t(y) && \text{for} \quad t \geqq y/c \end{aligned} \tag{6}$$

Since Eqs. (5) and (6) must yield the same result, we get

$$\begin{aligned} &E_y(t) = E_t(y) = 0 && \text{for} \quad t<y/c \\ &E_y(t) = ZH_0 t/\Delta T, \qquad E_t(y) = -ZH_0 y/c\,\Delta T && \text{for} \quad t \geqq y/c \end{aligned} \tag{7}$$

and

$$
\begin{aligned}
E_H(y, t) &= 0 && \text{for} \quad t < y/c \\
&= ZH_0(t/\Delta T - y/c\,\Delta T) && \text{for} \quad t \geqq y/c
\end{aligned}
\tag{8}
$$

We turn to the more general case $\sigma > 0$. Equations (2) and (2.1-27) yield:

$$
\begin{aligned}
E_H(y, t) &= e^{-\sigma t/\epsilon}\left(-\epsilon^{-1}\int \frac{\partial H_H}{\partial y} e^{\sigma t/\epsilon}\,dt + E_t(y)\right) \\
&= e^{-\sigma t/\epsilon} H_1\left(\epsilon^{-1}(s\sigma)^{1/2} e^{-y/L}\int (e^{\sigma t/\epsilon} - e^{(\sigma/\epsilon - 2a)t})dt \right. \\
&\quad \left. - \int \frac{\partial u}{\partial y} e^{\sigma t/\epsilon}\,dt\right) + E_t(y) \\
&= H_1\left((s/\sigma)^{1/2} e^{-y/L}[1 + (Z^2\sigma/s)e^{-2at}] \right. \\
&\quad - 2\pi \int_0^\infty [A_1(\kappa)(\epsilon\gamma_1 + \sigma)^{-1}\exp(\gamma_1 t) \\
&\quad \left. + A_2(\kappa)(\epsilon\gamma_2 + \sigma)^{-1}\exp(\gamma_2 t)]\kappa \cos 2\pi\kappa y\,d\kappa\right) + E_t(y)e^{-\sigma t/\epsilon}
\end{aligned}
\tag{9}
$$

From Eqs. (2.3-31) and (2.1-50) follows, with q_e and q_e' defined by Eq. (2.9-6):

$$
\begin{aligned}
\frac{2\pi A_1(\kappa)\kappa}{\epsilon\gamma_1 + \sigma} &= \frac{4\mu a}{b^2}(1 - q_e), \qquad a^2 > b^2c^2 \\
&= \frac{4\mu a}{b^2}(1 + jq_e'), \qquad a^2 < b^2c^2 \\
\frac{2\pi A_2(\kappa)\kappa}{\epsilon\gamma_2 + \sigma} &= \frac{4\mu a}{b^2}(1 + q_e), \qquad a^2 > b^2c^2 \\
&= \frac{4\mu a}{b^2}(1 - jq_e'), \qquad a^2 < b^2c^2
\end{aligned}
\tag{10}
$$

Equations (9) and (10) yield:

$$
\begin{aligned}
E_H(y, t) = H_1 Z\Bigg[&Z^{-1}\left(\frac{s}{\sigma}\right)^{1/2}\left(1 + \frac{Z^2\sigma}{s}e^{-2at}\right)\exp[-(s\sigma)^{1/2}y] \\
&- \frac{2a}{\pi c}e^{-at}\left(\int_0^{2\pi K}\{(1 - q_e)\exp[(a^2 - b^2c^2)^{1/2}t]\right. \\
&+ (1 + q_e)\exp[-(a^2 - b^2c^2)^{1/2}t]\}\frac{\cos\beta y}{b^2}\,d\beta
\end{aligned}
$$

$$+\int_{2\pi K}^{\infty}\{(1+jq'_e)\exp[j(b^2c^2-a^2)^{1/2}t]$$

$$+(1-jq'_e)\exp[-j(b^2c^2-a^2)^{1/2}t]\}\frac{\cos\beta y}{b^2}\,d\beta\bigg)\bigg]$$

$$+E_t(y)e^{-\sigma t/\epsilon} \tag{11}$$

We rewrite this equation with the help of hyperbolic and trigonometric functions:

$$E_H(y,t)=H_1Z\left\{Z^{-1}\left(\frac{s}{\sigma}\right)^{1/2}\left(1+\frac{Z^2\sigma}{s}e^{-2at}\right)\exp[-(s\sigma)^{1/2}y]\right.$$

$$-\frac{4a}{\pi c}e^{-at}\left[\int_0^{2\pi K}\left(\mathrm{ch}(a^2-b^2c^2)^{1/2}t\right.\right.$$

$$\left.-\frac{(\alpha-s/2\mu)\mathrm{sh}(a^2-b^2c^2)^{1/2}t}{(a^2-b^2c^2)^{1/2}}\right)\frac{\cos\beta y}{b^2}\,d\beta$$

$$+\int_{2\pi K}^{\infty}\left(\cos(b^2c^2-a^2)^{1/2}t-\frac{(\alpha-s/2\mu)\sin(b^2c^2-a^2)^{1/2}t}{(b^2c^2-a^2)^{1/2}}\right)$$

$$\left.\left.\times\frac{\cos\beta y}{b^2}\,d\beta\right]\right\}+E_t(y)e^{-\sigma t/\epsilon} \tag{12}$$

Next we make the transition $s\to 0$ for $\sigma>0$. The first term yields:

$$\lim_{s\to 0}Z^{-1}(s/\sigma)^{1/2}(1+Z^2\sigma s^{-1}e^{-2at})\exp[-(s\sigma)^{1/2}y]$$

$$=Z[(\sigma/s)^{1/2}-\sigma y]e^{-2\alpha t} \tag{13}$$

The calculation of the two integrals is facilitated by the observation that the only difference between the integrals in Eqs. (12) and (2.9-8) is the changed sign in front of the hyperbolic and trigonometric sine functions. Instead of I_{E3} of Eq. (2.9-12) we may thus define I_{H3} as the first integral in Eq. (12) for $s\to 0$:

$$I_{H3}(y,t)=\frac{2Z\sigma}{\pi}e^{-\alpha t}\left[\int_0^{Z\sigma/2}\left(\mathrm{ch}(\alpha^2-\beta^2c^2)^{1/2}t\right.\right.$$

$$\left.-\frac{\alpha\,\mathrm{sh}(\alpha^2-\beta^2c^2)^{1/2}t}{(\alpha^2-\beta^2c^2)^{1/2}}\right)\frac{d\beta}{\beta^2+s\sigma}$$

$$-2\int_0^{Z\sigma/2}\left(\mathrm{ch}(\alpha^2-\beta^2c^2)^{1/2}t\right.$$

$$\left.\left.-\frac{\alpha\,\mathrm{sh}(\alpha^2-\beta^2c^2)^{1/2}t}{(\alpha^2-\beta^2c^2)^{1/2}}\right)\frac{\sin^2(\beta y/2)}{\beta^2+s\sigma}\,d\beta\right] \tag{14}$$

We split this integral into three components:

$$I_{H3}(y, t) = -I_{41}(y, t) + I_{12}(t) - I_{13}(t) \tag{15}$$

where $I_{41}(y, t)$ is the second integral in Eq. (14), which remains finite for $s = 0$

$$I_{41}(y, t) = \frac{Z\sigma}{\pi} e^{-\alpha t} \int_0^{Z\sigma/2} \left(\operatorname{ch}(\alpha^2 - \beta^2 c^2)^{1/2} t - \frac{\alpha \operatorname{sh}(\alpha^2 - \beta^2 c^2)^{1/2} t}{(\alpha^2 - \beta^2 c^2)^{1/2}} \right) \left(\frac{\sin(\beta y/2)}{\beta/2} \right)^2 d\beta \tag{16}$$

and can be evaluated by computer. The other two components of Eq. (15) are defined by Eqs. (2.4-16) and (2.4-17). We write first $I_{12}(t) - I_{13}(t)$ according to Eqs. (2.4-35) and (2.4-36):

$$\begin{aligned} I_{12}(t) - I_{13}(t) = Z \Bigg\{ & \left[\left(\frac{\sigma}{s} \right)^{1/2} - \frac{2\sigma}{\pi d} \right] e^{-2\alpha t} \\ & + \frac{2\sigma}{\pi} e^{-\alpha t} \int_d^{Z\sigma/2} \left(\operatorname{ch}(\alpha^2 - \beta^2 c^2)^{1/2} t - \frac{\alpha \operatorname{sh}(\alpha^2 - \beta^2 c^2)^{1/2} t}{(\alpha^2 - \beta^2 c^2)^{1/2}} \right) \frac{d\beta}{\beta^2} \\ & + \frac{2\sigma}{\pi} e^{-2\alpha t} \int_0^d \frac{\operatorname{sh}^2(\beta^2 c^2 t/4\alpha)}{\beta^2} d\beta \\ & + \frac{2\sigma}{\pi} e^{-2\alpha t} \int_0^d \frac{\operatorname{sh}(\beta^2 c^2 t/2\alpha)}{\beta^2} d\beta \Bigg\}, \end{aligned}$$

$$d \ll Z\sigma/2 = \alpha/c \tag{17}$$

We observe that the last term becomes arbitrarily small for any value of t if d is made sufficiently small, and we ignore it. With Eqs. (15) and (16) we get:

$$\begin{aligned} I_{H3}(y, t) = Z \Bigg\{ & \left[\left(\frac{\sigma}{s} \right)^{1/2} - \frac{2\sigma}{\pi d} \right] e^{-2\alpha t} + \frac{2\sigma}{\pi} e^{-\alpha t} \int_d^{Z\sigma/2} \left(\operatorname{ch}(\alpha^2 - \beta^2 c^2)^{1/2} t \right. \\ & \left. - \frac{\alpha \operatorname{sh}(\alpha^2 - \beta^2 c^2)^{1/2} t}{(\alpha^2 - \beta^2 c^2)^{1/2}} \right) \frac{d\beta}{\beta^2} + \frac{2\sigma}{\pi} e^{-2\alpha t} \int_0^d \frac{\operatorname{sh}^2(\beta^2 c^2 t/4\alpha)}{\beta^2} d\beta \\ & - \frac{\sigma}{\pi} e^{-\alpha t} \int_0^{Z\sigma/2} \left(\operatorname{ch}(\alpha^2 - \beta^2 c^2)^{1/2} t - \frac{\alpha \operatorname{sh}(\alpha^2 - \beta^2 c^2)^{1/2} t}{(\alpha^2 - \beta^2 c^2)^{1/2}} \right) \\ & \times \left(\frac{\sin(\beta y/2)}{\beta/2} \right)^2 d\beta \Bigg\} \end{aligned} \tag{18}$$

The function $I_{H3}(y, t)$ without the term $Z(\sigma/s)^{1/2}e^{-2\alpha t}$ is denoted $-I'_{H3}(y, t)$:

$$-I'_{H3}(y, t) = I_{H3}(y, t) - Z(\sigma/s)^{1/2}e^{-2\alpha t} \tag{19}$$

We turn to the second integral in Eq. (12). It remains finite for $s \to 0$ if the electric conductivity σ is larger than zero:

$$I_{H4}(y, t) = \frac{2Z\sigma}{\pi} e^{-\alpha t} \int_{Z\sigma/2}^{\infty} \left(\cos(\beta^2 c^2 - \alpha^2)^{1/2} t - \frac{\alpha \sin(\beta^2 c^2 - \alpha^2)^{1/2} t}{(\beta^2 c^2 - \alpha^2)^{1/2}} \right) \frac{\cos \beta y}{\beta^2} \, d\beta \tag{20}$$

The electric field strength of Eq. (12) becomes:

$$\begin{aligned} E_H(y, t) &= H_1 Z\{[Z(\sigma/s)^{1/2} - Z\sigma y - Z(\sigma/s)^{1/2}]e^{-2\alpha t} \\ &\quad + I'_{H3}(y, t) - I_{H4}(y, t)\} + E_t(y)e^{-\sigma t/\epsilon} \\ &= H_1 Z[-Z\sigma y e^{-2\alpha t} + I'_{H3}(y, t) - I_{H4}(y, t)] + E_t(y)e^{-\sigma t/\epsilon} \end{aligned} \tag{21}$$

The two terms $Z(\sigma/s)^{1/2}$ and $-Z(\sigma/s)^{1/2}$ cancel and $E_H(y, t)$ remains finite for $s \to 0$.

We now substitute Eq. (2) into Eq. (2.1-25) instead of (2.1-27):

$$\begin{aligned} E_H(y, t) &= -\int \left(\mu \frac{\partial H_H}{\partial t} + sH_H \right) dy + E_y(t) \\ &= -H_1 \left([2\mu a e^{-2at} + s(1 - e^{-2at})] \int e^{-y/L} \, dy + \mu \int \frac{\partial u}{\partial t} dy + s \int u \, dy \right) + E_y(t) \\ &= H_1 \left[\left(\frac{s}{\sigma} \right)^{1/2} e^{-y/L} \left(1 + \frac{Z^2 \sigma}{s} e^{-2at} \right) - \mu \int \frac{\partial u}{\partial t} dy - s \int u \, dy \right] + E_y(t) \end{aligned} \tag{22}$$

From Eqs. (2.3-25) and (2.3-24) follows:

$$-\mu \int \frac{\partial u}{\partial t} dy = \frac{\mu}{2\pi} \int_0^{\infty} [A_1(\kappa)\gamma_1 \exp(\gamma_1 t) + A_2(\kappa)\gamma_2 \exp(\gamma_2 t)]\kappa^{-1} \cos 2\pi\kappa y \, d\kappa \tag{23}$$

$$-s \int u \, dy = \frac{s}{2\pi} \int_0^{\infty} [A_1(\kappa)\exp(\gamma_1 t) + A_2(\kappa)\exp(\gamma_2 t)]\kappa^{-1} \cos 2\pi\kappa y \, d\kappa \tag{24}$$

Insertion into Eq. (22) yields:

$$E_H(y,t) = H_1\Big((s/\sigma)^{1/2}e^{-y/L}(1 + Z^2\sigma s^{-1}e^{-2at}) + (2\pi)^{-1}\int_0^\infty [A_1(\kappa)(\mu\gamma_1 + s)\exp(\gamma_1 t) + A_2(\kappa)(\mu\gamma_2 + s)\exp(\gamma_2 t)]\beta^{-1}\cos\beta y\, d\beta\Big) + E_y(t) \quad (25)$$

One obtains with the help of Eqs. (2.3-31) and (2.1-50) with q_e and q_e' defined in Eq. (2.9-6):

$$\begin{aligned}
\frac{A_1(\kappa)(\mu\gamma_1 + s)}{\beta} &= \frac{8a(\mu\gamma_1 + s)}{b^2(\gamma_2 - \gamma_1)} = \frac{-4\mu a}{b^2}(1 - q_e), && a^2 > b^2c^2 \\
&= \frac{-4\mu a}{b^2}(1 + jq_e'), && a^2 < b^2c^2 \\
\frac{A_2(\kappa)(\mu\gamma_2 + s)}{\beta} &= \frac{8a(\mu\gamma_2 + s)}{b^2(\gamma_1 - \gamma_2)} = \frac{-4\mu a}{b^2}(1 + q_e), && a^2 > b^2c^2 \\
&= \frac{4\mu a}{b^2}(1 - jq_e'), && a^2 < b^2c^2
\end{aligned} \quad (26)$$

Substitution into Eq. (25) yields:

$$\begin{aligned}
E_H(y,t) = H_1 Z\Bigg[&Z^{-1}\left(\frac{s}{\sigma}\right)^{1/2}\left(1 + \frac{Z^2\sigma}{s}e^{-2at}\right)\exp[-(s\sigma)^{1/2}y] \\
&-\frac{2a}{\pi c}e^{-at}\Bigg(\int_0^{2\pi K}\{(1 - q_e)\exp[(a^2 - b^2c^2)^{1/2}t] \\
&+ (1 + q_e)\exp[-(a^2 - b^2c^2)^{1/2}t]\}\frac{\cos\beta y}{b^2}\,d\beta \\
&+ \int_{2\pi K}^\infty \{(1 + jq_e')\exp[j(b^2c^2 - a^2)^{1/2}t] \\
&+ (1 - jq_e')\exp[-j(b^2c^2 - a^2)^{1/2}t]\}\frac{\cos\beta y}{b^2}\,d\beta\Bigg)\Bigg] + E_y(t)
\end{aligned} \quad (27)$$

Comparison with Eq. (11) shows that the condition

$$E_t(y)e^{-\sigma t/\epsilon} = E_y(t) = E_{H0}e^{-\sigma t/\epsilon} \quad (28)$$

must be satisfied. The initial condition of Eq. (2.3-5) requires $E_{H0} = 0$.

We rewrite $H_H(y,t)$ of Eq. (2) for $s = 0$ in normalized form using Eqs.

(2.1-78) and (2.1-79):

$$H_H(\xi, \theta) = H_1[1 - e^{-2\theta} + u(\xi, \theta)] \tag{29}$$

$$u(\xi, \theta) = -\frac{4}{\pi} e^{-\theta} \left(\int_0^1 \frac{\operatorname{sh}(1-\eta^2)^{1/2}\theta}{(1-\eta^2)^{1/2}} \frac{\sin \xi\eta}{\eta} d\eta + \int_1^\infty \frac{\sin(\eta^2-1)^{1/2}\theta}{(\eta^2-1)^{1/2}} \frac{\sin \xi\eta}{\eta} d\eta \right) \tag{30}$$

Furthermore, we rewrite $E_H(y, t)$ of Eq. (21) for $E_t(y) = 0$:

$$E_H(\xi, \theta) = H_1 Z[-2\xi e^{-2\theta} + I'_{H3}(\xi, \theta) - I_{H4}(\xi, \theta)] \tag{31}$$

$$I'_{H3}(\xi, \theta) = \frac{2}{\pi} \left\{ 2 \left[\frac{1}{\delta} e^{-2\theta} - e^{-\theta} \int_\delta^1 \left(\operatorname{ch}(1-\eta^2)^{1/2}\theta - \frac{\operatorname{sh}(1-\eta^2)^{1/2}\theta}{(1-\eta^2)^{1/2}} \right) \frac{d\eta}{\eta^2} - e^{-2\theta} \int_0^\delta \frac{\operatorname{sh}^2(\eta^2\theta/4)}{\beta^2} d\beta \right] + e^{-\theta} \int_0^1 \left(\operatorname{ch}(1-\eta^2)^{1/2}\theta - \frac{\operatorname{sh}(1-\eta^2)^{1/2}\theta}{(1-\eta^2)^{1/2}} \right) \times \left(\frac{\sin(\xi\eta/2)}{\eta/2} \right)^2 d\eta \right\} \tag{32}$$

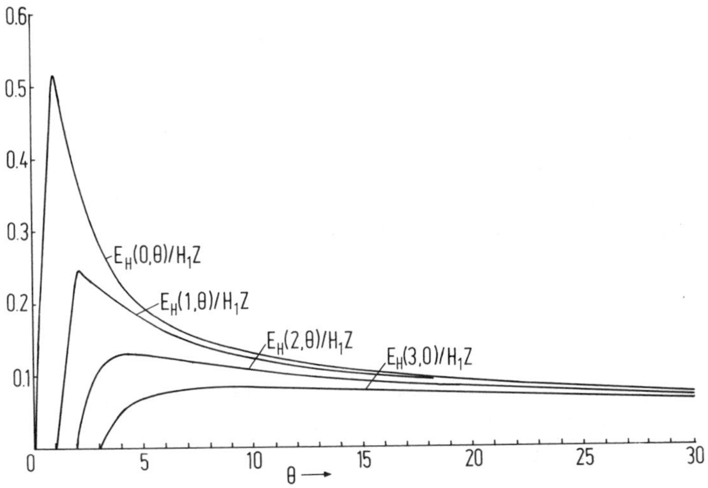

FIG. 2.10-1. Magntitude of the electric field strength due to excitation by a magnetic exponential ramp function for the locations $\xi = 0, 1, 2, 3$ in the time range $0 \leqq \theta \leqq 30$. This illustration is based on computer plots by R. Boules of University of Alexandria, Egypt, and M. Hussain of Kuwait University.

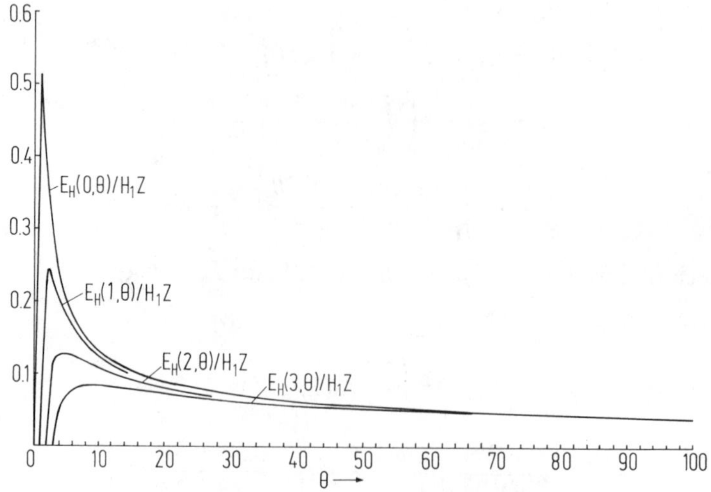

FIG. 2.10-2. Magnitude of the electric field strength due to excitation by a magnetic exponential ramp function for the locations $\xi = 0, 1, 2, 3$ in the time range $0 \leqq \theta \leqq 100$. This illustration is based on computer plots by M. Hussain of Kuwait University and R. Boules of University of Alexandria, Egypt.

$$I_{H4}(\xi, \theta) = \frac{4}{\pi} e^{-\theta} \int_1^{\infty} \left(\cos(\eta^2 - 1)^{1/2}\theta - \frac{\sin(\eta^2 - 1)^{1/2}\theta}{(\eta^2 - 1)^{1/2}} \right) \frac{\cos \xi\eta}{\eta^2} d\eta \quad (33)$$

$$\delta = 2d/Z\sigma \ll 1$$

The function $H_H(\xi, \theta)$ of Eq. (29) differs by a constant factor only from the function $E_E(\xi, \theta)$ of Eq. (2.3-37), for which plots are shown in Figs. 2.3-2 and 2.3-3. Plots of $E_H(\xi, \theta)$ are shown in Figs. 2.10-1 and 2.10-2. These plots should be compared with the plots of Fig. 2.6-1.

2.11 Boundary Conditions Using Ramp Functions

Let the two arbitrary excitation functions $E(0, t)S(t)$ and $H(0, t)S(t)$ of Eqs. (2.8-3) and (2.8-4) be represented by exponential ramp functions according to Eq. (1.3-33) rather than step functions:

$$E(0, t) = \tau \int_0^t \frac{d^2E(0, t'')}{dt''^2} S(t - t'')(1 - e^{-(t-t'')/\tau}) dt'' \quad (1)$$

$$H(0, t) = \tau \int_0^t \frac{d^2H(0, t'')}{dt''^2} S(t - t'')(1 - e^{-(t-t'')/\tau}) dt'' \quad (2)$$

These equations were derived for the exponential ramp function

$$QS(t)(1-e^{-t/\tau}) = (1-e^{-\Delta T/\tau})^{-1}S(t)(1-e^{-t/\tau}) \tag{3}$$

for which we also solved Maxwell's modified equations. Hence, we get from Eq. (1.3-38) the waves produced by $E(0, t)$ and $H(0, t)$ if we insert for $\tau h_0(t)$ in Eq. (1.3-38) the solutions E_E, H_E, H_H, and E_H from Eqs. (2.3-36), (2.9-20), (2.10-2), and (2.10-21). For E_E we get for $2\alpha\,\Delta T \ll 1$

$$E_E(y, t) = E_0/2\alpha\,\Delta T)[1-e^{-2\alpha t}+u(y, t)] \tag{4}$$

where $u(y, t)$ is defined by Eq. (2.3-35). For H_E we obtain from Eq. (2.9-20) with $H_t(y) = 0$ and

$$E_1 = E_0 Q = E_0/2\alpha\,\Delta T \tag{5}$$

the equation

$$H_E(y, t) = (E_0/2\alpha\,\Delta T)Z^{-1}[-Z\sigma y + I'_{E3}(y, t) - I_{E4}(y, t)] \tag{6}$$

where I'_{E3} and I_{E4} are defined by Eqs. (2.9-18) and (2.9-19).

We insert $E_E(y, t)$ for $h(t/\Delta T)$ in Eq. (1.3-37):

$$\begin{aligned}&\lim_{\Delta T\to dt} \Delta T(E_0/2\alpha\,\Delta T)[1-e^{-2\alpha t}+u(y, t)]\\ &\quad = E_0(2\alpha)^{-1}[1-e^{-2\alpha t}+u(y, t)]\end{aligned} \tag{7}$$

We recognize the two relations

$$\tau = 1/2\alpha \tag{8}$$

and

$$h_0(t) = E_0[1-e^{-2\alpha t}+u(y, t)] \tag{9}$$

From Eq. (1.3-38) follows:

$$E_E(y, t) = (2\alpha)^{-1}\int_0^t \frac{d^2E(0, t'')}{dt''^2}S(t-t'')[1-e^{-2\alpha(t-t'')}+u(y, t-t'')]dt'' \tag{10}$$

For $H_E(y, t)$ we get from Eq. (1.3-37):

$$\begin{aligned}&\lim_{\Delta T\to dt} \Delta T\, Z^{-1}E_0(2\alpha\,\Delta T)^{-1}[-Z\sigma y + I'_{E3}(y, t) - I_{E4}(y, t)]\\ &\quad = \tau E_0 Z^{-1}[-Z\sigma y + I'_{E3}(y, t) - I_{E4}(y, t)]\end{aligned} \tag{11}$$

Again we get $\tau = 1/2\alpha$. The magnetic field strength follows from Eq. (1.3-38):

$$\begin{aligned}H_E(y, t) = (2\alpha Z)^{-1}&\int_0^t \frac{d^2E(0, t'')}{dt''^2}S(t-t'')\\ &\times[-Z\sigma y + I'_{E3}(y, t-t'') - I_{E4}(y, t-t'')]dt''\end{aligned} \tag{12}$$

For excitation by a magnetic exponential ramp function we get from Eq. (2.10-2) for $s = 0$ and $2\alpha\,\Delta T \ll 1$:

$$H_H(y, t) = H_0(2\alpha\,\Delta T)^{-1}[1 - e^{-2\alpha t} + u(y, t)] \tag{13}$$

where $u(y, t)$ is defined by Eq. (2.3-35). For $E_H(y, t)$ we obtain from Eq. (2.10-21) with $E_t(y) = 0$

$$E_H(y, t) = H_0 Z(2\alpha\,\Delta T)^{-1}[-Z\sigma y e^{-2\alpha t} + I'_{H3}(y, t) - I_{H4}(y, t)] \tag{14}$$

where I'_{H3} and I_{H4} are defined by Eqs. (2.10-19) and (2.10-20). In analogy to Eqs. (10) and (12) we obtain for excitation with a magnetic field strength $\mathbf{H}(0, t)S(t)$ having a general time variation:

$$\begin{aligned} H_H(y, t) = {} & (2\alpha)^{-1} \int_0^t \frac{d^2H(0, t'')}{dt''^2} S(t - t'') \\ & \times [1 - e^{-2\alpha(t-t'')} + u(y, t - t'')]dt'' \end{aligned} \tag{15}$$

$$\begin{aligned} E_H(y, t) = {} & Z(2\alpha)^{-1} \int_0^t \frac{d^2H(0, t'')}{dt''^2} S(t - t'') \\ & \times [-Z\sigma y e^{-2\alpha(t-t'')} + I'_{H3}(y, t - t'') - I_{H4}(y, t - t'')]dt'' \end{aligned} \tag{16}$$

If both an electric and a magnetic field strength $\mathbf{E}(0, t)S(t)$ and $\mathbf{H}(0, t)S(t)$ with general time variation are applied, we obtain the sum of the magnitudes of the field strengths produced by them:

$$\begin{aligned} E(y, t) &= E_E(y, t) + E_H(y, t) \\ &= \frac{1}{2\alpha}\left(\int_0^t \frac{d^2E(0, t'')}{dt''^2} S(t - t'')[1 - e^{-2\alpha(t-t'')} + u(y, t - t'')]dt'' \right. \\ &\quad + Z \int_0^t \frac{d^2H(0, t'')}{dt''^2} S(t - t'')[-Z\sigma y e^{-2\alpha(t-t'')} + I'_{H3}(y, t - t'') \\ &\quad \left. - I_{H4}(y, t - t'')]dt'' \right) \end{aligned} \tag{17}$$

$$\begin{aligned} H(y, t) &= H_H(y, t) + H_E(y, t) \\ &= \frac{1}{2\alpha}\left(\int_0^t \frac{d^2H(0, t'')}{dt''^2} S(t - t'')[1 - e^{-2\alpha(t-t'')} + u(y, t - t'')]dt'' \right. \\ &\quad + \frac{1}{Z} \int_0^t \frac{d^2E(0, t'')}{dt''^2} S(t - t'')[-Z\sigma y + I'_{E3}(y, t - t'') \\ &\quad \left. - I_{E4}(y, t - t'')]dt'' \right) \end{aligned} \tag{18}$$

The value of these equations lies, of course, strictly in the simplicity provided for computer evaluation. Hence, we rewrite them into their discrete form using Eq. (1.3-40):

$$\begin{aligned}E(y, k) = {} & \sum_{i=0}^{k} [E(0, i+1) - (1 + e^{-\Delta T/\tau})E(0, i) \\ & + e^{-\Delta T/\tau}E(0, i-1)]S(k-i) \\ & \times [1 - e^{-(k-i)\Delta T/\tau} + u(y, k-i)] \\ & + Z \sum_{i=0}^{k} [H(0, i+1) - (1 + e^{-\Delta T/\tau})H(0, i) \\ & + e^{-\Delta T/\tau}H(0, i-1)]S(k-i) \\ & \times [-Z\sigma y e^{-(k-i)\Delta T/\tau} + I'_{H3}(y, k-i) - I_{H4}(y, k-i)] \end{aligned} \tag{19}$$

$$\begin{aligned}H(y, k) = {} & \sum_{i=0}^{k} [H(0, i+1) - (1 + e^{-\Delta T/\tau})H(0, i) \\ & + e^{-\Delta T/\tau}H(0, i-1)]S(k-i) \\ & \times [1 - e^{-(k-i)\Delta T/\tau} + u(y, k-i)] \\ & + Z^{-1} \sum_{i=0}^{k} [E(0, i+1) - (1 + e^{-\Delta T/\tau})E(0, i) \\ & + e^{-\Delta T/\tau}E(0, i-1)]S(k-i) \\ & \times [-Z\sigma y + I'_{E3}(y, k-i) - I_{E4}(y, k-i)], \end{aligned} \tag{20}$$

$$k\,\Delta T = t, \qquad \tau = 1/2\alpha$$

For the plane $y = 0$ we obtain from Eqs. (2.3-35), (2.10-19), (2.10-20), (2.9-18), and (2.9-19):

$$u(0, t) = 0 \tag{21}$$

$$\begin{aligned}I'_{H3}(0, t) = {} & \frac{2Z\sigma}{\pi}\left[\frac{1}{d}e^{-2\alpha t} - e^{-\alpha t}\int_{d}^{Z\sigma/2}\left(\operatorname{ch}(\alpha^2 - \beta^2c^2)^{1/2}t\right.\right. \\ & \left.\left. - \frac{\alpha\operatorname{sh}(\alpha^2 - \beta^2c^2)^{1/2}t}{(\alpha^2 - \beta^2c^2)^{1/2}}\right)\frac{d\beta}{\beta^2} - 2\int_0^d \frac{\operatorname{sh}^2(\beta^2c^2t/4\alpha)}{\beta^2}\,d\beta\right] \end{aligned} \tag{22}$$

$$\begin{aligned}I_{H4}(0, t) = {} & \frac{2Z\sigma}{\pi}e^{-\alpha t}\int_{Z\sigma/2}^{\infty}\left(\cos(\beta^2c^2 - \alpha^2)^{1/2}t\right. \\ & \left. - \frac{\alpha\sin(\beta^2c^2 - \alpha^2)^{1/2}t}{(\beta^2c^2 - \alpha^2)^{1/2}}\right)\frac{d\beta}{\beta^2} \end{aligned} \tag{23}$$

$$I'_{E3}(0,t) = \frac{2Z\sigma}{\pi}\left[\frac{1}{d} - e^{-\alpha t}\int_d^{Z\sigma/2}\left(\operatorname{ch}(\alpha^2-\beta^2c^2)^{1/2}t + \frac{\alpha\operatorname{sh}(\alpha^2-\beta^2c^2)^{1/2}t}{(\alpha^2-\beta^2c^2)^{1/2}}\right)\frac{d\beta}{\beta^2} - \int_0^d[\exp(-\beta^2c^2t/2\alpha)-1]\frac{d\beta}{\beta^2}\right] \tag{24}$$

$$I_{E4}(0,t) = \frac{2Z\sigma}{\pi}e^{-\alpha t}\int_{Z\sigma/2}^{\infty}\left(\cos(\beta^2c^2-\alpha^2)^{1/2}t + \frac{\alpha\sin(\beta^2c^2-\alpha^2)^{1/2}t}{(\beta^2c^2-\alpha^2)^{1/2}}\right)\frac{d\beta}{\beta^2} \tag{25}$$

$$d \ll Z\sigma/2 = \alpha/c$$

The impedance $Z(0, t) = E(0, t)/H(0, t)$ at the plane $y = 0$ becomes, with the help of Eq. (1.3-38):

$$\begin{aligned} Z(0,t) = {} & \left(E(0,t) + \frac{Z}{2\alpha}\int_0^t \frac{d^2H(0,t'')}{dt''^2}S(t-t'') \times [I'_{H3}(0,t-t'') - I_{H4}(0,t-t'')]dt''\right) \\ & \div \left(H(0,t) + \frac{1}{2Z\alpha}\int_0^t \frac{d^2E(0,t'')}{dt''^2}S(t-t'') \times [I'_{E3}(0,t-t'') - I_{E4}(0,t-t'')]dt''\right), \\ & Z = (\mu/\epsilon)^{1/2}, \qquad \tau = 1/2\alpha, \qquad \alpha = \sigma/2\epsilon \end{aligned} \tag{26}$$

3 Space–Time Variation of Excitation of Waves

3.1 Propagating Excitation of Waves

In Chapter 2 we have analyzed the waves produced in a medium excited by electric or magnetic step and exponential ramp functions. The exciting field strengths $\mathbf{E}(0, t)$ or $\mathbf{H}(0, t)$ were applied simultaneously at all points of the plane $y = 0$. We now consider the more general case where this excitation is not simultaneous but the points with coordinates x in the plane $y = 0$ are excited at the time $t = (x/c)\sin \vartheta$. Instead of the exciting field strengths $\mathbf{E}(0, t)$ or $\mathbf{H}(0, t)$ we thus have $\mathbf{E}[x, 0, t - (x/c)\sin \vartheta]$ and $\mathbf{H}[x, 0, t - (x/c)\sin \vartheta]$.

In order to show the reason for an investigation of such propagating boundary conditions consider Fig. 3.1-1. Let electrodes be placed at the locations $x = \ldots, x_{-2}, x_{-1}, 0, x_1, x_2, \ldots$. A planar wavefront $L(t)$ shall reach the line $x = 0, y = 0$ at the time t. The line $x = x_4, y = 0$ will be reached by the wavefront at the time $t + (x_4/c)\sin \vartheta$, if the wavefront propagates with velocity c and has the angle of incidence ϑ. If this wavefront triggers the excitation $E(0, 0, t)$ of the electrodes at $x = 0, y = 0$, it will trigger the excitation $E[x_i, 0, t - (x_i/c)\sin \vartheta]$ of the electrodes at $x = x_i, y = 0$. Our investigation of waves caused by a propagating excitation is thus an introduction to the study of reflected and transmitted waves caused by an incident planar wavefront with general time variation and angle of incidence ϑ. In particular, we will later study the case where the plane $y = 0$ separates a loss-free medium from a lossy medium.

We assume that a TEM wave that satisfies the modified Maxwell equations exists in the half-space $y \geqq 0$ of Fig. 3.1-1. Since the excitation in the plane $y = 0$ is only a function of x but not of z, it is clear that the field strengths should not vary with z:

$$\begin{aligned} \partial E_x/\partial z = \partial E_y/\partial z = \partial E_z/\partial z = 0 \\ \partial H_x/\partial z = \partial H_y/\partial z = \partial H_z/\partial z = 0 \end{aligned} \tag{1}$$

Writing the operator curl in cartesian coordinates and introducing Eq. (1) brings the modified Maxwell equations of Eqs. (2.1-2) and (2.1-3) into the form:

$$\partial H_y/\partial x - \partial H_x/\partial y = \epsilon\, \partial E_z/\partial t + \sigma E_z \tag{2}$$

$$\partial H_z/\partial y = \epsilon\, \partial E_x/\partial t + \sigma E_x \tag{3}$$

$$-\partial H_z/\partial x = \epsilon\, \partial E_y/\partial t + \sigma E_y \tag{4}$$

$$-\partial E_y/\partial x + \partial E_x/\partial y = \mu\, \partial H_z/\partial t + sH_z \tag{5}$$

$$-\partial E_z/\partial y = \mu\, \partial H_x/\partial t + sH_x \tag{6}$$

$$\partial E_z/\partial x = \mu\, \partial H_y/\partial t + sH_y \tag{7}$$

We now distinguish two types of waves, the perpendicularly polarized TEM wave and the parallel polarized TEM wave.

Assume that the electrodes in the plane $y = 0$ of Fig. 3.1-1 are excited so that an electric field strength is produced in the direction z

$$\mathbf{E} = E_z \mathbf{e}_z = E_z \mathbf{z}/z \tag{8}$$

as shown in Fig. 3.1-2. The notation $\mathbf{e}_z = \mathbf{z}/z$ indicates a unit vector in the direction of the z axis. For the magnetic field strength we assume an excitation in the direction of x and y but not in the direction z:

$$\mathbf{H} = H_x \mathbf{e}_x + H_y \mathbf{e}_y \tag{9}$$

Nothing in Fig. 3.1-2 suggests that the electric and magnetic field strengths of the excited wave will develop components E_x, E_y, and H_z during propagation. Hence, we will look for a solution of Eqs. (2)–(7) that satisfies for $y \geqq 0$ the conditions

$$E_x = E_y = 0 \tag{10}$$

$$H_z = 0 \tag{11}$$

We call this the perpendicularly polarized[1] TEM wave solution. Introducing Eqs. (10) and (11) into Eqs. (2)–(7), we obtain:

$$\partial H_y/\partial x - \partial H_x/\partial y = \epsilon\, \partial E_z/\partial t + \sigma E_z \tag{12}$$

$$-\partial E_z/\partial y = \mu\, \partial H_x/\partial t + sH_x \tag{13}$$

$$\partial E_z/\partial x = \mu\, \partial H_y/\partial t + sH_y \tag{14}$$

From Fig. 3.1-1 we recognize the relation

$$\lim_{\Delta x \to 0} \Delta y/\Delta x = dy/dx = \operatorname{tg} \vartheta \tag{15}$$

which holds for perpendicularly as well as parallel polarized waves. In addi-

[1] The perpendicularly polarized wave has an electric field strength **E** in the direction of z only, or perpendicularly to the *plane of incidence* $z =$ constant. The parallel polarized wave has an electric field strength **E** in the plane of incidence only; the component E_z is zero.

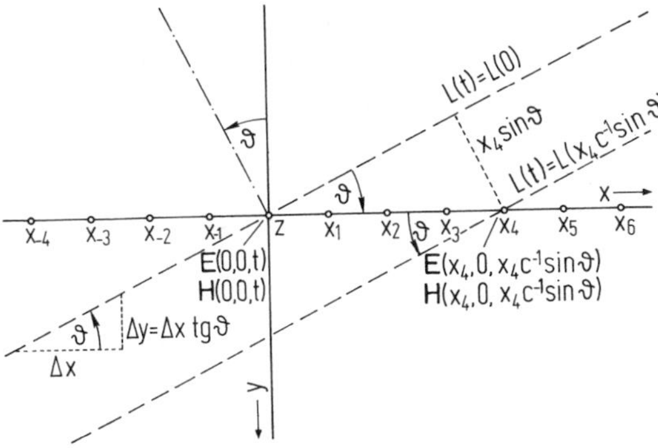

FIG. 3.1-1. General relations for the excitation of waves at the plane $y = 0$ by a wavefront $L(t)$ with angle of incidence ϑ.

tion we recognize from Fig. 3.1-2 the condition

$$H_y(x, y, t) = -H_x(x, y, t)\mathrm{tg}\,\vartheta \tag{16}$$

that holds for perpendicularly polarized waves only. With Eqs. (15) and (16) we may replace Eqs. (12) to (14) by two equations:

$$-(\partial H_x/\partial y)\cos^{-2}\vartheta = \mu\,\partial E_z/\partial t + \sigma E_z \tag{17}$$

$$-\partial E_z/\partial y = \mu\,\partial H_x/\partial t + sH_x \tag{18}$$

These two equations are the generalization of Eqs. (2.1-8) and (2.1-11) for $\vartheta \neq 0$.

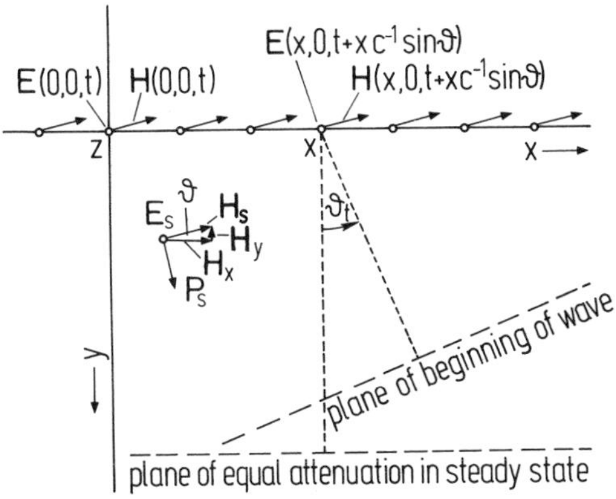

FIG. 3.1-2. Relations for the excitation of waves at the plane $y = 0$ by a perpendicularly polarized planar wavefront with angle of incidence ϑ.

We know from Fresnel's equations for sinusoidal waves in a conducting medium that one has to distinguish between planes of equal attenuation, which are parallel to the plane $y = 0$, and planes of equal phase, which have a different direction. For a wave excited by a step function we will expect the same plane of equal attenuation *in the steady-state limit* $t \to \infty$, as shown in Fig. 3.1-2. The concept of planes of equal phase is only applicable to sinusoidal waves. For a wave that begins at a certain time, we may define a *plane of the beginning of a wave.*

The transmission angle ϑ_t of the plane of the beginning of a wave can be readily determined if we make the *assumption* that the beginning of the wave propagates at all times with the velocity c_t. This appears to be a reasonable assumption, and we shall see in Chapter 5 that it is indeed justified if the signal-to-noise ratio is infinite and one has infinitely sensitive detection equipment. According to Fig. 3.1-3 the plane of propagating excitation $L(t)$ has the angle ϑ with the plane $y = 0$. It travels the distance $x_1 \sin \vartheta$ and reaches the point x_1 with a delay $(x_1/c) \sin \vartheta$ relative to the point $x = 0$. The wave excited in the point $x = 0$ will require the time $(x_1/c_t) \sin \vartheta_t$ to propagate the distance $x_1 \sin \vartheta_t$. To make the two times equal, the relation

$$c^{-1} \sin \vartheta = c_t^{-1} \sin \vartheta_t \tag{19}$$

must be satisfied. This is Snell's law of refraction if the propagating excitation is due to a planar wavefront in a medium 1 for $y < 0$, while the excited wave propagates in a medium 2 for $y \geqq 0$. We see that this law requires assumptions about the velocity c_t and the signal-to-noise ratio. It will turn out that the plane of the beginning of a wave in Fig. 3.1-3 in a lossy material is generally the curved wavefront shown, although there is a planar wavefront close to x_1 for step function excitation.

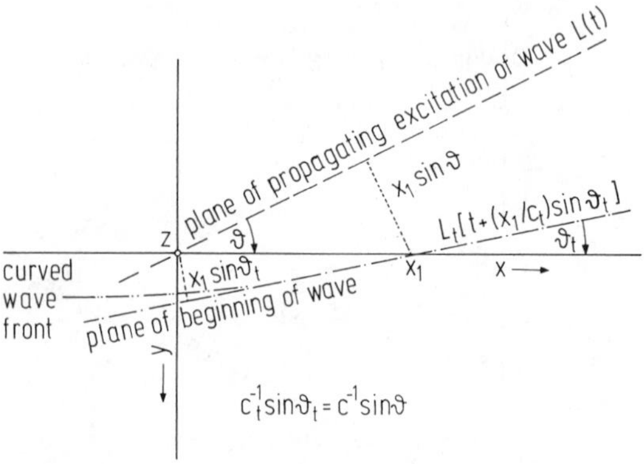

FIG. 3.1-3. Relation between the plane of excitation of wave $L(t)$ and the plane of beginning of wave $L_t(t)$ for infinite signal-to-noise ratio and infinitely sensitive detection equipment.

A parallel polarized TEM wave is defined by the excitation functions

$$\mathbf{H} = H_z\mathbf{e}_z \tag{20}$$

$$\mathbf{E} = E_x\mathbf{e}_x + E_y\mathbf{e}_y \tag{21}$$

In analogy to Eq. (11) we now look for solutions of Eqs. (2)–(7) that satisfy for $y \geqq 0$ the conditions

$$H_x = H_y = 0, \qquad E_z = 0 \tag{22}$$

Introduction of Eq. (22) into Eqs. (2)–(7) yields:

$$\partial H_z/\partial y = \epsilon\, \partial E_x/\partial t + \sigma E_x \tag{23}$$

$$-\partial H_z/\partial x = \epsilon\, \partial E_y/\partial t + \sigma E_y \tag{24}$$

$$-\partial E_y/\partial x + \partial E_x/\partial y = \mu\, \partial H_z/\partial t + sH_z \tag{25}$$

We have now an entirely different set of equations than Eqs. (12)–(14) holding for a perpendicularly polarized TEM wave. From Fig. 3.1-4 we recognize the condition

$$E_y(x, y, t) = -E_x(x, y, t)\operatorname{tg}\vartheta \tag{26}$$

instead of Eq. (16), while Eq. (15) remains unchanged. With Eqs. (15) and (26) we may rewrite Eqs. (23) to (25):

$$\partial H_z/\partial y = \epsilon\, \partial E_x/\partial t + \sigma E_x \tag{27}$$

$$(\partial E_x/\partial y)\cos^{-2}\vartheta = \mu\, \partial H_z/\partial t + sH_z \tag{28}$$

These two equations are the generalization of Eqs. (2.1-9) and (2.1-10) for $\vartheta \neq 0$.

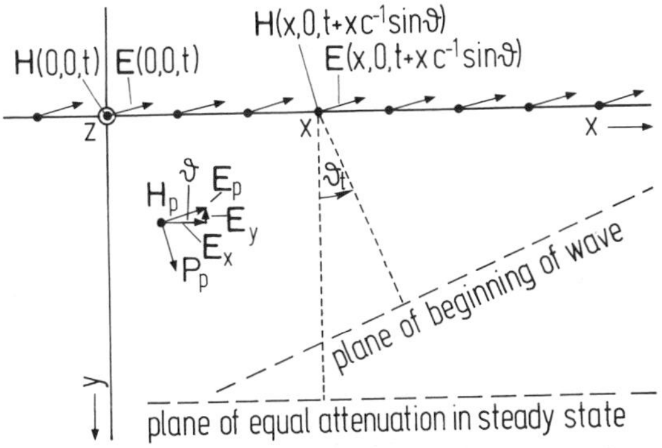

FIG. 3.1-4. Excitation of waves at the plane $y = 0$ by a parallel polarized planar wavefront having the angle of incidence ϑ.

3.2 Perpendicularly Polarized Electric Step Function

We want to find solutions of Eqs. (3.1-17) and (3.1-18) that satisfy certain boundary and initial conditions. In order to obtain an equation that contains E_z only, we differentiate Eq. (3.1-17) with respect to t and Eq. (3.1-18) with respect to y:

$$-(\partial^2 H_x/\partial y\,\partial t)\cos^{-2}\vartheta = \epsilon\,\partial^2 E_z/\partial t^2 + \sigma\,\partial E_z/\partial t \qquad (1)$$

$$-\partial^2 E_z/\partial y^2 = \mu\,\partial^2 H_x/\partial y\,\partial t + s\,\partial H_x/\partial y \qquad (2)$$

From Eqs. (1), (2), and (3.1-17) follows:

$$(\partial^2 E_z/\partial y^2)\cos^{-2}\vartheta - \mu\epsilon\,\partial^2 E_z/\partial t^2 - (\mu\sigma + \epsilon s)\partial E_z/\partial t - s\sigma E_z = 0 \quad (3)$$

This is the generalization of Eq. (2.1-17) for $E = E_z$ according to Eq. (2.1-12), if ϑ is not equal to zero. If E_z is found from this equation one obtains H_x from either Eq. (3.1-17) or (3.1-18) in analogy to Eqs. (2.1-18) and (2.1-23):

$$H_x = H_{xEs} = -\cos^2\vartheta \int\left(\epsilon\frac{\partial E_{zEs}}{\partial t} + \sigma E_{zEs}\right)dy + H_y(t) \qquad (4)$$

$$H_{xEs} = e^{-st/\mu}\left(-\mu^{-1}\int \frac{\partial E_{zEs}}{\partial y}\,e^{st/\mu}\,dt + H_t(y)\right) \qquad (5)$$

$$E_{zEs} = E_z$$

We have introduced a new notation with the subscripts Es, where E stands for "electric excitation" and s for "perpendicular[1] polarization." Later on we will use the subscripts Ep for electric excitation with parallel polarization, Hs for magnetic excitation with perpendicular polarization, and Hp for magnetic excitation with parallel polarization. To simplify notation we will use the subscripts Es, Ep, Hs, and Hp only when strictly required, which usually means in the final result of the calculation but not at all intermediate stages.

The integration constants $H_y(t)$ and $H_t(y)$ in Eqs. (4) and (5) are determined by the requirement that both equations must yield the same result.

We have written $H_x = H_{xEs}$ in Eqs. (4) and (5) in order to avoid momentarily the question whether H_{xEs} is a function of y and t according to Eqs. (3)–(5) or of x, y, and t as suggested by Figs. 3.1-1 and 3.1-2. According to Fig. 3.1-1 an advance of the wavefront $L(0)$ by ct makes its intersection with

---

[1] Since the English words perpendicular and parallel both start with p it is a common practice to use the letter s from the German word *senkrecht* for perpendicular (Reitz *et al.*, 1979). A great variety of notations exists, ranging from special symbols for perpendicular and parallel (Stratton, 1941; Kraus and Carver, 1973), to no distinguishing notation (Adler *et al.*, 1960; Kelso, 1964).

the x axis move from $x = 0$ for $t = 0$ to

$$x = ct/\sin \vartheta \tag{6}$$

for a general value of t. Hence, x is not an independent variable, and it does not show up in Eqs. (3.1-17) and (3.1-18) for this reason. We could, of course, have chosen t as the dependent variable, or even $y = ct/\cos \vartheta$, but we try to maintain the similarity with the equations in Section 2.1. For reasons soon to be seen we will write the variable x from here on, e.g., $H_{xEs} = H_{xEs}(x, y, t)$.

Instead of deriving a differential equation for E_z from Eqs. (3.1-17) and (3.1-18), we may derive one for H_x. This is not a trivial repetition of what we just did. Equations (3)–(5) give us a solution if the boundary conditions are defined for E_z. Since the boundary conditions for H_x may be defined independently, we need the differential equation for H_x to obtain a solution that holds for any physically possible set of boundary conditions.

Differentiation of Eq. (3.1-17) with respect to y and of Eq. (3.1-18) with respect to t

$$-(\partial^2 H_x/\partial y^2)\cos^{-2}\vartheta = \epsilon\, \partial^2 E_z/\partial t\, \partial y + \sigma\, \partial E_z/\partial y \tag{7}$$

$$-\partial^2 E_z/\partial y\, \partial t = \mu\, \partial^2 H_x/\partial t^2 + s\, \partial H_x/\partial t \tag{8}$$

permits one with the help of Eq. (3.1-18) to eliminate E_z from Eq. (7):

$$(\partial^2 H_x/\partial y^2)\cos^{-2}\vartheta - \mu\epsilon\, \partial^2 H_x/\partial t^2 - (\mu\sigma + \epsilon s)\partial H_x/\partial t - s\sigma H_x = 0 \tag{9}$$

This equation has the same form as Eq. (3). If $H_x = H_{xHs}$ is found from Eq. (9) one obtains E_z from either Eq. (3.1-17) or (3.1-18) in analogy to Eqs. (2.1-25) and (2.1-27):

$$E_z = E_{zHs}(x, y, t) = -\int \left(\mu \frac{\partial H_{xHs}}{\partial t} + sH_x\right) dy + E_y(t) \tag{10}$$

$$E_z = E_{zHs}(x, y, t) = e^{-\sigma t/\epsilon}\left(-\epsilon^{-1}\cos^{-2}\vartheta \int \frac{\partial H_{xHs}}{\partial y} e^{\sigma t/\epsilon}\, dt + E_t(y)\right) \tag{11}$$

$$H_x = H_{xHs}$$

The integration constants $E_y(t)$ and $E_t(y)$ are again defined by the requirement that Eqs. (10) and (11) yield the same result for $E_{zHs}(x, y, t)$.

We turn to the boundary conditions. Consider many electrodes in the plane $y = 0$ of Fig. 3.1-2 which apply an electric field strength $\mathbf{E}_z$ in this plane as a function of time:

$$\begin{aligned} E_z(x, 0, t) = E_0 S(t - xc^{-1}\sin\vartheta) &= 0 \qquad \text{for} \quad t < xc^{-1}\sin\vartheta \\ &= E_0 \qquad \text{for} \quad t \geqq xc^{-1}\sin\vartheta \end{aligned} \tag{12}$$

In Section 2.1 the boundary condition of Eq. (2.1-28) required that a step function with amplitude E_0 be applied to all points of the plane $y=0$ at the time $t=0$. The boundary condition of Eq. (12), on the other hand, requires that such a step function be applied to the points on the line x in the plane $y=0$ at the time $t=xc^{-1}\sin\vartheta$. This is the time at which the wavefront $L(t)$ in Fig. 3.1-1 reaches the coordinate x, if it reaches $x=0$ at the time $t=0$.

At the plane $y\rightarrow\infty$ we have the further boundary condition

$$E_z(x,\infty,t)=\text{finite} \tag{13}$$

Let E_z and H_x initially be zero along the line x, $y>0$ at the time $t=xc^{-1}\sin\vartheta$. We have then the initial conditions

$$E_z(x,y,xc^{-1}\sin\vartheta)=H_x(x,y,xc^{-1}\sin\vartheta)=0 \tag{14}$$

If these relations are satisfied for all coordinates $y>0$, their derivatives with respect to y must also be zero:

$$\partial E_z/\partial y=\partial H_x/\partial y=0 \qquad \text{for} \quad t=xc^{-1}\sin\vartheta, \qquad y>0 \tag{15}$$

Equations (14) and (15) also imply the initial conditions

$$\partial E_z/\partial t=\partial H_x/\partial t=0 \tag{16}$$

for $y>0$ and $t=xc^{-1}\sin\vartheta$ according to Eqs. (3.1-17) and (3.1-18).

We assume that the solution of Eq. (3) can be written in the form

$$E_z(x,y,t)=E_{zEs}(x,y,t)=E_0[w(x,y,t)+F(x,y)] \tag{17}$$

First we insert $F(x,y)$ into Eq. (3):

$$(\partial^2F/\partial y^2)\cos^{-2}\vartheta-s\sigma F=0 \tag{18}$$

Consider the following solution of this equation, keeping in mind that F is a function of x but any derivatives of F with respect to x must be zero:

$$\begin{aligned} F(x,y)&=A_{00}e^{-(y\cos\vartheta)/L}+A_{01}e^{(y\cos\vartheta)/L} && \text{for} \quad x\leqq ct/\sin\vartheta \\ &=0 && \text{for} \quad x>ct/\sin\vartheta \\ L&=(s\sigma)^{-1/2} \end{aligned} \tag{19}$$

In order to meet the boundary conditions of Eqs. (12) and (13) one must choose $A_{00}=1$ and $A_{01}=0$:

$$\begin{aligned} F(x,y)&=e^{-(y\cos\vartheta)/L} && \text{for} \quad x\leqq ct/\sin\vartheta \\ &=0 && \text{for} \quad x>ct/\sin\vartheta \end{aligned} \tag{20}$$

This function is shown in Fig. 3.2-1.

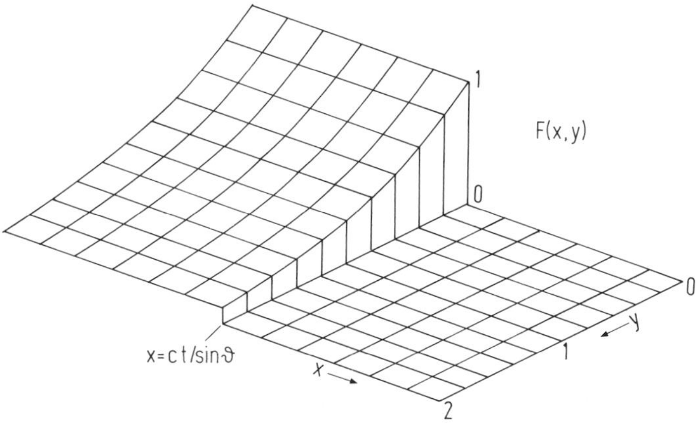

FIG. 3.2-1. The function $F(x,y)$ according to Eq. (20).

For the calculation of $w(x, y, t)$ of Eq. (17) we observe that the introduction of the function $F(x, y)$ transforms the boundary conditions of Eq. (12),

$$\begin{aligned} E_{zEs}(x, 0, t) &= E_0 w(x, 0, t) = 0 && \text{for} \quad t < xc^{-1} \sin\vartheta \\ &= E_0[w(x, 0, t) + 1] = E_0 && \text{for} \quad t \geqq xc^{-1} \sin\vartheta \end{aligned} \tag{21}$$

which implies the homogeneous boundary condition

$$w(x, 0, t) = 0 \tag{22}$$

for any value of x and t.

Equation (13) yields:

$$w(x, \infty, t) = \text{finite} \tag{23}$$

The initial conditions of Eqs. (14) and (16) assume the form:

$$w(x, y, xc^{-1} \sin\vartheta) = -F(x, y) \tag{24}$$

$$\partial w(x, y, t)/\partial t = 0 \qquad \text{for} \quad t = xc^{-1} \sin\vartheta \tag{25}$$

For the calculation of $w(x, y, t)$ we substitute Eq. (17) into Eq. (3). The terms containing $F(x, y)$ drop out due to Eq. (18), and one obtains for $w(x, y, t)$ the same differential equation as for E_z:

$$(\partial^2 w/\partial y^2)\cos^{-2}\vartheta - \mu\epsilon\, \partial^2 w/\partial t^2 - (\mu\sigma + \epsilon s)\partial w/\partial t - s\sigma w = 0 \tag{26}$$

A particular solution $w_\kappa(x, y, t)$ is obtained by the separation of variables using Bernoulli's product method in the following form:

$$w_\kappa(x, y, t) = \varphi(y)\psi(t - xc^{-1} \sin\vartheta) \tag{27}$$

Insertion into Eq. (26) yields

$$\varphi^{-1}(\partial^2\varphi/\partial y^2)\cos^{-2}\vartheta = \mu\epsilon\psi^{-1}\,\partial^2\psi/\partial t^2 + (\mu\,\sigma + \epsilon s)\psi^{-1}\,\partial\psi/\partial t + s\sigma = -(2\pi\kappa)^2 \qquad (28)$$

and one obtains two ordinary differential equations:

$$\partial^2\varphi/\partial y^2 + (2\pi\kappa\cos\vartheta)^2\varphi = 0 \qquad (29)$$

$$\partial^2\psi/\partial t^2 + c^2(\mu\sigma + \epsilon s)\partial\psi/\partial t + [(2\pi c\kappa)^2 + s\sigma c^2]\psi = 0 \qquad (30)$$

Their solutions are

$$\varphi(y) = A_{10}\sin(2\pi\kappa y\cos\vartheta) + A_{11}\cos(2\pi\kappa y\cos\vartheta) \qquad (31)$$

$$\psi(t - xc^{-1}\sin\vartheta) = A_{20}\exp[\gamma_1(t - xc^{-1}\sin\vartheta] + A_{21}\exp[\gamma_2(t - xc^{-1}\sin\vartheta)] \qquad (32)$$

where γ_1 and γ_2 have the same values as in Eq. (2.1-50).

The boundary condition of Eq. (22) requires $A_{11} = 0$ in Eq. (31). The particular solution $w_\kappa(x, y, t)$ thus becomes:

$$w_\kappa(x, y, t) = [A_1\exp(\gamma_1 t') + A_2\exp(\gamma_2 t')]\sin(2\pi\kappa y\cos\vartheta), \qquad (33)$$

$$t' = t - xc^{-1}\sin\vartheta \geqq 0$$

A more general solution is obtained by making A_1 and A_2 functions of κ, and integrating over all possible values of the wavenumber κ:

$$w(x, y, t) = \int_0^\infty [A_1(\kappa)\exp(\gamma_1 t') + A_2(\kappa)\exp(\gamma_2 t')]\sin(2\pi\kappa y\cos\vartheta)d\kappa \qquad (34)$$

The functions $A_1(\kappa)$ and $A_2(\kappa)$ must be chosen so that the initial conditions of Eqs. (24) and (25) are satisfied. To this end we need the time derivative of $w(x, y, t)$:

$$\frac{\partial w}{\partial t} = \int_0^\infty [A_1(\kappa)\gamma_1\exp(\gamma_1 t') + A_2(\kappa)\gamma_2\exp(\gamma_2 t')]\sin(2\pi\kappa y\cos\vartheta)d\kappa \qquad (35)$$

Insertion of Eqs. (34) and (35) into Eqs. (24) and (25) yields for $t = xc^{-1}\sin\vartheta$ or $t' = 0$:

$$\int_0^\infty [A_1(\kappa) + A_2(\kappa)]\sin(2\pi\kappa y\cos\vartheta)d\kappa = -e^{-(y\cos\vartheta)/L} \qquad (36)$$

$$\int_0^\infty [A_1(\kappa)\gamma_1 + A_2(\kappa)\gamma_2]\sin(2\pi\kappa y\cos\vartheta)d\kappa = 0 \qquad (37)$$

Equations (36) and (37) are not quite Fourier sine transforms, but they can be brought into that form by the substitution

$$\kappa \cos \vartheta = \kappa', \qquad d\kappa = \cos^{-1} \vartheta \, d\kappa' \tag{38}$$

which brings:

$$\int_0^\infty [A_1(\kappa') + A_2(\kappa')] \sin 2\pi\kappa' y \, d\kappa' = -\cos \vartheta \, e^{-(y \cos \vartheta)/L} \tag{39}$$

$$\int_0^\infty [A_1(\kappa')\gamma_1 + A_2(\kappa')\gamma_2] \sin 2\pi\kappa' y \, d\kappa' = 0 \tag{40}$$

A comparison of Eq. (39) with Eqs. (2.1-54) and (2.1-60) yields

$$A_1(\kappa') + A_2(\kappa') = -(8\pi\kappa' \cos \vartheta)/[s\sigma \cos^2 \vartheta + (2\pi\kappa')^2] \tag{41}$$

or

$$A_1(\kappa) + A_2(\kappa) = -8\pi\kappa/[s\sigma + (2\pi\kappa)^2] = -8\pi\kappa/b^2 \tag{42}$$

From Eqs. (40), (2.1-55), and (2.1-58) follows:

$$A_1(\kappa)\gamma_1 + A_2(\kappa)\gamma_2 = 0 \tag{43}$$

From Eqs. (42) and (43) we obtain the same values for $A_1(\kappa)$ and $A_2(\kappa)$ as in Eq. (2.1-62), since we have already seen that the values of γ_1 and γ_2 are the same here as in Section 2.1. Hence, $w(x, y, t)$ of Eq. (34) has the same form as $w(y, t)$ of Eq. (2.1-64), except that t is replaced by $t' = t - xc^{-1} \sin \vartheta$ and $\sin 2\pi\kappa y$ by $\sin(2\pi\kappa y \cos \vartheta)$:

$$\begin{aligned} w(x, y, t) = w(y \cos \vartheta, t') = -\frac{2}{\pi} e^{-at'} \Bigg[\int_0^{2\pi K} \Bigg(& \operatorname{ch}(a^2 - b^2c^2)^{1/2} t' \\ & + \frac{a \operatorname{sh}(a^2 - b^2c^2)^{1/2} t'}{(a^2 - b^2c^2)^{1/2}} \Bigg) \frac{\sin(\beta y \cos \vartheta)}{b^2} \beta \, d\beta \\ + \int_{2\pi K}^\infty \Bigg(& \cos(b^2c^2 - a^2)^{1/2} t' \\ & + \frac{a \sin(b^2c^2 - a^2)^{1/2} t'}{(b^2c^2 - a^2)^{1/2}} \Bigg) \frac{\sin(\beta y \cos \vartheta)}{b^2} \beta \, d\beta \Bigg], \end{aligned}$$

$$t' = t - xc^{-1} \sin \vartheta \geqq 0, \quad \beta = 2\pi\kappa, \quad 2\pi K = (a^2/c^2 - s\sigma)^{1/2} \tag{44}$$

The solution for the electric field strength $\mathbf{E}_{zEs}(x, y, t)$ follows from Eqs. (17) and (20):

$$\begin{aligned} E_{zEs}(x, y, t') &= 0 && \text{for} \quad t' < 0 \\ &= E_0[e^{-(y \cos \vartheta)/L} + w(y \cos \vartheta, t')] && \text{for} \quad t' \geqq 0 \end{aligned} \tag{45}$$

This equation is reduced to Eq. (2.1-65) for $\vartheta = 0$. As a further check consider the case of vanishing conductances $\sigma = s = 0$. For $s = 0$ we get:

$$\begin{aligned} E_{zEs}(x, y, t') &= 0 && \text{for } t' < 0 \\ &= E_0(1 + w(y\cos\vartheta, t')] && \text{for } t' \geqq 0 \end{aligned}$$

$$\begin{aligned} w(y\cos\vartheta, t') = -\frac{2}{\pi} e^{-\alpha t'} \Bigg[\int_0^{Z\sigma/2} \Bigg(&\operatorname{ch}(\alpha^2 - \beta^2 c^2)^{1/2} t' \\ &+ \frac{\alpha \operatorname{sh}(\alpha^2 - \beta^2 c^2)^{1/2} t'}{(\alpha^2 - \beta^2 c^2)^{1/2}} \Bigg) \\ &\times \frac{\sin(\beta y \cos\vartheta)}{\beta} d\beta + \int_{Z\sigma/2}^{\infty} \Bigg(\cos(\beta^2 c^2 - \alpha^2)^{1/2} t' \\ &+ \frac{\alpha \sin(\beta^2 c^2 - \alpha^2)^{1/2} t'}{(\beta^2 c^2 - \alpha^2)^{1/2}} \Bigg) \frac{\sin(\beta y \cos\vartheta)}{\beta} d\beta \Bigg] \end{aligned} \tag{46}$$

The transition $\sigma \to 0$ eliminates the first integral and makes α equal to zero:

$$\begin{aligned} E_{zEs}(x, y, t) &= 0 \qquad \text{for } t < xc^{-1}\sin\vartheta \\ &= E_0\Big[1 - \frac{2}{\pi}\int_0^{\infty} \frac{\cos[\beta c(t - xc^{-1}\sin\vartheta)]\sin(\beta y\cos\vartheta)}{\beta} d\beta\Big] \\ &\qquad \text{for } t \geqq xc^{-1}\sin\vartheta \end{aligned} \tag{47}$$

With

$$\begin{aligned} &2\cos[\beta c(t - xc^{-1}\sin\vartheta)]\sin(\beta y\cos\vartheta) \\ &\quad = \sin[\beta(y\cos\vartheta + x\sin\vartheta - ct)] + \sin[\beta(y\cos\vartheta - x\sin\vartheta + ct)] \end{aligned} \tag{48}$$

and Eq. (2.1–70) we get:

$$\begin{aligned} E_{zEs}(x, y, t) &= E_0 && \text{for } y = 0,\ ct - x\sin\vartheta = 0 \\ &= 0 && \text{for } y > 0,\ ct - x\sin\vartheta < y\cos\vartheta \\ &= E_0/2 && \text{for } y > 0,\ ct - x\sin\vartheta = y\cos\vartheta \\ &= E_0 && \text{for } y \geqq 0,\ ct - x\sin\vartheta > y\cos\vartheta \end{aligned} \tag{49}$$

For a discussion of Eq. (49) refer to Fig. 3.2-2. A wavefront $L(0)$ at the time $t = 0$ propagates the distance ct during the time t to become the wavefront $L(t)$. An arbitrary point x in the plane $y = 0$ is chosen. The distances $x\sin\vartheta$ and $ct - x\sin\vartheta$ are shown for this point. For the point x, y_1 the distance $y_1\cos\vartheta$ is smaller than $ct - x\sin\vartheta$; the field strength is thus $\mathbf{E}_0$ in this point.

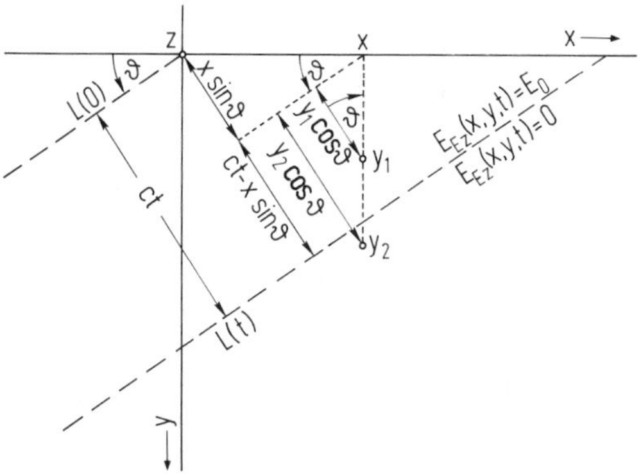

FIG. 3.2-2. The function $E_{zEs}(x,y,t)$ defined by Eq. (49).

On the other hand, for the point x, y_2 the distance $y_2 \cos \vartheta$ is larger than $ct - x \sin \vartheta$, and the field strength is zero in this point. Along the line $L(t)$ the field strength equals $\mathbf{E}_0/2$; we have pointed out before that this is due to the fact that the Fourier transform converges to the median $E_0/2$ if a function has the two values E_0 and 0 in a point.

With the substitutions

$$\begin{aligned}
&\eta = \beta c/\alpha, \qquad \theta' = \alpha t' = \alpha(t - xc^{-1} \sin \vartheta), \\
&\xi' = \alpha c^{-1} y \cos \vartheta = (Z\sigma y/2)\cos \vartheta \\
&\beta y \cos \vartheta = \eta\xi', \qquad d\beta = \alpha c^{-1}\, d\eta, \qquad d\beta/\beta = d\eta/\eta, \\
&\eta = 1 \qquad \text{for} \quad \beta = Z\sigma/2
\end{aligned} \tag{50}$$

one may rewrite Eq. (46) in a normalized form:

$$\begin{aligned}
E_{zEs}(\xi', \theta') &= 0 && \text{for} \quad \theta' < 0 \\
&= E_0[1 + w(\xi', \theta')] && \text{for} \quad \theta' \geqq 0
\end{aligned} \tag{51}$$

$$\begin{aligned}
w(\xi', \theta') = -\frac{2}{\pi} e^{-\theta'} \Bigg[&\int_0^1 \left(\operatorname{ch}(1 - \eta^2)^{1/2}\theta' + \frac{\operatorname{sh}(1 - \eta^2)^{1/2}\theta'}{(1 - \eta^2)^{1/2}} \right) \frac{\sin \eta\xi'}{\eta}\, d\eta \\
+ &\int_1^\infty \left(\cos(\eta^2 - 1)^{1/2}\theta' + \frac{\sin(\eta^2 - 1)^{1/2}\theta'}{(\eta^2 - 1)^{1/2}} \right) \frac{\sin \eta\xi'}{\eta}\, d\eta \Bigg]
\end{aligned}$$

Note that this is formally the same equation as Eq. (2.1-80). The difference is only in the different definition of θ' and ξ' compared with θ and ξ.

Let us consider the magnetic field strength for the special case $\sigma = 0$. We have seen before that one may choose $s = 0$ in this case at the beginning rather than the end of the calculation. We obtain from Eq. (4) and (47):

$$H_{xEs}(x, y, t) = -\epsilon \cos^2 \vartheta \int \frac{\partial E_{zEs}}{\partial t} dy + H_y(t)$$

$$= 2E_0 \epsilon c \pi^{-1} \cos^2 \vartheta \int_0^\infty \left(\sin[\beta c(t - xc^{-1} \sin \vartheta)] \right.$$

$$\left. \times \int \sin(\beta y \cos \vartheta) dy \right) d\beta + H_y(t)$$

$$= \frac{2E_0}{Z\pi} \cos \vartheta$$

$$\times \int_0^\infty \frac{\sin[\beta c(t - xc^{-1} \sin \vartheta)] \cos(\beta y \cos \vartheta)}{\beta} d\beta + H_y(t) \quad (52)$$

From Eqs. (5) and (47) we get:

$$H_{xEs}(x, y, t) = -\mu^{-1} \int \frac{\partial E_{zEs}}{\partial y} dt + H_t(y)$$

$$= 2E_0 \pi^{-1} \mu^{-1} \cos \vartheta \int_0^\infty \left(\int \cos[\beta c(t - xc^{-1} \sin \vartheta)] dt \right)$$

$$\times \cos(\beta y \cos \vartheta) d\beta + H_t(y)$$

$$= \frac{2E_0}{Z\pi} \cos \vartheta$$

$$\times \int_0^\infty \frac{\sin[\beta c(t - xc^{-1} \sin \vartheta)] \cos(\beta y \cos \vartheta)}{\beta} d\beta + H_t(y) \quad (53)$$

Since Eqs. (52) and (53) must yield the same value we obtain

$$H_y(t) = H_t(y) = H_c = 0 \quad (54)$$

and

$$H_{xEs}(x, y, t) = \frac{E_0}{Z\pi} \cos \vartheta \left(\int_0^\infty \frac{\sin[\beta(y \cos \vartheta + ct - x \sin \vartheta)]}{\beta} d\beta \right.$$

$$\left. - \int_0^\infty \frac{\sin[\beta(y \cos \vartheta - ct + x \sin \vartheta)]}{\beta} d\beta \right) \quad (55)$$

For $\vartheta = 0$ this equation becomes Eq. (2.1-84).

If we use Eq. (2.1-70) we obtain:

$$
\begin{aligned}
H_{xEs}(x, y, t) &= 0 && \text{for } y=0, && ct - x\sin\vartheta = 0\\
&= 0 && \text{for } y>0, && ct - x\sin\vartheta < y\cos\vartheta\\
&= (E_0/2Z)\cos\vartheta && \text{for } y>0, && ct - x\sin\vartheta = y\cos\vartheta\\
&= (E_0/Z)\cos\vartheta && \text{for } y>0, && ct - x\sin\vartheta > y\cos\vartheta
\end{aligned} \tag{56}
$$

3.3 Parallel Polarized Electric Step Function

We consider now the excitation of a wave by a parallel polarized electric step function according to Fig. 3.1-4. Equations (3.1-27) and (3.1-28) have to be solved. We differentiate Eq. (3.1-28) with respect to y and Eq. (3.1-27) with respect to t, and obtain after some substitutions a differential equation for E_x:

$$(\partial^2 E_x/\partial y^2)\cos^{-2}\vartheta - \mu\epsilon\,\partial^2 E_x/\partial t^2 - (\mu\sigma + \epsilon s)\partial E_x/\partial t - s\sigma E_x = 0 \tag{1}$$

This equation for E_x has the same form as Eq. (3.2-3) for E_z. The magnetic field strength H_z follows from Eqs. (3.1-27) and (3.1-28) in analogy to Eqs. (2.1-18) and (2.1-23):

$$H_z = H_{zEp}(x, y, t) = \int \left(\epsilon \frac{\partial E_{xEp}}{\partial t} + \sigma E_{xEp}\right) dy + H_y(t) \tag{2}$$

$$H_{zEp}(x, y, t) = e^{-st/\mu}\left(\mu^{-1}\cos^{-2}\vartheta \int \frac{\partial E_{xEp}}{\partial y} e^{st/\mu}\,dt + H_t(y)\right) \tag{3}$$

$$E_{xEp} = E_x$$

The integration constants are again determined by the requirement that both equations must yield the same result. The variables x and t are connected by the relation

$$x = ct/\sin\vartheta \tag{4}$$

as in Section 3.2.

The solution of Eqs. (3.1-27) and (3.1-28) for H_z rather than for E_x yields the differential equation

$$(\partial^2 H_z/\partial y^2)\cos^{-2}\vartheta - \mu\epsilon\,\partial^2 H_z/\partial t^2 - (\mu\sigma + \epsilon s)\partial H_z/\partial t - s\sigma H_z = 0 \tag{5}$$

while the following two equations are obtained for E_x:

$$E_x = E_{xHp}(x, y, t) = \cos^2\vartheta \int \left(\mu \frac{\partial H_{zHp}}{\partial t} + sH_{zHp}\right) dy + E_y(t) \tag{6}$$

$$E_{xHp}(x, y, t) = e^{-\sigma t/\epsilon}\left(\epsilon^{-1}\int \frac{\partial H_{zHp}}{\partial y}\, e^{\sigma t/\epsilon}\, dt + E_t(y)\right) \tag{7}$$

$$H_z = H_{zHp}$$

Turning to the boundary conditions, we consider many electrodes in the plane $y = 0$ of Fig. 3.1-4 which apply an electric field strength $\mathbf{E}_x$, and a second set of electrodes that apply the electric field strength $\mathbf{E}_y$ according to Eq. (3.1-26), in this plane as a function of time:

$$\begin{aligned} E_x(x, 0, t) = E_0 S(t - xc^{-1}\sin\vartheta) &= 0 \qquad \text{for} \quad t < xc^{-1}\sin\vartheta \\ &= E_0 \qquad \text{for} \quad t \geqq xc^{-1}\sin\vartheta \end{aligned} \tag{8}$$

At the plane $y \to \infty$ we have as always the boundary condition

$$E_x(x, \infty, t) = \text{finite} \tag{9}$$

If E_x and H_z are initially zero along the line x, $y > 0$ at the time $t = xc^{-1}\sin\vartheta$, we have the initial conditions

$$E_x(x, y, xc^{-1}\sin\vartheta) = H_z(x, y, xc^{-1}\sin\vartheta) = 0 \tag{10}$$

The derivatives of E_x and H_z with respect to y must also be zero if the conditions of Eq. (10) are satisfied for all coordinates $y > 0$:

$$\partial E_x/\partial y = \partial H_z/\partial y = 0 \qquad \text{for} \quad t = xc^{-1}\sin\vartheta,\ y > 0 \tag{11}$$

Equations (10) and (11) imply the initial conditions

$$\partial E_x/\partial t = \partial H_z/\partial t = 0 \tag{12}$$

for $y > 0$ and $t = xc^{-1}\sin\vartheta$ according to Eqs. (3.1-27) and (3.1-28).

We write the solution of Eq. (1) in the form

$$E_x(x, y, t) = E_{xEp}(x, y, t) = E_0[w(x, y, t) + F(x, y)] \tag{13}$$

Since the differential equation, the boundary conditions, and the initial conditions are the same as in Section 3.2, we may immediately write the result in analogy to Eq. (3.2-45),

$$\begin{aligned} E_{xEp}(x, y, t) &= 0 \qquad &&\text{for} \quad t' < 0 \\ &= E_0[e^{-(y\cos\vartheta)/L} + w(y\cos\vartheta, t')] \qquad &&\text{for} \quad t' \geqq 0 \end{aligned} \tag{14}$$

where $w(y\cos\vartheta, t')$ is given by Eq. (3.2-44), and obtain for $s = 0$

$$E_{xEp}(x, y, t) = E_0[1 + w(y\cos\vartheta, t')] \qquad \text{for} \quad t' \geqq 0 \tag{15}$$

where $w(y\cos\vartheta, t')$ is now defined by Eq. (3.2-46).

3.4 Magnetic Field Strength Due to Electric Step Function

We calculate first the magnetic field strength $H_{xEs}(x, y, t)$ according to Eqs. (3.2-4) and (3.2-5) for electric step function excitation with perpendicular polarization. Equation (3.2-45) is inserted into Eq. (3.2-5):

$$H_{xEs} = E_0 e^{-st/\mu}\left[-\mu^{-1}\int\left(-(s\sigma)^{1/2}\cos\vartheta e^{-(y\cos\vartheta)/L} + \frac{\partial w}{\partial y}\right)e^{st/\mu}\,dt + H_t(y)\right]$$

$$= E_0\left(\frac{\sigma}{s}\right)^{1/2}\cos\vartheta\, e^{-(y\cos\vartheta)/L} + e^{-st/\mu}\left(-E_0\mu^{-1}\int\frac{\partial w}{\partial y}e^{st/\mu}\,dt + H_t(y)\right) \tag{1}$$

For $w(x, y, t)$ of Eq. (3.2-44) we use the form of Eq. (2.1-52), replacing t by t' and y by $y\cos\vartheta$

$$w(x, y, t) = \int_0^\infty [A_1(\kappa)\exp(\gamma_1 t') + A_2(\kappa)\exp(\kappa_2 t')]\sin(2\pi\kappa y\cos\vartheta)d\kappa \tag{2}$$

and obtain

$$\frac{\partial w}{\partial y} = 2\pi\cos\vartheta\int_0^\infty [A_1(\kappa)\exp(\gamma_1 t') + A_2(\kappa)\exp(\gamma_2 t')] \times \cos(2\pi\kappa y\cos\vartheta)\kappa\, d\kappa$$

$$e^{-st/\mu}\int\frac{\partial w}{\partial y}e^{st/\mu}\,dt = 2\pi\cos\vartheta\int_0^\infty [A_1(\kappa)(\gamma_1 + s/\mu)^{-1}\exp(\gamma_1 t') + A_2(\kappa)(\gamma_2 + s/\mu)^{-1}\exp(\gamma_2 t')]\cos(2\pi\kappa y\cos\vartheta)\kappa\, d\kappa \tag{3}$$

The magnetic field strength becomes:

$$H_{xEs}(x, y, t) = E_0 Z^{-1}\cos\vartheta\left[Z\left(\frac{\sigma}{s}\right)^{1/2}e^{-(y\cos\vartheta)/L} - \frac{1}{2\pi}\int_0^\infty [A_1(\kappa)(\gamma_1 + s/\mu)^{-1}\exp(\gamma_1 t') + A_2(\kappa)(\gamma_2 + s/\mu)^{-1}\exp(\gamma_2 t)]\beta c\cos(2\pi\kappa y\cos\vartheta)d\beta\right] + H_t(y)e^{-st/\mu} \tag{4}$$

A comparison of this equation with Eq. (2.4-4) shows that the only difference is the factor $\cos\vartheta$ in front, as well as the substitution of t' and $y\cos\vartheta$ for t and y. We may thus write $H_{xEs}(x, y, t)$ in analogy to Eq. (2.4-8):

$$
\begin{aligned}
H_{xEs}(x, y, t) = E_0 Z^{-1} \cos\vartheta \Bigg\{ & Z\left(\frac{\sigma}{s}\right)^{1/2} \exp[-(s\sigma)^{1/2} y \cos\vartheta] \\
& - \frac{2Z\sigma}{\pi} e^{-at'} \Bigg[\int_0^{2\pi K} \bigg(\operatorname{ch}(a^2 - b^2c^2)^{1/2} t' \\
& + \frac{(2a\alpha - b^2c^2)\operatorname{sh}(a^2 - b^2c^2)^{1/2} t'}{2\alpha(a^2 - b^2c^2)^{1/2}} \bigg) \\
& \times \frac{\cos(\beta y \cos\vartheta)}{b^2}\, d\beta + \int_{2\pi K}^{\infty} \bigg(\cos(b^2c^2 - a^2)^{1/2} t' \\
& + \frac{(2a\alpha - b^2c^2)\sin(b^2c^2 - a^2)^{1/2} t'}{2\alpha(b^2c^2 - a^2)^{1/2}} \bigg) \frac{\cos(\beta y \cos\vartheta)}{b^2}\, d\beta \Bigg] \Bigg\} \\
& + H_t(y) e^{-st/\mu} \qquad (5)
\end{aligned}
$$

We make now the transition $s \to 0$ for $\sigma > 0$. Using the results of Eqs. (2.4-9) to (2.4-41) we substitute $E_0 Z^{-1}\cos\vartheta$, $y\cos\vartheta$, t' for $E_0 Z^{-1}$, y, t in Eqs. (2.4-38) to (2.4-41):

$$
\begin{aligned}
H_{xEs}(x, y, t) = E_0 Z^{-1} \cos\vartheta [& -Z\sigma y \cos\vartheta + I'_{E1}(y\cos\vartheta, t') \\
& - I_{E2}(y\cos\vartheta, t')] + H_t(y\cos\vartheta) \qquad (6)
\end{aligned}
$$

$$
\begin{aligned}
I'_{E1}(y\cos\vartheta, t') = \frac{2Z\sigma}{\pi} \Bigg[& \frac{1}{d} - e^{-\alpha t'} \int_d^{Z\sigma/2} \bigg(\operatorname{ch}(\alpha^2 - \beta^2c^2)^{1/2} t' \\
& + \frac{\alpha \operatorname{sh}(\alpha^2 - \beta^2c^2)^{1/2} t'}{(\alpha^2 - \beta^2c^2)^{1/2}} \bigg) \frac{d\beta}{\beta} \\
& - \int_0^d [\exp(-\beta^2c^2 t'/2\alpha) - 1] \frac{d\beta}{\beta^2} \Bigg] \\
& + \frac{2c}{\pi} e^{-\alpha t'} \int_0^{Z\sigma/2} \frac{\operatorname{sh}(\alpha^2 - \beta^2c^2)^{1/2} t'}{(\alpha^2 - \beta^2c^2)^{1/2}}\, d\beta \\
& + \frac{Z\sigma}{\pi} e^{-\alpha t'} \int_0^{Z\sigma/2} \bigg(\operatorname{ch}(\alpha^2 - \beta^2c^2)^{1/2} t' \\
& + \frac{(2\alpha^2 - \beta^2c^2)\operatorname{sh}(\alpha^2 - \beta^2c^2)^{1/2} t'}{2\alpha(\alpha^2 - \beta^2c^2)^{1/2}} \bigg) \\
& \times \left(\frac{\sin[(\beta y/2)\cos\vartheta]}{\beta/2} \right)^2 d\beta \qquad (7)
\end{aligned}
$$

$$I_{E2}(y\cos\vartheta, t') = \frac{2Z\sigma}{\pi} e^{-\alpha t'} \int_{Z\sigma/2}^{\infty} \left(\cos(\beta^2c^2 - \alpha^2)^{1/2} t' + \frac{(2\alpha^2 - \beta^2c^2)\sin(\beta^2c^2 - \alpha^2)^{1/2} t'}{2\alpha(\beta^2c^2 - \alpha^2)^{1/2}} \right) \frac{\cos(\beta y \cos\vartheta)}{\beta^2} d\beta, \quad (8)$$

$d \ll Z\sigma/2$

We turn to Eq. (3.2-4). Insertion of Eq. (3.2-45) yields:

$$\begin{aligned} H_{xEs}(x, y, t) &= -E_0 \cos^2\vartheta \left(\epsilon \int \frac{\partial w}{\partial t} dy + \sigma \int e^{-(y\cos\vartheta)/L} dy + \sigma \int w\, dy \right) + H_y(t) \\ &= E_0 \left[\left(\frac{\sigma}{s} \right)^{1/2} \cos\vartheta\, e^{-(y\cos\vartheta)/L} - \cos^2\vartheta \left(\epsilon \int \frac{\partial w}{\partial t} dy + \sigma \int w\, dy \right) \right] + H_y(t) \end{aligned} \quad (9)$$

From Eq. (2) follows:

$$\int w\, dy = -(2\pi\cos\vartheta)^{-1} \int_0^{\infty} [A_1(\kappa)\exp(\gamma_1 t') + A_2(\kappa)\exp(\gamma_2 t')] \times \kappa^{-1} \cos(2\pi\kappa y \cos\vartheta) d\kappa \quad (10)$$

$$\int \frac{\partial w}{\partial t} dy = -(2\pi\cos\vartheta)^{-1} \int_0^{\infty} [A_1(\kappa)\gamma_1 \exp(\gamma_1 t') + A_2(\kappa)\gamma_2 \exp(\gamma_2 t')]\kappa^{-1} \cos(2\pi\kappa y\cos\vartheta) d\kappa \quad (11)$$

The magnetic field strength becomes:

$$H_{xEs}(x, y, t) = E_0 Z^{-1} \cos\vartheta \left[Z \left(\frac{\sigma}{s} \right)^{1/2} e^{-(y\cos\vartheta)/L} - \frac{1}{2\pi} \int_0^{\infty} [-A_1(\kappa)(\gamma_1 + 2\alpha)\exp(\gamma_1 t') - A_2(\kappa)(\gamma_2 + 2\alpha)\exp(\gamma_2 t')] \frac{\cos(\beta y\cos\vartheta)}{\beta c} d\beta \right] + H_y(t) \quad (12)$$

This equation equals Eq. (2.4-43) if $E_0Z^{-1}\cos\vartheta$, $y\cos\vartheta$, t' are substituted there for E_0Z^{-1}, y, t. We may thus draw the same conclusions as in Eq.

(2.4-45) for $H_t(y)$ and $H_y(t)$ in Eqs. (5) and (12):

$$H_t(y)e^{-st/\mu} = H_y(t) = H_{E0}e^{-st/\mu} \tag{13}$$

The initial condition $H_{xEs}(x, y, 0)$ requires $H_{E0} = 0$.

We rewrite Eqs. (6) to (8) for $H_{E0} = 0$ in a normalized form using Eq. (3.2-50):

$$H_{xEs}(\xi', \theta') = E_0 Z^{-1} (\cos \vartheta)[-2\xi' + I'_{E1}(\xi', \theta') - I_{E2}(\xi', \theta')] \tag{14}$$

$$\begin{aligned} I'_{E1}(\xi', \theta') = \frac{2}{\pi}\Bigg\{ & 2\left[\frac{1}{\delta} - e^{-\theta'} \int_0^1 \left(\mathrm{ch}(1-\eta^2)^{1/2}\theta' \right.\right. \\ & \left.\left. + \frac{\mathrm{sh}(1-\eta^2)^{1/2}}{(1-\eta^2)^{1/2}}\right)\frac{d\eta}{\eta^2} - \int_0^\delta [\exp(-\eta^2\theta'/2) - 1]\frac{d\eta}{\eta^2}\right] \\ & + e^{-\theta'}\left[\left[\int_0^1 \frac{\mathrm{sh}(1-\eta^2)^{1/2}\theta'}{(1-\eta^2)^{1/2}}\, d\eta \right.\right. \\ & + \int_0^1 \left(\mathrm{ch}(1-\eta^2)^{1/2}\theta' + \frac{(2-\eta^2)\mathrm{sh}(1-\eta^2)^{1/2}\theta'}{2(1-\eta^2)^{1/2}}\right) \\ & \left.\left. \times \left(\frac{\sin(\xi'\eta/2)}{\eta/2}\right)^2 d\eta\right]\right]\Bigg\} \end{aligned} \tag{15}$$

$$\begin{aligned} I_{E2}(\xi', \theta') = \frac{4}{\pi} e^{-\theta'} & \int_1^\infty \left(\cos(\eta^2-1)^{1/2}\theta' + \frac{(2-\eta^2)\sin(\eta^2-1)^{1/2}\theta'}{2(\eta^2-1)^{1/2}}\right) \\ & \times \frac{\cos \xi'\eta}{\eta^2}\, d\eta, \end{aligned} \tag{16}$$

$$\delta = 2d/Z\sigma \ll 1, \qquad \xi' = \alpha c^{-1} y \cos \vartheta, \qquad \theta' = \alpha(t - xc^{-1} \sin \vartheta)$$

For $\vartheta = 0$ one obtains Eqs. (2.4-46) to (2.4-48).

Let us turn to the magnetic field strength due to a parallel polarized electric step function. According to Eqs. (3.3-15) and (3.2-46) the electric field strength $\mathbf{E}_{xEp}$ of the parallel polarized wave varies like the field strength $\mathbf{E}_{zEs}$ of the perpendicularly polarized wave. We may thus substitute E_{zEs} for E_{xEp} in Eqs. (3.3-2) and (3.3-3). Furthermore, Eq. (3.2-5) is changed into Eq. (3.3-3) if we multiply all terms except $H_t(y)$ with $-\cos^{-2}\vartheta$ in addition to changing E_{zEs} to E_{xEp}. We obtain from Eq. (5):

$$\begin{aligned} H_{zEp}(x, y, t) = -E_0 Z^{-1} (\cos^{-1} \vartheta) & \left\{ Z\left(\frac{\sigma}{s}\right)^{1/2} \exp[-(s\sigma)^{1/2} y \cos \vartheta] \right. \\ & - \frac{2Z\sigma}{\pi} e^{-\alpha t'}\left[\int_0^{2\pi K}\left(\mathrm{ch}(a^2 - b^2c^2)^{1/2} t' \right.\right. \end{aligned}$$

$$+\frac{(2a\alpha - b^2c^2)\mathrm{sh}(a^2 - b^2c^2)^{1/2}t'}{2\alpha(a^2 - b^2c^2)^{1/2}}\Bigg)$$

$$\times\frac{\cos(\beta y\cos\vartheta)}{b^2}\,d\beta + \int_{2\pi K}^{\infty}\Bigg(\cos(b^2c^2 - a^2)^{1/2}t'$$

$$+\frac{(2a\alpha - b^2c^2)\sin(b^2c^2 - a^2)^{1/2}t'}{2\alpha(b^2c^2 - a^2)^{1/2}}\Bigg)$$

$$\times\frac{\cos(\beta y\cos\vartheta)}{b^2}\,d\beta\Bigg]\Bigg\} + H_t(y)e^{-st/\mu} \tag{17}$$

Similarly, Eq. (3.2-4) is transformed into Eq. (3.3-2) if all terms except $H_y(t)$ are multiplied by $-\cos^{-2}\vartheta$ and E_{xEp} is substituted for E_{zEs}. We get thus from Eq. (12):

$$H_{zEp}(x, y, t) = -E_0Z^{-1}(\cos^{-1}\vartheta)\Bigg[Z\left(\frac{\sigma}{s}\right)^{1/2}e^{-(y\cos\vartheta)/L}$$

$$-\frac{1}{2\pi}\int_0^{\infty}[-A_1(\kappa)(\gamma_1 + 2\alpha)\exp(\gamma_1 t')$$

$$-A_2(\kappa)(\gamma_2 + 2\alpha)\exp(\gamma_2 t')]\,\frac{\cos(\beta y\cos\vartheta)}{\beta c}\,d\beta\Bigg] + H_y(t) \tag{18}$$

Equations (17) and (18) have the same factor $-E_0Z^{-1}\cos^{-1}\vartheta$ in front. The terms multiplied by this factor are equal just as they were for Eqs. (12) and (5), and one again gets the condition

$$H_t(y)e^{-st/\mu} = H_y(t) = H_{E0}e^{-st/\mu} \tag{19}$$

for the equality of Eqs. (17) and (18). Note that H_{xEs} of Eq. (5) and H_{zEp} of Eq. (17) differ only in the factors $\cos\vartheta$ and $-\cos^{-1}\vartheta$. Hence, we get for $s \to 0$

$$H_{zEp}(x, y, t) = -E_0Z^{-1}(\cos^{-1}\vartheta)[-Z\sigma y\cos\vartheta + I'_{E1}(y\cos\vartheta, t')$$

$$-I_{E2}(y\cos\vartheta, t')] + H_{E0} \tag{20}$$

where I'_{E1} and I_{E2} are defined by Eqs. (7) and (8). The constant H_{E0} must be zero to satisfy the initial conditions.

The normalized version of H_{zEp}, using Eq. (3.2-50), has the form

$$H_{zEp}(\xi', \theta') = -E_0Z^{-1}(\cos^{-1}\vartheta)[-2\xi' + I'_{E1}(\xi', \theta') - I_{E2}(\xi', \theta')] \tag{21}$$

where I'_{E1} and I_{E2} are defined by Eqs. (15) and (16).

3.5 Polarized Magnetic Step Function Excitation

A perpendicularly polarized wave always means a wave whose electric field strength is perpendicular to the plane of incidence and whose magnetic field strength is thus in the plane of incidence. When we want to excite such a wave with a magnetic step function we must excite the field strengths $\mathbf{H}_x$ and $\mathbf{H}_y$ in the plane of incidence, and not the field strength $\mathbf{H}_z$ perpendicular to the plane of incidence. For the calculation of a perpendicularly polarized wave caused by excitation with a magnetic step function we start with Eq. (3.2-9) for $H_x = H_{xHs}$, obtain $H_y = H_{yHs}$ from Eq. (3.1-16), and $E_z = E_{zHs}$ from Eqs. (3.2-10) and (3.2-11).

The boundary conditions for a magnetic step function with propagating excitation follow from analogy with Eq. (3.2-12):

$$\begin{aligned} H_x(x, 0, t) = H_0 S(t - xc^{-1} \sin \vartheta) &= 0 \qquad \text{for} \quad t < xc^{-1} \sin \vartheta \\ &= H_0 \qquad \text{for} \quad t \geqq xc^{-1} \sin \vartheta \end{aligned} \tag{1}$$

At the plane $y \to \infty$ we have the boundary condition

$$H_x(x, \infty, t) = \text{finite} \tag{2}$$

The differential equations for E_z and H_x, Eqs. (3.2-3) and (3.2-9), are the same; the boundary conditions of Eqs. (3.2-12), (3.2-13), (1), and (2) are the same, and the initial conditions of Eqs. (3.2-14) to (3.2-16) are used again. We thus obtain the solution by substituting $H_x = H_{xHs}$ and H_0 for E_{zEs} and E_0 in Eq. (3.2-45):

$$\begin{aligned} H_{xHs}(x, y, t) &= 0 \qquad &\text{for} \quad t' < 0 \\ &= H_0[e^{-(y\cos\vartheta)/L} + w(y\cos\vartheta, t')] \qquad &\text{for} \quad t' \geqq 0 \end{aligned} \tag{3}$$

where $w(y \cos \vartheta, t')$ is defined by Eq. (3.2-44). For $s = 0$ we obtain from Eq. (3)

$$H_{xHs}(x, y, t) = H_0[1 + w(y \cos \vartheta, t')] \qquad \text{for} \quad t' > 0 \tag{4}$$

where $w(y \cos \vartheta, t')$ is now defined in Eq. (3.2-46).

The electric field strength E_{zHs} is obtained by first comparing Eqs. (3.2-5) and (3.2-11). We obtain Eq. (3.2-11) by exchanging E_{zEs} with H_{xHs}, H_t with E_t, μ with ϵ, and s with σ in Eq. (3.2-5); furthermore, all terms except E_t are multiplied by $\cos^{-2} \vartheta$. Let us observe that an exchange of ϵ and μ transforms Z into Z^{-1}. With these changes we get from Eqs. (3.4-5) and (3.4-19):

$$\begin{aligned} E_{zHs}(x, y, t) = H_0 Z \cos^{-1} \vartheta \Bigg\{ & Z^{-1} \left(\frac{s}{\sigma}\right)^{1/2} \exp[-(s\sigma)^{1/2} y \cos \vartheta] \\ & - \frac{2s}{Z\pi} e^{-at'} \Bigg[\int_0^{2\pi K} \Bigg(\operatorname{ch}(a^2 - b^2c^2)^{1/2} t' \\ & + \frac{(2a\alpha' - b^2c^2) \operatorname{sh}(a^2 - b^2c^2)^{1/2} t'}{2\alpha'(a^2 - b^2c^2)^{1/2}} \Bigg) \end{aligned}$$

$$\times \frac{\cos(\beta y \cos \vartheta)}{b^2} d\beta + \int_{2\pi K}^{\infty} \left(\cos(b^2c^2 - a^2)^{1/2} t' \right.$$
$$\left. + \frac{(2a\alpha' - b^2c^2)\sin(b^2c^2 - a^2)^{1/2}t'}{2\alpha'(b^2c^2 - a^2)^{1/2}} \right) \frac{\cos(\beta y \cos \vartheta)}{b^2} d\beta \Bigg] \Bigg\}$$
$$+ E_{H0} e^{-\sigma t/\epsilon}, \qquad \alpha' = s/2\mu \tag{5}$$

We choose $E_{H0} = 0$ to satisfy the initial condition $E_{zHs}(x, y, 0) = 0$, and make the transition $s \to 0$, which differs from the calculation following Eq. (3.4-5):

$$E_{zHs}(x, y, t) = H_0 Z \cos^{-1} \vartheta [I_{H1}(y \cos \vartheta, t') + I_{Hs}(y \cos \vartheta, t')] \tag{6}$$

$$I_{H1}(y \cos \vartheta, t') = \frac{2}{\pi} e^{-\alpha t'} \int_0^{Z\sigma/2} \frac{\operatorname{sh}(\alpha^2 - \beta^2 c^2)^{1/2} t'}{(\alpha^2/c^2 - \beta^2)^{1/2}} \cos(\beta y \cos \vartheta) d\beta$$

$$I_{H2}(y \cos \vartheta, t') = \frac{2}{\pi} e^{-\alpha t'} \int_{Z\sigma/2}^{\infty} \frac{\sin(\beta^2 c^2 - \alpha^2)^{1/2} t'}{(\beta^2 - \alpha^2/c^2)^{1/2}} \cos(\beta y \cos \vartheta) d\beta$$

The transition $\sigma \to 0$ is possible:

$$E_{zHs}(x, y, t) = \frac{2H_0 Z}{\pi \cos \vartheta} \int_0^{\infty} \frac{\sin[\beta c(t - xc^{-1} \sin \vartheta)] \cos(\beta y \cos \vartheta)}{\beta} d\beta \tag{7}$$

The integral is the same as in Eq. (3.2-53).

We turn to the parallel polarized wave excited by a propagating magnetic step function. The magnetic field strength $\mathbf{H}_z$ is now defined by Eq. (3.3-5), while the electric field strengths $\mathbf{E}_{xHp}$ and $\mathbf{E}_{yHp} = -e_y \mathbf{E}_{xHp} \operatorname{tg} \vartheta$ follow from Eqs. (3.3-6), (3.3-7), and (3.1-26).

The boundary condition for excitation by a magnetic step function H_z follow from analogy to Eqs. (1) and (2):

$$\begin{aligned} H_z(x, 0, t) = H_0 S(t - xc^{-1} \sin \vartheta) &= 0 && \text{for} \quad t < xc^{-1} \sin \vartheta \\ &= H_0 && \text{for} \quad t \geqq xc^{-1} \sin \vartheta \end{aligned} \tag{8}$$

$$H_z(x, \infty, t) = \text{finite} \tag{9}$$

The initial conditions for $H_{zHp} = H_z$ are the same as for E_x according to Eqs. (3.3-10)–(3.3-12). The solution for H_{zHp} thus follows from Eq. (3.3-14)

$$\begin{aligned} H_{zHp} &= 0 && \text{for} \quad t' < 0 \\ &= H_0[e^{-(y \cos \vartheta)/L} + w(y \cos \vartheta, t')] && \text{for} \quad t' \geqq 0 \end{aligned} \tag{10}$$

where $w(y \cos \vartheta, t')$ is given by Eq. (3.2-44). For $s = 0$ we obtain from Eq. (10)

$$H_{zHp}(x, y, t) = H_0[1 + w(y \cos \vartheta, t')] \qquad \text{for} \quad t' \geqq 0 \tag{11}$$

and $w(y \cos \vartheta, t')$ is now defined by Eq. (3.2-46).

The electric field strength follows from a comparison of Eqs. (3.3-3) and (3.3-7). One must exchange E_{xHp} with H_{zEp}, H_t with E_t, μ with ϵ, s with σ, and multiply all terms except E_t by $\cos^2 \vartheta$. We thus get from Eqs. (3.4-17) and (3.4-19):

$$\begin{aligned}
E_{xHp}(x, y, t) = &-H_0 Z (\cos \vartheta) \left\{ Z^{-1} \left(\frac{s}{\sigma}\right)^{1/2} \exp[-(s\sigma)^{1/2} y \cos \vartheta] \right. \\
&- \frac{2s}{Z\pi} e^{-\alpha' t'} \left[\int_0^{2\pi K} \left(\operatorname{ch}(a^2 - b^2c^2)^{1/2} t' \right. \right. \\
&+ \left. \frac{(2a\alpha' - b^2c^2) \operatorname{sh}(a^2 - b^2c^2)^{1/2} t'}{2\alpha'(a^2 - b^2c^2)^{1/2}} \right) \\
&\times \frac{\cos(\beta y \cos \vartheta)}{b^2} \, d\beta + \int_{2\pi K}^{\infty} \left(\cos(b^2c^2 - a^2)^{1/2} t' \right. \\
&+ \left. \left. \left. \frac{(2a\alpha' - b^2c^2) \sin(b^2c^2 - a^2)^{1/2} t'}{2\alpha'(b^2c^2 - a^2)^{1/2}} \right) \frac{\cos(\beta y \cos \vartheta)}{b^2} \, d\beta \right] \right\} \\
&+ E_{H0} e^{-\sigma t/\epsilon}
\end{aligned} \tag{12}$$

In the limit $s = 0$ we obtain with $E_{H0} = 0$

$$E_{xHp}(x, y, t) = -ZH_0 (\cos \vartheta)[I_{H1}(y \cos \vartheta, t') + I_{H2}(y \cos \vartheta, t')] \tag{13}$$

where I_{H1} and I_{H2} are defined by Eq. (6). For $\sigma = 0$ we obtain from Eq. (13):

$$E_{xHp}(x, y, t) = -\frac{2H_0 Z}{\pi} \cos \vartheta \int_0^{\infty} \frac{\sin[\beta c(t - xc^{-1} \sin \vartheta)] \cos(\beta y \cos \vartheta)}{\beta} \, d\beta \tag{14}$$

3.6 General Propagating Boundary Conditions

In Sections 3.2 and 3.4 we calculated the electric and magnetic field strengths $\mathbf{E}_{zEs}(x, y, t)$ and $\mathbf{H}_{xEs}(x, y, t)$ of a perpendicularly polarized wave caused by an electric field strength $\mathbf{E}_z(x, 0, t)$ with the time variation of a step function $S(t - xc^{-1} \sin \vartheta)$ in the y plane

$$E_z(x, 0, t) = E_0 S(t - xc^{-1} \sin \vartheta) = E_0 S(t') \tag{1}$$

In Section 3.5 we calculated the field strength $E_{zHs}(x, y, t)$ and H_{xHs} of a perpendicularly polarized wave caused by a magnetic field strength

$$H_x(x, 0, t) = H_0 S(t - xc^{-1} \sin \vartheta) = H_0 S(t') \tag{2}$$

with the time variation of a step function in the y plane. We generalize now these boundary conditions as in Section 2.8 and assume that E_z as well as H_x are zero for $t < xc^{-1} \sin \vartheta$, but may have any time variation for $t \geqq xc^{-1} \sin \vartheta$ that permits a representation by a superposition of step functions according to Fig. 1.3-3 and Eqs. (1.3-1) or (1.3-7):

$$E_z(0, 0, t') = \int_0^{t'} \frac{dE_z(0, 0, t'')}{dt''} S(t' - t'')dt'' \tag{3}$$

$$H_x(0, 0, t') = \int_0^{t''} \frac{dH_x(0, 0, t'')}{dt''} S(t' - t'')dt'' \tag{4}$$

No convergence problems are caused if we consider a vanishing magnetic conductivity only, and we restrict ourselves to this simpler case. Electric and magnetic field strengths due to excitation by a propagating electric step function for a perpendicularly polarized wave are given by Eqs. (3.2-46) and (3.4-6) with $H_t = 0$:

$$E_{zEs}(x, y, t) = E_0[1 + w(y \cos \vartheta, t')] \tag{5}$$

$$\begin{aligned} H_{xEs}(x, y, t) = E_0 Z^{-1} (\cos \vartheta)[&-Z\sigma y \cos \vartheta + I'_{E1}(y \cos \vartheta, t') \\ &- I_{E2}(y \cos \vartheta, t')], \qquad t' = t - xc^{-1} \sin \vartheta \geqq 0 \end{aligned} \tag{6}$$

where w, I'_{E1}, and I_{E2} are given by Eqs. (3.2-46), (3.4-7), and (3.4-8).

The terms in brackets in Eqs. (5) and (6) are the changed step function $S(t)g(t)$ used in Eqs. (1.3-8) and (1.3-9). The electric field strength $\mathbf{E}_z(0, 0, t)$ $S(t)$ with general time variation applied at the line $x = 0$, $y = 0$ and propagating along the plane $y = 0$ according to the relation $x = ct/\sin \vartheta$ will produce the following magnitudes of electric and magnetic field strength along the line x, y:

$$E_{zEs}(x, y, t) = \int_0^{t'} \frac{dE_z(0, 0, t'')}{dt''} S(t' - t'')[1 + w(y \cos \vartheta, t' - t'')]dt'' \tag{7}$$

$$\begin{aligned} H_{xEs}(x, y, t) = Z^{-1} \cos \vartheta \int_0^{t'} \frac{dE_z(0, 0, t'')}{dt''} S(t' - t'')[&-Z\sigma y \cos \vartheta \\ + I'_{E1}(y \cos \vartheta, t' - t'') - I_{E2}(y \cos \vartheta, t' - t'')]dt''& \end{aligned} \tag{8}$$

Electric and magnetic field strengths due to the excitation by a propagating magnetic step function for a perpendicularly polarized wave are given by Eqs. (3.5-4) and (3.5-6) for $E_{H0} = 0$:

$$H_{xHs}(x, y, t) = H_0[1 + w(y \cos \vartheta, t')] \tag{9}$$

$$E_{zHs}(x, y, t) = H_0 Z \cos^{-1} \vartheta[I_{H1}(y \cos \vartheta, t') + I_{H2}(y \cos \vartheta, t')] \tag{10}$$

A magnetic field strength $\mathbf{H}_x(0, 0, t)S(t)$ with general time variation applied at the line $x = 0$, $y = 0$ and propagating along the plane $y = 0$ according to the relation $x = ct/\sin\vartheta$ will produce the following magnitudes of electric and magnetic field strength along the line x, y in analogy to Eqs. (7) and (8):

$$H_{xHs}(x, y, t) = \int_0^{t'} \frac{dH_x(0, 0, t'')}{dt''} S(t' - t'')[1 + w(y\cos\vartheta, t' - t'')]dt'' \qquad (11)$$

$$E_{zHs}(x, y, t) = \frac{Z}{\cos\vartheta} \int_0^{t'} \frac{dH_x(0, 0, t'')}{dt''} S(t' - t'') \times [I_{H1}(y\cos\vartheta, t' - t'') + I_{H2}(y\cos\vartheta, t' - t'')]dt'' \qquad (12)$$

If both an electric and a magnetic field strength with general time variation $\mathbf{E}_z(0, 0, t)S(t)$ and $\mathbf{H}_x(0, 0, t)S(t)$ are applied at the line $x = 0$, $y = 0$ and propagate along the plane $y = 0$ according to the relation $x = ct/\sin\vartheta$, we obtain the sum of the magnitudes of the field strengths produced by them:

$$\begin{aligned} E_{zs}(x, y, t) &= E_{zEs}(x, y, t) + E_{zHs}(x, y, t) \\ &= \int_0^{t'} \frac{dE_z(0, 0, t'')}{dt''} S(t' - t'')[1 + w(y\cos\vartheta, t' - t'')]dt'' \\ &\quad + \frac{Z}{\cos\vartheta} \int_0^{t'} \frac{dH_x(0, 0, t'')}{dt''} S(t' - t'') \\ &\quad \times [I_{H1}(y\cos\vartheta, t' - t'') + I_{H2}(y\cos\vartheta, t' - t'')]dt'' \end{aligned} \qquad (13)$$

$$\begin{aligned} H_{xs}(x, y, t) &= H_{xHs}(x, y, t) + H_{xEs}(x, y, t) \\ &= \int_0^{t'} \frac{dH_x(0, 0, t'')}{dt''} S(t' - t'')[1 + w(y\cos\vartheta, t' - t'')]dt'' \\ &\quad + \frac{\cos\vartheta}{Z} \int_0^{t'} \frac{dE_z(0, 0, t'')}{dt''} S(t' - t'')[-Z\sigma y\cos\vartheta \\ &\quad + I'_{E1}(y\cos\vartheta, t' - t'') - I_{E2}(y\cos\vartheta, t' - t'')]dt'' \end{aligned} \qquad (14)$$

The values of E_{zs} and H_{xs} in the plane $y = 0$ will be of particular interest later on. We obtain from Eqs. (3.2-46), (3.5-6), (3.4-7), and (3.4-8):

$$w(0, t') = 0 \qquad (15)$$

$$I_{H1}(0, t') = \frac{2}{\pi} e^{-\alpha t'} \int_0^{Z\sigma/2} \frac{\mathrm{sh}(\alpha^2 - \beta^2 c^2)^{1/2} t'}{(\alpha^2/c^2 - \beta^2)^{1/2}} d\beta \qquad (16)$$

$$I_{H2}(0, t') = \frac{2}{\pi} e^{-\alpha t'} \int_{Z\sigma/2}^{\infty} \frac{\sin(\beta^2 c^2 - \alpha^2)^{1/2} t'}{(\beta^2 - \alpha^2/c^2)^{1/2}} d\beta \qquad (17)$$

$$I'_{E1}(0,t') = \frac{2Z\sigma}{\pi}\left\{\frac{1}{d} - e^{-\alpha t'}\int_d^{Z\sigma/2}\left(\operatorname{ch}(\alpha^2-\beta^2c^2)^{1/2}t' + \frac{\alpha\operatorname{sh}(\alpha^2-\beta^2c^2)^{1/2}t'}{(\alpha^2-\beta^2c^2)^{1/2}}\right)\frac{d\beta}{\beta} - \int_0^d\left[\exp\left(-\frac{\beta^2c^2t'}{2\alpha}\right)-1\right]\frac{d\beta}{\beta^2}\right\} + \frac{2c}{\pi}e^{-\alpha t'}\int_0^{Z\sigma/2}\frac{\operatorname{sh}(\alpha^2-\beta^2c^2)^{1/2}t'}{(\alpha^2-\beta^2c^2)^{1/2}}\,d\beta \tag{18}$$

$$I_{E2}(0,t') = \frac{2Z\sigma}{\pi}e^{-\alpha t'}\int_{Z\sigma/2}^{\infty}\left(\cos(\beta^2c^2-\alpha^2)^{1/2}t' + \frac{(2\alpha^2+\beta^2c^2)\sin(\beta^2c^2-\alpha^2)^{1/2}t'}{2\alpha(\beta^2c^2-\alpha^2)^{1/2}}\right)\frac{d\beta}{\beta^2} \tag{19}$$

A comparison with Eqs. (2.8-15)–(2.8–19) shows that t has been replaced by $t' = t - xc^{-1}\sin\vartheta$, while Eqs. (13) and (14) differ from Eqs. (2.8-13) and (2.8-14) by the interchange of $Z\cos^{-1}\vartheta$ and $Z^{-1}\cos\vartheta$ with Z and Z^{-1}.

The impedance $Z_s(0, t')$ at the plane $y = 0$ becomes:

$$Z_s(0,t') = \frac{E_{zs}(x,0,t)}{H_{xs}(x,0,t)} = \frac{E_{zs}(0,0,t')}{H_{xs}(0,0,t')}$$
$$= \left(E_z(0,0,t') + Z\cos^{-1}\vartheta\int_0^{t'}\frac{dH_x(0,0,t'')}{dt''}S(t'-t'') \times [I_{H1}(0,t'-t'') + I_{H2}(0,t'-t'')]dt''\right)$$
$$\div\left(H_x(0,0,t') + Z^{-1}\cos\vartheta\int_0^{t'}\frac{dE_z(0,0,t'')}{dt''}S(t'-t'') \times [I'_{E1}(0,t'-t'') - I_{E2}(0,t'-t'')]dt''\right) \tag{20}$$

Compared with Eq. (2.8-20) we have factors $\cos\vartheta$ as well as $\cos^{-1}\vartheta$, and t is replaced by $t' = t - xc^{-1}\sin\vartheta$.

For a parallel polarized wave excited by an electric step function we get from Eqs. (3.3-15) and (3.4-20) for $H_{E0} = 0$:

$$E_{xEp}(x,y,t) = E_0[1 + w(y\cos\vartheta, t')], \qquad t' \geqq 0 \tag{21}$$

$$H_{zEp}(x,y,t) = -E_0Z^{-1}\cos^{-1}\vartheta[-Z\sigma y\cos\vartheta + I'_{E1}(y\cos\vartheta,t') - I_{E2}(y\cos\vartheta,t')] \tag{22}$$

If the time variation of the excitation function is not that of a step function $E_0S(t)$ but has the general form $E_x(0, 0, t)S(t)$, we obtain for E_{xEp} and H_{zEp} in analogy to Eqs. (7) and (8):

$$E_{xEp}(x, y, t) = \int_0^{t'} \frac{dE_x(0, 0, t'')}{dt''} S(t' - t'') \times [1 + w(y \cos \vartheta, t' - t'')]dt'' \tag{23}$$

$$H_{zEp}(x, y, t) = -\frac{1}{Z \cos \vartheta} \int_0^{t'} \frac{dE_x(0, 0, t'')}{dt''} S(t' - t'')[-Z\sigma y \cos \vartheta + I'_{E1}(y \cos \vartheta, t' - t'') - I_{E2}(y \cos \vartheta, t' - t'')]dt'' \tag{24}$$

Excitation of a parallel polarized wave by a magnetic step function yields the field strengths

$$H_{zHp}(x, y, t) = H_0[1 + w(y \cos \vartheta, t')], \qquad t' \geqq 0 \tag{25}$$

$$E_{xHp}(x, y, t) = -H_0 Z \cos \vartheta [I_{H1}(y \cos \vartheta, t') + I_{H2}(y \cos \vartheta, t')] \tag{26}$$

for $E_{H0} = 0$ according to Eqs. (3.5-11) and (3.5-13). A magnetic excitation function $H_z(0, 0, t)S(t)$ with general time variation instead of a step function $H_0S(t)$ yields in analogy to Eqs. (11) and (12):

$$H_{zHp}(x, y, t) = \int_0^{t'} \frac{dH_z(0, 0, t'')}{dt''} S(t' - t'') \times [1 + w(y \cos \vartheta, t' - t'')]dt'' \tag{27}$$

$$E_{xHp}(x, y, t) = -Z \cos \vartheta \int_0^{t'} \frac{dH_z(0, 0, t'')}{dt''} S(t' - t'') \times [I_{H1}(y \cos \vartheta, t' - t'') + I_{H2}(y \cos \vartheta, t' - t'')]dt'' \tag{28}$$

The field strengths produced by both an electric and a magnetic excitation with arbitrary time variation $E_x(0, 0, t)$ and $H_z(0, 0, t)$ follow in analogy to Eqs. (13) and (14):

$$\begin{aligned} E_{xp}(x, y, t) &= E_{xEp}(x, y, t) + E_{xHp}(x, y, t) \\ &= \int_0^{t'} \frac{dE_x(0, 0, t'')}{dt''} S(t' - t'')[1 + w(y \cos \vartheta, t' - t'')]dt'' \\ &\quad - Z \cos \vartheta \int_0^{t'} \frac{dH_z(0, 0, t'')}{dt''} S(t' - t'') \\ &\quad \times [I_{H1}(y \cos \vartheta, t' - t'') + I_{H2}(y \cos \vartheta, t' - t'')]dt'' \end{aligned} \tag{29}$$

$$H_{zp}(x, y, t) = H_{zHp}(x, y, t) + H_{zEp}(x, y, t)$$
$$= \int_0^{t'} \frac{dH_z(0, 0, t'')}{dt''} S(t' - t'')[1 + w(y \cos \vartheta, t' - t'')]dt''$$
$$- \frac{1}{Z \cos \vartheta} \int_0^{t'} \frac{dE_x(0, 0, t'')}{dt''} S(t' - t'')[-Z\sigma y \cos \vartheta$$
$$+ I'_{E1}(y \cos \vartheta, t' - t'') - I_{E2}(y \cos \vartheta, t' - t'')]dt'' \quad (30)$$

The functions w, I_{H1}, I_{H2}, I'_{E1}, and I_{E2} are the same as in Eqs. (13) and (14). For $y = 0$ we get again the relations of Eqs. (15)–(19). The impedance $Z_p(0, t')$ at the plane $y = 0$ becomes:

$$Z_p(0, t') = \frac{E_{xp}(x, 0, t)}{H_{zp}(x, 0, t)} = \frac{E_{xp}(0, 0, t')}{H_{zp}(0, 0, t')}$$
$$= \left(E_x(0, 0, t') - Z \cos \vartheta \int_0^{t'} \frac{dH_z(0, 0, t'')}{dt''} S(t' - t'') \right.$$
$$\left. \times [I_{H1}(y \cos \vartheta, t' - t'') + I_{H2}(y \cos \vartheta, t' - t'')]dt'' \right)$$
$$\div \left(H_z(0, 0, t') - \frac{1}{Z \cos \vartheta} \int_0^{t'} \frac{dE_x(0, 0, t'')}{dt''} S(t' - t'') \right.$$
$$\left. \times [I'_{E1}(y \cos \vartheta, t' - t'') - I_{E2}(y \cos \vartheta, t' - t'')]dt'' \right) \quad (31)$$

A comparison with Eq. (20) shows that the two impedances Z_s and Z_p differ by factors $-\cos^2 \vartheta$ and $-\cos^{-2} \vartheta$ in front of the integrals.

3.7 Perpendicularly Polarized Electric Ramp Function

Consider Eq. (3.2-3), which determines the electric field strength $\mathbf{E}_z$ due to an electric excitation:

$$(\partial^2 E_z/\partial y^2) \cos^{-2} \vartheta - \mu\epsilon\, \partial^2 E_z/\partial t^2 - (\mu\sigma + \epsilon s)\partial E_z/\partial t - s\sigma E_z = 0 \quad (1)$$

Instead of the boundary condition of a propagating step function given in Eq. (3.2-12) we use now a propagating exponential ramp function in analogy to Eqs. (2.3-2) and (2.3-3):

$$\begin{aligned} E_z(x, 0, t) = E_0 r(t') &= 0 && \text{for } t' < 0 \\ &= E_0 Q(1 - e^{-t'/\tau}) = E_1(1 - e^{-t'/\tau}) && \text{for } t' \geqq 0 \end{aligned} \quad (2)$$

$$t' = t - xc^{-1} \sin \vartheta$$

For $y \to \infty$ we have the usual boundary condition

$$E_z(x, \infty, t) = \text{finite} \tag{3}$$

The initial conditions of Eqs. (3.2-14) and (3.2-16) remain unchanged for $t' = 0$:

$$E_z(x, y, xc^{-1} \sin \vartheta) = H_x(x, y, xc^{-1} \sin \vartheta) = 0 \tag{4}$$

$$\partial E_z/\partial t = \partial H_x/\partial t = 0 \qquad \text{for} \quad y > 0,\ t = xc^{-1} \sin \vartheta \tag{5}$$

We assume that the solution of Eq. (1) can be written in the form

$$E_z(x, y, t) = E_{zEs}(x, y, t) = E_0 Q[u(x, y, t) + (1 - e^{-t'/\tau})F(x, y)], \qquad t' \geqq 0 \tag{6}$$

Insertion of the term $E_0 Q(1 - e^{-t'/\tau})F(x, y)$ into Eq. (1) yields:

$$(1 - e^{-t'/\tau})(\partial^2 F/\partial y^2)\cos^{-2} \vartheta + \tau^{-2}\mu\epsilon e^{-t'/\tau}F - \tau^{-1}(\mu\sigma + \epsilon s)e^{-t'/\tau}F - s\sigma(1 - e^{-t'/\tau})F = 0 \tag{7}$$

Since F is not a function of t', the terms with different functions of t' must vanish independently. We get an equation with the first and the last term of Eq. (7):

$$(\partial^2 F/\partial y^2)\cos^{-2} \vartheta - s\sigma F = 0 \tag{8}$$

and a second equation with the two remaining terms:

$$\tau^{-1}(\tau^{-1}\mu\epsilon - \mu\sigma - \epsilon s)e^{-t'/\tau} = 0 \tag{9}$$

Equation (8) is the same as Eq. (3.2-18) which had the solution

$$\begin{aligned} F(x, y) &= e^{-(y \cos \vartheta)/L} && \text{for} \quad x \leqq ct/\sin \vartheta \\ &= 0 && \text{for} \quad x > ct/\sin \vartheta \end{aligned} \tag{10}$$

Equation (9) equals Eq. (2.3-12) except for the replacement of t by t'. The solution of primary interest is not affected by this replacement:

$$\tau^{-1} = c^2(\mu\sigma + \epsilon s) = 2a \tag{11}$$

The introduction of the function $F(x, y)$ transforms the boundary condition of Eq. (2),

$$\begin{aligned} E_z(x, 0, t) = E_{zEs}(x, 0, t) = E_1 u(x, 0, t) &= 0 && \text{for} \quad t < xc^{-1} \sin \vartheta \\ = E_1[u(x, 0, t) + 1] &= E_1 && \text{for} \quad t \geqq xc^{-1} \sin \vartheta \end{aligned} \tag{12}$$

which implies the homogeneous boundary condition

$$u(x, 0, t) = 0 \tag{13}$$

for any value of x and t.

Equation (3) yields:

$$u(x, \infty, t) = \text{finite} \tag{14}$$

The initial conditions of Eqs. (4) and (5) assume for $t' = 0$, $y \geqq 0$ the form

$$u(x, y, xc^{-1} \sin \vartheta) = 0 \tag{15}$$

$$\partial u(x, y, t)/\partial t + 2ae^{-(y \cos \vartheta)/L} = 0 \tag{16}$$

For the calculation of $u(x, y, t)$ we insert this function into Eq. (1):

$$(\partial^2 u/\partial y^2)\cos^{-2} \vartheta - \mu\epsilon\, \partial^2 u/\partial t^2 - (\mu\sigma + \epsilon s)\partial u/\partial t - s\sigma u = 0 \tag{17}$$

The calculation of u proceeds as that of w in Section 3.2 from Eq. (3.2-26)–Eqs. (3.2-34) and (3.2-35):

$$u(x, y, t) = \int_0^\infty [A_1(\kappa)\exp(\gamma_1 t') + A_2(\kappa)\exp(\gamma_2 t')]\sin(2\pi\kappa y \cos \vartheta)d\kappa \tag{18}$$

$$\frac{\partial u}{\partial t} = \int_0^\infty [A_1(\kappa)\gamma_1 \exp(\gamma_1 t') + A_2(\kappa)\gamma_2 \exp(\gamma_2 t')]\sin(2\pi\kappa y \cos \vartheta)d\kappa \tag{19}$$

Insertion of Eqs. (18) and (19) into Eqs. (15) and (16) yields for $t' = 0$:

$$\int_0^\infty [A_1(\kappa) + A_2(\kappa)]\sin(2\pi\kappa y \cos \vartheta)d\kappa = 0 \tag{20}$$

$$\int_0^\infty [A_1(\kappa)\gamma_1 + A_2(\kappa)\gamma_2]\sin(2\pi\kappa y \cos \vartheta)d\kappa = -2ae^{-(y \cos \vartheta)/L} \tag{21}$$

These two equations should be compared with Eqs. (2.3-26) and (2.3-27) as well as (3.2-36) and (3.2-37).

Equation (20) yields

$$A_1(\kappa) = -A_2(\kappa) \tag{22}$$

while Eq. (21) is brought by the substitution

$$\kappa \cos \vartheta = \kappa', \qquad d\kappa = \cos^{-1} \vartheta\, d\kappa' \tag{23}$$

into the form

$$\int_0^\infty [A_1(\kappa')\gamma_1 + A_2(\kappa')\gamma_2]\sin 2\pi\kappa'\, d\kappa = -(2a \cos \vartheta)e^{-(y \cos \vartheta)/L} \tag{24}$$

which yields according to Eqs. (2.3-27)–(2.3-30)

$$A_1(\kappa')\gamma_1 + A_2(\kappa')\gamma_2 = -16a\pi\kappa' b^{-2}\cos\vartheta \tag{25}$$

or

$$A_1(\kappa)\gamma_1 + A_2(\kappa)\gamma_2 = -16a\pi\kappa/b^2 \tag{26}$$

Since Eqs. (22) and (26) are equal to Eqs. (2.3-28) and (2.3-30), we get for $A_1(\kappa)$ and $A_2(\kappa)$ the values of Eq. (2.3-31). Furthermore, $u(x, y, t)$ follows from Eq. (2.3-33) by substitution of t' for t and $y\cos\vartheta$ for y:

$$u(x, y, t) = -\frac{4ae^{-at'}}{\pi}\left(\int_0^{2\pi K}\frac{\mathrm{sh}(a^2-b^2c^2)^{1/2}t'}{(a^2-b^2c^2)^{1/2}}\frac{\beta\sin(\beta y\cos\vartheta)}{b^2}\,d\beta + \int_{2\pi K}^{\infty}\frac{\sin(b^2c^2-a^2)^{1/2}t'}{(b^2c^2-a^2)^{1/2}}\frac{\beta\sin(\beta y\cos\vartheta)}{b^2}\,d\beta\right) \tag{27}$$

The solution $E_{zEs}(x, y, t)$ follows from Eq. (6) with the help of Eq. (10) and $Q = (1 - e^{-2a\,\Delta T})^{-1}$:

$$E_{zEs}(x, y, t) = E_1[(1 - e^{-2at'})e^{-(y\cos\vartheta)/L} + u(x, y, t)]$$
$$E_1 = E_0 Q = E_0(1 - e^{-2a\,\Delta T})^{-1} \tag{28}$$

At this point we may make the transition $s \to 0$:

$$u(y\cos\vartheta, t') = -\frac{4\alpha e^{-\alpha t'}}{\pi}\left(\int_0^{Z\sigma/2}\frac{\mathrm{sh}(\alpha^2-\beta^2c^2)^{1/2}t'}{(\alpha^2-\beta^2c^2)^{1/2}} \times\frac{\sin(\beta y\cos\vartheta)}{\beta}\,d\beta + \int_{Z\sigma/2}^{\infty}\frac{\sin(\beta^2c^2-\alpha^2)^{1/2}t'}{(\beta^2c^2-\alpha^2)^{1/2}} \times\frac{\sin(\beta y\cos\vartheta)}{\beta}\,d\beta\right) \tag{29}$$

$$E_{zEs}(x, y, t) = E_1[1 - e^{-2\alpha t'} + u(y\cos\vartheta, t')] \tag{30}$$

Equations (29) and (30) may be written in a more compact normalized form for $\sigma \neq 0$ by means of the substitutions of Eqs. (3.2-50):

$$E_{zEs}(\xi', \theta') = E_1[1 - e^{-2\theta'} + u(\xi', \theta')], \qquad \theta' \geqq 0 \tag{31}$$

$$u(\xi', \theta') = -\frac{4}{\pi}e^{-\theta'}\left(\int_0^1\frac{\mathrm{sh}(1-\eta^2)^{1/2}\theta'}{(1-\eta^2)^{1/2}}\frac{\sin\xi'\eta}{\eta}\,d\eta + \int_1^{\infty}\frac{\sin(\eta^2-1)^{1/2}\theta'}{(\eta^2-1)^{1/2}}\frac{\sin\xi'\eta}{\eta}\,d\eta\right)$$

These two equations are formally the same as Eqs. (2.3-37) and (2.3-38); only the definition of ξ and ξ' as well as θ and θ' differs.

Consider the loss-free case $\sigma = 0$. Using Eqs. (2.3-39) to (2.3-41) we obtain from Eq. (30):

$$E_{zEs}(x, y, t) = E_0\left(\frac{t'}{\Delta T} - \frac{2}{\pi c \Delta T}\int_0^\infty \frac{\sin \beta ct' \sin(\beta y \cos \vartheta)}{\beta^2}\, d\beta\right) \tag{32}$$

With the help of Eq. (2.3-44) we get:

$$\begin{aligned}\int_0^\infty \frac{\sin \beta ct' \sin(\beta y \cos \vartheta)}{\beta^2}\, d\beta &= \pi ct/2 && \text{for} \quad 0 \leqq ct' \leqq y \cos \vartheta \\ &= (\pi y/2)\cos \vartheta && \text{for} \quad 0 \leqq y \cos \vartheta \leqq ct'\end{aligned} \tag{33}$$

$$t' = t - xc^{-1}\sin\vartheta$$

Substitution into Eq. (32) brings:

$$\begin{aligned}E_{zEs}(x, y, t) &= 0 && \text{for} \quad t \leqq (x/c)\sin \vartheta + (y/c)\cos \vartheta \\ &= E_0\{t/\Delta T - [(x/c\,\Delta T)\sin \vartheta \\ &\quad + (y/c\,\Delta T)\cos \vartheta]\} && \text{for} \quad t \geqq (x/c)\sin \vartheta + (y/c)\cos \vartheta\end{aligned} \tag{34}$$

For a discussion of this formula refer to Fig. 3.7-1a. It shows a wavefront $L(t) = L(0)$ propagating at the time $t = 0$ through the point $x = 0$, $y = 0$, and an arbitrary point x, y. Let $L(0)$ denote the wavefront where the transition from $E_{zEs} = 0$ to a linear increase with time takes place. The time variation of $E_{zEs}(x, y, t)$ in the point x, y is shown in Fig. 3.7-1b. For

$$ct < x \sin \vartheta + y \cos \vartheta$$

the wavefront $L(t)$ has not yet reached the point x, y and E_{zEs} is zero. At the time

$$ct = x \sin \vartheta + y \cos \vartheta$$

the wavefront $L(t)$ becomes $L[(x/c)\sin \vartheta + (y/c)\cos \vartheta]$ and intersects the point x, y. From this time on, E_{zEs} in the point x, y will increase linearly with time, as shown in Fig. 3.7-1b.

For the calculation of the magnetic field strength $\mathbf{H}_{xEs}$ we start with Eq. (3.2-5):

$$H_{xEs}(x, y, t) = e^{-st/\mu}\left(-\mu^{-1}\int \frac{\partial E_{zEs}}{\partial y} e^{st/\mu}\, dt + H_t(y)\right) \tag{35}$$

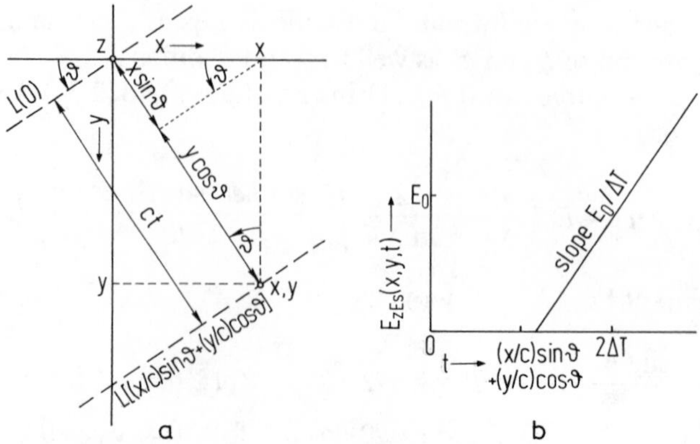

FIG. 3.7-1. Propagation of a linear ramp function as the limit of an exponential ramp function for $\sigma = 0$ using propagating excitation in the plane $y = 0$.

Substitution of Eq. (28) yields:

$$H_{xEs}(x, y, t) = E_1 e^{-st/\mu}\Big(\mu^{-1}(s\sigma)^{1/2} \cos \vartheta e^{-(y\cos\vartheta)/L} \int (e^{st/\mu} - e^{(2ax/c)\sin\vartheta} e^{-(2a-s/\mu)t})dt - \mu^{-1}\int \frac{\partial u}{\partial y} e^{st/\mu} + H_t(y)\Big)$$

$$= E_1\left(\frac{\sigma}{s}\right)^{1/2} \cos\vartheta\left(1 + \frac{s}{Z^2\sigma}\right) e^{-2at'} e^{-(y\cos\vartheta)/L} + e^{st/\mu}\left(-E_1\mu^{-1}\int \frac{\partial u}{\partial y} e^{st/\mu}\, dt + H_t(y)\right) \quad (36)$$

The term $e^{-st/\mu}\int (\partial u/\partial y)e^{st/\mu}\, dt$ was evaluated in Eq. (3.4-3) for w rather than for u:

$$H_{xEs}(x, y, t) = E_1 Z^{-1} \cos\vartheta\Big[Z\left(\frac{\sigma}{s}\right)^{1/2}\left(1 + \frac{s}{Z^2\sigma} e^{-2at'}\right) e^{-(y\cos\vartheta)/L} - \frac{1}{2\pi}\int_0^\infty [A_1(\kappa)(\gamma_1 + s/\mu)^{-1} \exp(\gamma_2 t') + A_2(\kappa)(\gamma_2 + s/\mu)^{-1} \exp(\gamma_2 t)] \times \beta c \cos(2\pi\kappa y \cos\vartheta)d\beta\Big] + H_t(y)e^{-st/\mu} \quad (37)$$

A comparison with Eq. (2.9-5) shows the additional factor $\cos\vartheta$, as well as the replacement of t and y by t' and $y \cos \vartheta$. We obtain thus from Eq. (2.9-8):

$$\begin{aligned}
H_{xEs}(x, y, t) = E_1 Z^{-1} \cos \vartheta \Bigg\{ & Z\left(\frac{\sigma}{s}\right)^{1/2} \left(1 + \frac{s}{Z^2\sigma} e^{-2at'}\right) \\
& \times \exp[-(s\sigma)^{1/2} y \cos \vartheta] \\
& - \frac{4a}{\pi c} e^{-at'} \Bigg[\int_0^{2\pi K} \Bigg(\operatorname{ch}(a^2 - b^2c^2)^{1/2} t' \\
& + \frac{(\alpha - s/2\mu)\operatorname{sh}(a^2 - b^2c^2)^{1/2} t'}{(a^2 - b^2c^2)^{1/2}} \Bigg) \\
& \times \frac{\cos(\beta y \cos \vartheta)}{b^2} d\beta + \int_{2\pi K}^{\infty} \Bigg(\cos(b^2c^2 - a^2)^{1/2} t' \\
& + \frac{(\alpha - s/2\mu)\sin(b^2c^2 - a^2)^{1/2} t'}{(b^2c^2 - a^2)^{1/2}} \Bigg) \\
& \times \frac{\cos(\beta y \cos \vartheta)}{b^2} d\beta \Bigg] \Bigg\} + H_t(y) e^{-st'/\mu}
\end{aligned} \tag{38}$$

For $s \to 0$ we get with the same change from y and t to $y \cos \vartheta$ and t' from Eqs. (2.9-20), (2.9-18), and (2.9-19) with $H_t(y) = 0$:

$$\begin{aligned}
H_{xEs}(x, y, t) = E_1 Z^{-1} \cos \vartheta [& -Z\sigma y \cos \vartheta + I'_{E3}(y \cos \vartheta, t') \\
& - I_{E4}(y \cos \vartheta, t')]
\end{aligned} \tag{39}$$

$$\begin{aligned}
I'_{E3}(y \cos \vartheta, t') = \frac{2Z\sigma}{\pi} \Bigg[& \frac{1}{d} - e^{-\alpha t'} \int_d^{Z\sigma/2} \Bigg(\operatorname{ch}(\alpha^2 - \beta^2 c^2)^{1/2} t' \\
& + \frac{\alpha \operatorname{sh}(\alpha^2 - \beta^2c^2)^{1/2} t'}{(\alpha^2 - \beta^2c^2)^{1/2}} \Bigg) \frac{d\beta}{\beta^2} \\
& - \int_0^d [\exp(-\beta^2c^2t'/2\alpha) - 1] \frac{d\beta}{\beta^2} \\
& + \tfrac{1}{2} e^{-\alpha t'} \int_0^{Z\sigma/2} \Bigg(\operatorname{ch}(\alpha^2 - \beta^2c^2)^{1/2} t' \\
& + \frac{\alpha \operatorname{sh}(\alpha^2 - \beta^2c^2)^{1/2} t'}{(\alpha^2 - \beta^2c^2)^{1/2}} \Bigg) \\
& \times \left(\frac{\sin(2^{-1}\beta y \cos \vartheta)}{\beta/2} \right)^2 d\beta \Bigg]
\end{aligned} \tag{40}$$

$$I_{E4}(y\cos\vartheta, t') = \frac{2Z\sigma}{\pi} e^{-\alpha t'} \int_{Z\sigma/2}^{\infty} \left(\cos(\beta^2 c^2 - \alpha^2)^{1/2} t' + \frac{\alpha \sin(\beta^2 c^2 - \alpha^2)^{1/2}}{(\beta^2 c^2 - \alpha^2)^{1/2}} \right) \frac{\cos(\beta y \cos\vartheta)}{\beta^2} d\beta \tag{41}$$

Using the substitutions of Eq. (3.2-50) we get the following normalized expressions for $\sigma \neq 0$:

$$H_{xEs}(\xi', \theta') = E_1 Z^{-1} \cos\vartheta[-2\xi' + I'_{E3}(\xi', \theta') - I_{E4}(\xi', \theta')] \tag{42}$$

$$\begin{aligned} I'_{E3}(\xi', \theta') = \frac{2}{\pi} \Bigg\{ 2\Bigg[\frac{1}{\delta} - e^{-\theta'} \int_{\delta}^{1} \left(\mathrm{ch}(1-\eta^2)^{1/2}\theta' + \frac{\mathrm{sh}(1-\eta^2)^{1/2}\theta'}{(1-\eta^2)^{1/2}} \right) \\ \times \frac{d\eta}{\eta^2} - \int_0^{\infty} [\exp(-\eta^2\theta'/2) - 1] \frac{d\eta}{\eta^2} \Bigg] \\ + e^{-\theta'} \int_0^1 \left(\mathrm{ch}(1-\eta^2)^{1/2}\theta' + \frac{\mathrm{sh}(1-\eta^2)^{1/2}\theta'}{(1-\eta^2)^{1/2}} \right) \\ \times \left(\frac{\sin(\xi'\eta/2)}{\eta/2} \right)^2 d\eta \Bigg\} \end{aligned} \tag{43}$$

$$I_{E4}(\xi', \theta') = \frac{4}{\pi} e^{-\theta'} \int_1^{\infty} \left(\cos(\eta^2 - 1)^{1/2}\theta' + \frac{\sin(\eta^2-1)^{1/2}\theta'}{(\eta^2-1)^{1/2}} \right) \frac{\cos \xi'\eta}{\eta^2} d\eta \tag{44}$$

$$\delta = 2d/Z\sigma \ll 1$$

3.8 Parallel Polarized Electric Ramp Function

For excitation by a parallel polarized exponential ramp function we have to start with Eq. (3.3-1) for $E_{xEp} = E_x$:

$$(\partial^2 E_x/\partial y^2)\cos^{-2}\vartheta - \mu\epsilon\, \partial^2 E_x/\partial t^2 - (\mu\sigma + \epsilon s)\partial E_x/\partial t - s\sigma E_x = 0 \tag{1}$$

The boundary condition of a propagating exponential ramp function follows in analogy to Eq. (3.7-2):

$$\begin{aligned} E_x(x, 0, t) &= E_0 r(t') = 0 && \text{for } t' < 0 \\ &= E_0 Q(1 - e^{-t'/\tau}) = E_1(1 - e^{-t'/\tau}) && \text{for } t' \geqq 0 \end{aligned} \tag{2}$$

For $y \to \infty$ we have as always:

$$E_x(x, \infty, t) = \text{finite} \tag{3}$$

The initial conditions are the same as in Eqs. (3.7-4) and (3.7-5), except that

E_z and H_x are replaced by E_x and H_z:

$$E_x(x, y, xc^{-1}\sin\vartheta) = H_z(x, y, xc^{-1}\sin\vartheta) = 0 \tag{4}$$

$$\partial E_x/\partial t = \partial H_z/\partial t = 0 \qquad \text{for} \quad y > 0, \quad t = xc^{-1}\sin\vartheta \tag{5}$$

From the analogy of Eqs. (1)–(5) with Eqs. (3.7-1)–(3.7-5) we deduce that $E_x = E_{xEp}$ can be written in the form of Eq. (3.7-28):

$$E_{xEp}(x, y, t) = E_1[(1 - e^{-2at'})e^{-(y\cos\vartheta)/L} + u(x, y, t)] \tag{6}$$

where $u(x, y, t)$ is given be Eq. (3.7-27). For $s \to 0$ we get from Eq. (3.7-30)

$$E_{xEp}(x, y, t) = E_1[1 - e^{-2\alpha t'} + u(y\cos\vartheta, t')], \qquad t' \geqq 0 \tag{7}$$

with $u(y\cos\vartheta, t')$ defined by Eq. (3.7-29).

We turn to the magnetic field strength produced by a parallel polarized electric exponential ramp function. Equations (3.3-2) and (3.3-3) give this field strength:

$$H_{zEp}(x, y, t) = \int \left(\epsilon \frac{\partial E_{xEp}}{\partial t} + E_{xEp}\right) dy + H_y(t) \tag{8}$$

$$H_{zEp}(x, y, t) = e^{-st/\mu}\left(\mu^{-1}\cos^{-2}\vartheta \int \frac{\partial E_{xEp}}{\partial y} e^{st/\mu}\, dt + H_t(y)\right) \tag{9}$$

According to Eqs. (6) and (3.7-28) the electric field strength E_{xEp} of the parallel polarized wave varies like the field strength E_{zEs} of the perpendicularly polarized wave. We may thus substitute E_{zEs} for E_{xEp} in Eqs. (8) and (9). Furthermore, Eq. (3.7-35) is changed into Eq. (9) if we multiply all terms except $H_t(y)$ with $-\cos^{-2}\vartheta$ in addition to changing E_{zEs} to E_{xEp}. With these changes we obtain from Eq. (3.7-38):

$$\begin{aligned} H_{zEp}(x, y, t) = -E_1 Z^{-1}\cos^{-1}\vartheta \Bigg\{ & Z\left(\frac{\sigma}{s}\right)^{1/2}\left(1 + \frac{s}{Z^2\sigma}e^{-2at'}\right) \\ & \times \exp[-(s\sigma)^{1/2} y\cos\vartheta] \\ & - \frac{4a}{\pi c}e^{-at'}\Bigg[\int_0^{2\pi K}\Bigg(\operatorname{ch}(a^2 - b^2c^2)^{1/2}t' \\ & + \frac{(\alpha - s/2\mu)\operatorname{sh}(a^2 - b^2c^2)^{1/2}t'}{(a^2 - b^2c^2)^{1/2}}\Bigg) \\ & \times \frac{\cos(\beta y\cos\vartheta)}{b^2}\, d\beta + \int_{2\pi K}^{\infty}\Bigg(\cos(b^2c^2 - a^2)^{1/2}t' \\ & + \frac{(\alpha - s/2\mu)\sin(b^2c^2 - a^2)^{1/2}t'}{(b^2c^2 - a^2)^{1/2}}\Bigg) \\ & \times \frac{\cos(\beta y\cos\vartheta)}{b^2}\, d\beta\Bigg]\Bigg\} + H_t(y)e^{-st/\mu} \end{aligned} \tag{10}$$

For $s \to 0$ we get in analogy to the transition from Eq. (3.7-38) to (3.7-39)

$$H_{zEp}(x, y, t) = -E_1 Z^{-1} \cos^{-1} \vartheta[-Z\sigma \cos \vartheta + I'_{E3}(y \cos \vartheta, t') - I_{E4}(y \cos \vartheta, t')] \quad (11)$$

where I'_{E3} and I_{E4} are defined by Eqs. (3.7-40) and (3.7-41).

Using the normalized notation of Eq. (3.2-50) we may rewrite Eqs. (7) and (11):

$$E_{xEp}(\xi', \theta') = E_1[1 - e^{-2\theta'} + u(\xi', \theta')], \qquad \theta' \geqq 0 \quad (12)$$

$$H_{zEp}(\xi', \theta') = -E_1 Z^{-1} \cos^{-1} \vartheta[-2\xi' + I'_{E3}(\xi', \theta') - I_{E4}(\xi', \theta')] \quad (13)$$

The functions $u(\xi', \theta')$, $I'_{E3}(\xi', \theta')$, and $I_{E4}(\xi', \theta')$ are defined in Eqs. (3.7-31), (3.7-43), and (3.7-44).

3.9 Polarized Magnetic Ramp Function Excitation

Instead of exciting the perpendicular and the parallel polarized waves by exponential ramp functions E_z and E_x we may use magnetic exponential ramp functions H_x and H_z. For the perpendicularly polarized wave we use the boundary conditions of Eqs. (3.7-2) and (3.7-3) with E_z replaced by H_x:

$$\begin{aligned} H_x(x, 0, t) &= H_0 r(t') = 0 && \text{for} \quad t' < 0 \\ &= H_0 Q(1 - e^{-t'/\tau}) = H_1(1 - e^{-t'/\tau}) && \text{for} \quad t' \geqq 0 \end{aligned} \quad (1)$$

$$H_x(x, \infty, t) = \text{finite} \quad (2)$$

The initial conditions of Eqs. (3.7-4) and (3.7-5) remain unchanged. The differential equation for H_x, given by Eq. (3.2-9), is the same as for E_z, given by Eq. (3.7-1). Hence, the solution for H_{xHs} is given by Eq. (3.7-28) if E_{zEs} and E_0 are replaced by H_{xHs} and H_0:

$$H_{xHs}(x, y, t) = H_1[(1 - e^{-2at'})e^{-(y \cos \vartheta)/L} + u(y \cos \vartheta, t')] \quad (3)$$

For $s = 0$ we get from Eq. (3.7-30)

$$H_{xHs}(x, y, t) = H_1[1 - e^{-2\alpha t'} + u(y \cos \vartheta, t')] \quad (4)$$

where $u(y \cos \vartheta, t')$ is defined by Eq. (3.7-29).

The electric field strength E_{zHs} follows by insertion of Eq. (3) into Eq. (3.2-11):

$$E_{zHs}(x, y, t) = e^{-\sigma t/\epsilon}\left(-\epsilon^{-1} \cos^{-2} \vartheta \int \frac{\partial H_{xHs}}{\partial y} e^{\sigma t/\epsilon}\, dt + E_t(y)\right) \quad (5)$$

A comparison of Eqs. (3) and (3.7-28) as well as Eqs. (5) and (3.7-35) shows that they are transformed into each other if we exchange E_{zEs}, E_0, H_{xEs}, s, μ, H_t with H_{xHs}, H_0, E_{zHs}, σ, ϵ, E_t; furthermore, all terms except E_t are multiplied with $\cos^{-2}\vartheta$. Hence, we get from Eq. (3.7-38) with $\alpha' - \sigma/2\epsilon = -(\alpha - s/2\mu)$:

$$E_{zHs}(x, y, t) = H_1 Z \cos^{-1}\vartheta \Bigg\{ Z^{-1}\left(\frac{s}{\sigma}\right)^{1/2}\left(1 + \frac{Z^2\sigma}{s}e^{-2at'}\right) \times \exp[-(s\sigma)^{1/2} y\cos\vartheta] - \frac{4a}{\pi c}e^{-at'}\Bigg[\int_0^{2\pi K}\Bigg(\text{ch}(a^2 - b^2c^2)^{1/2}t' - \frac{(\alpha - s/2\mu)\text{sh}(a^2 - b^2c^2)^{1/2}t'}{(a^2 - b^2c^2)^{1/2}}\Bigg)\frac{\cos(\beta y\cos\vartheta)}{b^2}d\beta + \int_{2\pi K}^{\infty}\Bigg(\cos(b^2c^2 - a^2)^{1/2}t' - \frac{(\alpha - s/2\mu)\sin(b^2c^2 - a^2)^{1/2}t'}{(b^2c^2 - a^2)^{1/2}}\Bigg) \times \frac{\cos(\beta y\cos\vartheta)}{b^2}d\beta\Bigg]\Bigg\} + E_t(y)e^{-\sigma t'/\epsilon} \tag{6}$$

This is the generalization of Eq. (2.10-12) for $\vartheta \neq 0$. We note that the substitution of $y\cos\vartheta$ and t' for y and t as well as multiplication by $\cos^{-1}\vartheta$ transforms Eq. (2.10-12) into Eq. (6). For $s = 0$ we get with these changes from Eq. (2.10-21) for $E_t = 0$:

$$E_{zHs}(x, y, t) = H_1 Z \cos^{-1}\vartheta[-Z\sigma e^{-2\alpha t'} y\cos\vartheta + I'_{H3}(y\cos\vartheta, t') - I_{H4}(y\cos\vartheta, t')] \tag{7}$$

where I'_{H3} and I_{H4} follow from Eqs. (2.10-19) and (2.10-20):

$$I'_{H3}(y\cos\vartheta, t') = \frac{2Z\sigma}{\pi}\Bigg[\frac{1}{d}e^{-2\alpha t'} - e^{-\alpha t'}\int_d^{Z\sigma/2}\Bigg(\text{ch}(\alpha^2 - \beta^2c^2)^{1/2}t' - \frac{\alpha\,\text{sh}(\alpha^2 - \beta^2c^2)^{1/2}t'}{(\alpha^2 - \beta^2c^2)^{1/2}}\Bigg)\frac{d\beta}{\beta^2} - 2\int_0^d \frac{\text{sh}^2(\beta^2c^2t/4\alpha)}{\beta^2}d\beta + \tfrac{1}{2}e^{-\alpha t'}\int_0^{Z\sigma/2}\Bigg(\text{ch}(\alpha^2 - \beta^2c^2)^{1/2}t' + \frac{\alpha\,\text{sh}(\alpha^2 - \beta^2c^2)^{1/2}t'}{(\alpha^2 - \beta^2c^2)^{1/2}}\Bigg) \times \left(\frac{\sin(2^{-1}\beta y\cos\vartheta)}{\beta/2}\right)^2 d\beta\Bigg] \tag{8}$$

$$I_{H4}(y\cos\vartheta, t') = \frac{2Z\sigma}{\pi} e^{-\alpha t'} \int_{Z\sigma/2}^{\infty} \left(\cos(\beta^2c^2 - \alpha^2)^{1/2}t' - \frac{\alpha\sin(\beta^2c^2 - \alpha^2)^{1/2}t'}{(\beta^2c^2 - \alpha^2)^{1/2}} \right) \frac{\cos(\beta y\cos\vartheta)}{\beta^2} d\beta \tag{9}$$

For a parallel polarized wave with excitation by a magnetic exponential ramp function we have the boundary conditions

$$\begin{aligned} H_z(x, 0, t) = H_0 r(t') &= 0 && \text{for} \quad t' < 0 \\ &= H_1(1 - e^{-t'/\tau}) && \text{for} \quad t' \geqq 0 \end{aligned} \tag{10}$$

$$H_z(x, \infty, t) = \text{finite} \tag{11}$$

and the initial conditions of Eqs. (3.8-4) as well as (3.8-5). We obtain the same solution as in Eq. (3)

$$H_{zHp}(x, y, t) = H_1[(1 - e^{-2at'})e^{-(y\cos\vartheta)/L} + u(y\cos\vartheta, t')] \tag{12}$$

and we get for $s = 0$

$$H_{zHp}(x, y, t) = H_1[1 - e^{-2\alpha t'} + u(y\cos\vartheta, t')] \tag{13}$$

where $u(y\cos\vartheta, t')$ is defined by Eq. (3.7-29).

The electric field strength follows from a comparison of Eqs. (3.3-3) and (3.3-7). One must exchange E_{xHp} with H_{zEp}, H_t with E_t, μ with ϵ, s with σ, and multiply all terms except E_t by $\cos^2\vartheta$. We thus get from Eq. (3.8-10) with $\alpha' - \sigma/2\epsilon = -(\alpha - s/2\mu)$:

$$\begin{aligned} E_{xHp}(x, y, t) = -H_1 Z\cos\vartheta \Bigg\{ & Z^{-1} \left(\frac{s}{\sigma}\right)^{1/2} \left(1 + \frac{Z^2\sigma}{s} e^{-2at'}\right) \\ & \times \exp[-(s\sigma)^{1/2} y\cos\vartheta] \\ & - \frac{4a}{\pi c} e^{-at'} \Bigg[\int_0^{2\pi K} \Bigg(\operatorname{ch}(a^2 - b^2c^2)^{1/2}t' \\ & - \frac{(\alpha - s/2\mu)\operatorname{sh}(a^2 - b^2c^2)^{1/2}t'}{(a^2 - b^2c^2)^{1/2}} \Bigg) \\ & \times \frac{\cos(\beta y\cos\vartheta)}{b^2} d\beta + \int_{2\pi k}^{\infty} \Bigg(\cos(b^2c^2 - a^2)^{1/2}t' \\ & - \frac{(\alpha - s/2\mu)\sin(b^2c^2 - a^2)^{1/2}t'}{(b^2c^2 - a^2)^{1/2}} \Bigg) \frac{\cos(\beta y\cos\vartheta)}{b^2} d\beta \Bigg] \Bigg\} \\ & + E_{H0} e^{-\sigma t'/\epsilon} \end{aligned} \tag{14}$$

This is the same result as Eq. (2.10-12), except that t is replaced by t', y by $y \cos \vartheta$, and the whole equation except $E_t(y)$ is multiplied by $-\cos \vartheta$. Hence, we get in analogy to Eq. (2.10-21) for $s = 0$:

$$E_{xHp}(x, y, t) = -H_1 Z (\cos \vartheta)[-Z\sigma e^{-2\alpha t'} y \cos \vartheta + I'_{H3}(y \cos \vartheta, t') - I_{H4}(y \cos \vartheta, t')] \quad (15)$$

In accordance with Eq. (2.10-28) we have chosen $E_{H0} = 0$. The functions I'_{H3} and I_{H4} are defined by Eqs. (8) and (9).

We still rewrite Eqs. (4), (7), (13), (15), (8), and (9) into normalized form using the notation of Eq. (3.2-50):

$$H_{xHs}(\xi', \theta') = H_1[1 - e^{-2\theta'} + u(\xi', \theta')] \quad (16)$$

$$E_{zHs}(\xi', \theta') = H_1 Z (\cos^{-1} \vartheta)[-2\xi' e^{-2\theta} + I'_{H3}(\xi', \theta') - I_{H4}(\xi', \theta')] \quad (17)$$

$$H_{zHp}(\xi', \theta') = H_1[1 - e^{-2\theta'} + u(\xi', \theta')] \quad (18)$$

$$E_{xHp}(\xi', \theta') = -H_1 Z (\cos \vartheta)[-2\xi' e^{-2\theta'} + I'_{H3}(\xi', \theta') - I_{H4}(\xi', \theta')] \quad (19)$$

$$\begin{aligned} I'_{H3}(\xi', \theta') = \frac{2}{\pi} \Bigg\{ & 2\left[\frac{1}{\delta} - e^{-\theta'} \int_\delta^1 \left(\operatorname{ch}(1-\eta^2)^{1/2}\theta' - \frac{\operatorname{sh}(1-\eta^2)^{1/2}\theta'}{(1-\eta^2)^{1/2}} \right) \frac{d\eta}{\eta^2} - 2 \int_0^\delta \frac{\operatorname{sh}^2(\eta^2\theta'/4)}{\eta^2} d\eta \right] \\ & + e^{-\theta'} \int_0^1 \left(\operatorname{ch}(1-\eta^2)^{1/2}\theta' + \frac{\operatorname{sh}(1-\eta^2)^{1/2}\theta'}{(1-\eta^2)^{1/2}} \right) \\ & \times \left(\frac{\sin(\eta\xi'/2)}{\eta/2} \right)^2 d\eta \Bigg\} \end{aligned} \quad (20)$$

$$\begin{aligned} I_{H4}(\xi', \theta') = \frac{4}{\pi} e^{-\theta'} \int_1^\infty & \left(\cos(\eta^2 - 1)^{1/2}\theta' \right. \\ & \left. - \frac{\sin(\eta^2-1)^{1/2}\theta'}{(\eta^2-1)^{1/2}} \right) \frac{\cos \eta\xi'}{\eta^2} d\eta \end{aligned} \quad (21)$$

$$\delta = 2d/Z\sigma \ll 1$$

3.10 Propagating Boundary Conditions Using Ramp Functions

In Section 3.6 we investigated the excitation of electromagnetic waves by electric and magnetic field strengths with general time variation represented by a superposition of step functions. From a comparison of Eqs. (1.3-7) and (1.3-33) we see that Eqs. (3.6-3) and (3.6-4) have to be rewritten in the following form for exponential ramp functions:

$$E_z(0, 0, t') = \tau \int_0^{t'} \frac{d^2E_z(0, 0, t'')}{dt''^2} S(t' - t'')(1 - e^{-(t'-t'')/\tau})dt'' \quad (1)$$

$$H_x(0, 0, t') = \tau \int_0^{t'} \frac{d^2H_x(0, 0, t'')}{dt''^2} S(t' - t'')(1 - e^{-(t'-t'')/\tau})dt'' \quad (2)$$

For a perpendicularly polarized electric field strength with the time variation of an exponential ramp function we get from Eqs. (3.7-30) and (3.7-28) for $2\alpha\,\Delta T \ll 1$

$$E_{zEs}(x, y, t) = (E_0/2\alpha\,\Delta T)[1 - e^{-2\alpha t'} + u(y \cos\vartheta, t')] \quad (3)$$

$$E_0/2\alpha\,\Delta T = E_1 \quad (4)$$

where $u(y \cos\vartheta, t')$ is defined by Eq. (3.7-29).

The magnetic field strength H_{xEs} follows from Eq. (3.7-39), with I'_{E3} and I_{E4} defined by Eqs. (3.7-40) and (3.7-41):

$$H_{xEs}(x, y, t) = E_1 Z^{-1}(\cos\vartheta)[-Z\sigma y \cos\vartheta + I'_{E3}(y\cos\vartheta, t') - I_{E4}(y\cos\vartheta, t')] \quad (5)$$

A comparison of Eqs. (1)–(5) with Eqs. (2.11-1), (2.11-2), (2.11-4), and (2.11-6) shows that Eqs. (2.11-10) and (2.11-12) assume the following form:

$$E_{zEs}(x, y, t) = \frac{1}{2\alpha} \int_0^{t'} \frac{d^2E_z(0, 0, t'')}{dt''^2} S(t' - t'')[1 - e^{-2\alpha(t'-t'')} + u(y\cos\vartheta, t' - t'')]dt'' \quad (6)$$

$$H_{xEs}(x, y, t) = \frac{\cos\vartheta}{2Z\alpha} \int_0^{t'} \frac{d^2E_z(0, 0, t'')}{dt''^2} S(t' - t'')[-Z\sigma y\cos\vartheta + I'_{E3}(y\cos\vartheta, t' - t'') - I_{E4}(y\cos\vartheta, t' - t'')]dt'' \quad (7)$$

The functions I'_{E3} and I_{E4} are defined by Eqs. (3.7-40) and (3.7-41).

Electric and magnetic field strengths due to excitation by a propagating magnetic exponential ramp function for a perpendicularly polarized wave are given by Eqs. (3.9-4) and (3.9-7). For $\alpha\,\Delta T \ll 1$ we get:

$$H_{xHs}(x, y, t) = (H_0/2\alpha\,\Delta T)[1 - e^{-2\alpha t'} + u(y\cos\vartheta, t')] \quad (8)$$

$$H_0/2\alpha\,\Delta T = H_1$$

$$E_{zHs}(x, y, t) = H_1 Z(\cos^{-1}\vartheta)[-Z\sigma e^{-2\alpha t'} y\cos\vartheta + I'_{H3}(y\cos\vartheta, t') - I_{H4}(y\cos\vartheta, t')] \quad (9)$$

The functions I'_{H3} and I_{H4} are defined by Eqs. (3.9-8) and (3.9-9).

A magnetic field strength $\mathbf{H}_x(0, 0, t)S(t)$ with general time variation applied at the line $x = 0$, $y = 0$ and propagating along the plane $y = 0$ according to the relation $x = ct/\sin\vartheta$ will produce the following magnitudes of electric and magnetic field strengths along the line x, y in analogy to Eqs. (6) and (7):

$$H_{xHs}(x, y, t) = \frac{1}{2\alpha}\int_0^{t'} \frac{d^2H_x(0, 0, t'')}{dt''^2} S(t' - t'')[1 - e^{-2\alpha(t'-t'')} + u(y\cos\vartheta, t' - t'')]dt'' \tag{10}$$

$$E_{zHs}(x, y, t) = \frac{Z}{2\alpha\cos\vartheta}\int_0^{t'} \frac{d^2H_x(0, 0, t'')}{dt''^2} S(t' - t'') \times [-Z\sigma e^{-2\alpha(t'-t'')} y\cos\vartheta + I'_{H3}(y\cos\vartheta, t' - t'') - I_{H4}(y\cos\vartheta, t' - t'')]dt'' \tag{11}$$

If both an electric and a magnetic field strength with general time variation, $\mathbf{E}_z(0, 0, t)S(t)$ and $\mathbf{H}_x(0, 0, t)S(t)$, are applied at the line $x = 0$, $y = 0$ and propagate along the plane $y = 0$ according to the relation $x = ct/\sin\vartheta$, we obtain the sum of the magnitudes of the field strengths produced by them:

$$\begin{aligned} E_{zs}(x, y, t) &= E_{zEs}(x, y, t) + E_{zHs}(x, y, t) \\ &= \frac{1}{2\alpha}\int_0^{t'} \frac{d^2E_z(0, 0, t'')}{dt''^2} S(t' - t'')[1 - e^{-2\alpha(t'-t'')} \\ &\quad + u(y\cos\vartheta, t' - t'')]dt'' \\ &\quad + \frac{Z}{2\alpha\cos\vartheta}\int_0^{t'} \frac{d^2H_x(0, 0, t'')}{dt''^2} S(t' - t'') \\ &\quad \times [-Z\sigma e^{-2\alpha(t'-t'')} y\cos\vartheta \\ &\quad + I'_{H3}(y\cos\vartheta, t' - t'') - I_{H4}(y\cos\vartheta, t' - t'')]dt'' \end{aligned} \tag{12}$$

$$\begin{aligned} H_{xs}(x, y, t) &= H_{xHs}(x, y, t) + H_{xEs}(x, y, t) \\ &= \frac{1}{2\alpha}\int_0^{t'} \frac{d^2H_x(0, 0, t'')}{dt''^2} S(t' - t'')[1 - e^{-2\alpha(t'-t'')} \\ &\quad + u(y\cos\vartheta, t' - t'')]dt'' \\ &\quad + \frac{\cos\vartheta}{2Z\alpha}\int_0^{t'} \frac{d^2E_z(0, 0, t'')}{dt''^2} S(t' - t'')[-Z\sigma y\cos\vartheta \\ &\quad + I'_{E3}(y\cos\vartheta, t' - t'') - I_{E4}(y\cos\vartheta, t' - t'')]dt'' \end{aligned} \tag{13}$$

For parallel polarization we get from Eqs. (3.8-7) and (3.8-11):

$$E_{xEp}(x, y, t) = (E_0/2\alpha\,\Delta T)[1 - e^{-2\alpha t'} + u(y\cos\vartheta, t')] \tag{14}$$

$$E_0/2\alpha\,\Delta T = E_1$$

$$H_{zEp}(x, y, t) = -E_1 Z^{-1}\cos^{-1}\vartheta[-Z\sigma y\cos\vartheta + I'_{E3}(y\cos\vartheta, t') - I_{E4}(y\cos\vartheta, t')] \tag{15}$$

Furthermore, we get from Eqs. (3.9-13) and (3.9-15):

$$H_{zHp}(x, y, t) = (H_0/2\alpha\,\Delta T)[1 - e^{-2\alpha t'} + u(y\cos\vartheta, t')] \tag{16}$$

$$H_0/2\alpha\,\Delta T = H_1$$

$$E_{xHp}(x, y, t) = -H_1 Z\cos\vartheta[-Z\sigma e^{-\sigma t'/\epsilon} y\cos\vartheta + I'_{H3}(y\cos\vartheta, t') - I_{H4}(y\cos\vartheta, t')] \tag{17}$$

Equations (6), (7), (10), and (11) for field strengths with general time variation are rewritten for parallel polarization by replacing $E_z(0, 0, t'')$ and $H_x(0, 0, t'')$ by $E_x(0, 0, t'')$ and $H_z(0, 0, t'')$; furthermore, Eq. (7) is multiplied by $\cos^{-2}\vartheta$ and Eq. (11) by $\cos^2\vartheta$:

$$E_{xEp}(x, y, t) = \frac{1}{2\alpha}\int_0^{t'} \frac{d^2E_x(0, 0, t'')}{dt''^2} S(t' - t'')[1 - e^{-2\alpha(t'-t'')} + u(y\cos\vartheta, t' - t'')]dt'' \tag{18}$$

$$H_{zEp}(x, y, t) = -\frac{1}{2\alpha Z\cos\vartheta}\int_0^{t'} \frac{d^2E_x(0, 0, t'')}{dt''^2} S(t' - t'')[-Z\sigma y\cos\vartheta + I'_{E3}(y\cos\vartheta, t'-t'') - I_{E4}(y\cos\vartheta, t' - t'')]dt'' \tag{19}$$

$$H_{zHp}(x, y, t) = \frac{1}{2\alpha}\int_0^{t'} \frac{d^2H_z(0, 0, t'')}{dt''^2} S(t - t'')[1 - e^{-2\alpha(t'-t'')} + u(y\cos\vartheta, t'-t'')]dt'' \tag{20}$$

$$E_{xHp}(x, y, t) = -\frac{Z\cos\vartheta}{2\alpha}\int_0^{t'} \frac{d^2H_z(0, 0, t'')}{dt''^2} S(t' - t'') \times [-Z\sigma e^{-2\alpha(t'-t'')} y\cos\vartheta + I'_{H3}(y\cos\vartheta, t' - t'') - I_{H4}(y\cos\vartheta, t' - t'')]dt'' \tag{21}$$

If both electric and magnetic field strengths are applied for excitation, we obtain the following sums for the produced field strengths:

$$E_{xp}(x, y, t) = E_{xEp}(x, y, t) + E_{xHp}(x, y, t) \tag{22}$$

$$H_{zp}(x, y, t) = H_{zHp}(x, y, t) + H_{zEp}(x, y, t) \tag{23}$$

4 Reflection and Transmission at Boundaries

4.1 Nonsinusoidal Waves in Nonconducting Media

For the study of reflection and transmission of nonsinusoidal waves at the boundary between two nonconducting media, we follow the approach used by Fresnel[1] for sinusoidal waves. Figure 4.1-1 shows a plane wave in medium 1 coming from the upper left and reaching the boundary with a medium 2. Both media shall be described by constants of permeability μ_1, μ_2, and permittivity ϵ_1, ϵ_2. The conductivity of both media shall be zero. The time variation of electric and magnetic field strengths is the same, and the amplitude of the magnetic field strength equals E_0/Z if the amplitude of the electric field strength equals E_0; the wave impedance Z equals either $\sqrt{\mu_1/\epsilon_1}$ or $\sqrt{\mu_2/\epsilon_2}$.

We use the three unit vectors $\mathbf{e}_x$, $\mathbf{e}_y$, $\mathbf{e}_z$ in the direction of the x, y, and z axis in Fig. 4.1-1. The electric field strengths $\mathbf{E}_\mathrm{i}$, $\mathbf{E}_\mathrm{r}$, and $\mathbf{E}_\mathrm{t}$ of the incident, reflected, and transmitted wave shall have the direction of $\mathbf{e}_z$, which means the waves are polarized perpendicularly to the plane of incidence $z = 0$. The waves propagate in the direction of the propagation vectors $\mathbf{r}_\mathrm{i}$, $\mathbf{r}_\mathrm{r}$, and $\mathbf{r}_\mathrm{t}$. These are the same directions as those of the three Poynting vectors $\mathbf{P}_\mathrm{i}$, $\mathbf{P}_\mathrm{r}$, and $\mathbf{P}_\mathrm{t}$. Finally, we define three unit vectors $\mathbf{e}_\mathrm{i}$, $\mathbf{e}_\mathrm{r}$, and $\mathbf{e}_\mathrm{t}$ in the direction of the magnetic field strengths $\mathbf{H}_\mathrm{i}$, $\mathbf{H}_\mathrm{r}$, and $\mathbf{H}_\mathrm{t}$. These unit vectors may be written as follows:

$$\begin{aligned}
\mathbf{e}_\mathrm{i} &= \mathbf{e}_x \cos \vartheta_\mathrm{i} - \mathbf{e}_y \sin \vartheta_\mathrm{i} \\
\mathbf{e}_\mathrm{r} &= -\mathbf{e}_x \cos \vartheta_\mathrm{r} - \mathbf{e}_y \sin \vartheta_\mathrm{r} \qquad (1) \\
\mathbf{e}_\mathrm{t} &= \mathbf{e}_x \cos \vartheta_\mathrm{t} - \mathbf{e}_y \sin \vartheta_\mathrm{t}
\end{aligned}$$

The incident planar wave propagates in the direction of the propagation vector $\mathbf{r}_\mathrm{i}$. It is described by a solution of the one-dimensional wave equation, which yields the two solutions $f(c_1 t - \mathbf{r}_\mathrm{i})$ and $g(c_1 t + \mathbf{r}_\mathrm{i})$. We discard the second solution since we want only the solution representing a wave propagating toward increasing values of $\mathbf{r}_\mathrm{i}$. Instead of writing $f(c_1 t - \mathbf{r}_\mathrm{i})$ we may also write $f(t - \mathbf{r}_\mathrm{i}/c_1)$. This distinction may appear trivial, but it is not. The notation $f(c_1 t - \mathbf{r}_\mathrm{i})$ states that the wave propagates a certain distance r_i in a time t that depends on the velocity c_1, while the notation $f(t - \mathbf{r}_\mathrm{i}/c_1)$ states

[1] Augustin Jean Fresnel, engineer, 1788–1827; born in Broglie, France.

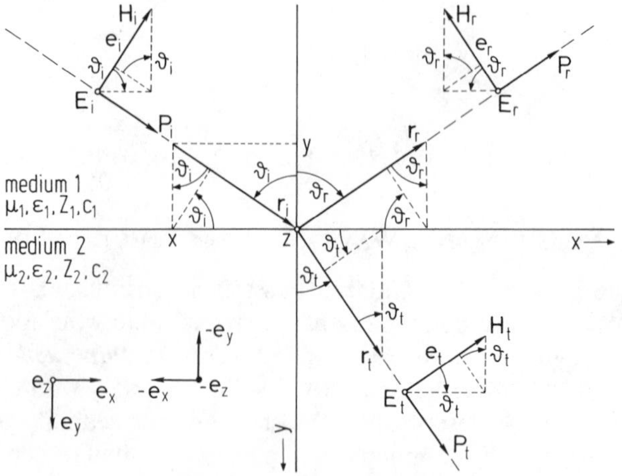

FIG. 4.1-1. Reflection and transmission of a wave that is polarized perpendicularly to the plane of incidence.

that the wave propagates in a certain time t a distance r_i that depends on the velocity c_1. If we write the solution of the wave equation generally in the form $f(t - \mathbf{r}/c)$ we use the same time unit in any medium the wave propagates through, and we avoid thus the need to discuss clocks and time measurements.

The propagation vectors $\mathbf{r}_i$, $\mathbf{r}_r$, and $\mathbf{r}_t$ of Fig. 4.1-1 may be written in terms of the unit vectors $\mathbf{e}_x$ and $\mathbf{e}_y$:

$$\begin{aligned}\mathbf{r}_i &= r(\mathbf{e}_x \sin \vartheta_i + \mathbf{e}_y \cos \vartheta_i)\\ \mathbf{r}_r &= r(\mathbf{e}_x \sin \vartheta_r - \mathbf{e}_y \cos \vartheta_r)\\ \mathbf{r}_t &= r(\mathbf{e}_x \sin \vartheta_t + \mathbf{e}_y \cos \vartheta_t)\end{aligned} \tag{2}$$

$$(\mathbf{r}_i \cdot \mathbf{r}_i)^{1/2} = (\mathbf{r}_r \cdot \mathbf{r}_r)^{1/2} = (\mathbf{r}_t \cdot \mathbf{r}_t)^{1/2} = r \tag{3}$$

The choice of the same constant r for all three propagation vectors indicates that the same scale is used for distance measurements in medium 1 and 2.

Using Eqs. (1) and (2) we may write the electric and magnetic field strength of the incident wave:

$$\begin{aligned}\mathbf{E}_i &= \mathbf{e}_z E_0 f(t - \mathbf{r}_i/c_1)\\ &= \mathbf{e}_z E_0 f[t - (\mathbf{e}_x \sin \vartheta_i + \mathbf{e}_y \cos \vartheta_i) r/c_1]\end{aligned} \tag{4}$$

$$\begin{aligned}\mathbf{H}_i &= \mathbf{e}_i E_0 Z_1^{-1} f(t - \mathbf{r}_i/c_1)\\ &= (\mathbf{e}_x \cos \vartheta_i - \mathbf{e}_y \sin \vartheta_i) E_0 Z_1^{-1} f[t - (\mathbf{e}_x \sin \vartheta_i + \mathbf{e}_y \cos \vartheta_i) r/c_1]\end{aligned} \tag{5}$$

$$c_1 = (\mu_1 \epsilon_1)^{-1/2}, \qquad Z_1 = (\mu_1/\epsilon_1)^{1/2} \tag{6}$$

We define a reflection coefficient ρ_s and a transmission coefficient τ_s that shall have the same value for any point x along the boundary $y=0$ in Fig. 4.1-1 and for any time t:

$$\rho_s = E_r/E_i \tag{7}$$

$$\tau_s = E_t/E_i \tag{8}$$

$$E_i = |\mathbf{E}_i|, \qquad E_r = |\mathbf{E}_r|, \qquad E_t = |\mathbf{E}_t| \tag{9}$$

Next we assume that the time variation of the reflected and the transmitted wave is given by $f(t - \mathbf{r}_r/c_1)$ and $f(t - \mathbf{r}_t/c_2)$, where f denotes the same function as $f(t - \mathbf{r}_i/c_1)$ of the incident wave. This assumption is suggested by the fact that both the reflected and the transmitted wave are solutions of the one-dimensional wave equation, but it will only be justified by the success in satisfying the boundary conditions at the boundary of medium 1 and 2. We obtain thus for the field strengths of the reflected and the transmitted wave:

$$\begin{aligned}\mathbf{E}_r &= \mathbf{e}_z \rho_s E_0 f(t - \mathbf{r}_r/c_1)\\ &= \mathbf{e}_z \rho_s E_0 f[t - (\mathbf{e}_x \sin \vartheta_r - \mathbf{e}_y \cos \vartheta_r) r/c_1]\end{aligned} \tag{10}$$

$$\begin{aligned}\mathbf{H}_r &= \mathbf{e}_r \rho_s E_0 Z_1^{-1} f(t - \mathbf{r}_r/c_1)\\ &= -(\mathbf{e}_x \cos \vartheta_r + \mathbf{e}_y \sin \vartheta_r)\rho_s E_0 Z_1^{-1}\\ &\quad \times f[t - (\mathbf{e}_x \sin \vartheta_r - \mathbf{e}_y \cos \vartheta_r) r/c_1]\end{aligned} \tag{11}$$

$$\begin{aligned}\mathbf{E}_t &= \mathbf{e}_z \tau_s E_0 f(t - \mathbf{r}_t/c_2)\\ &= \mathbf{e}_z \tau_s E_0 f[t - (\mathbf{e}_x \sin \vartheta_t + \mathbf{e}_y \cos \vartheta_t) r/c_2]\end{aligned} \tag{12}$$

$$\begin{aligned}\mathbf{H}_t &= \mathbf{e}_t \tau_s E_0 Z_2^{-1} f(t - \mathbf{r}_t/c)\\ &= (\mathbf{e}_x \cos \vartheta_t - \mathbf{e}_y \sin \vartheta_t)\tau_s E_0 Z_2^{-1}\\ &\quad \times f[t - (\mathbf{e}_x \sin \vartheta_t + \mathbf{e}_y \cos \vartheta_t) r/c_2]\end{aligned} \tag{13}$$

$$c_2 = (\mu_2 \epsilon_2)^{-1/2}, \qquad Z_2 = (\mu_2/\epsilon_2)^{1/2} \tag{14}$$

Two boundary conditions must be met at the boundary $y=0$ in Fig. 4.1-1; these boundary conditions are as much a part of the physical content of Maxwell's theory as Maxwell's equations themselves:

(a) The tangential components ($\mathbf{e}_z$) of the electric field strengths in both media must be equal at $y=0$.

(b) The tangential components ($\mathbf{e}_x$) of the magnetic field strengths in both media must be equal at $y=0$.

In order to see which part of the incident electric field strength $\mathbf{E}_i$ is the tangential component to the plane $y=0$, or generally $y=$ constant, refer to Fig. 4.1-2. A wavefront of $\mathbf{E}_i$ according to Eq. (4) is defined by the function

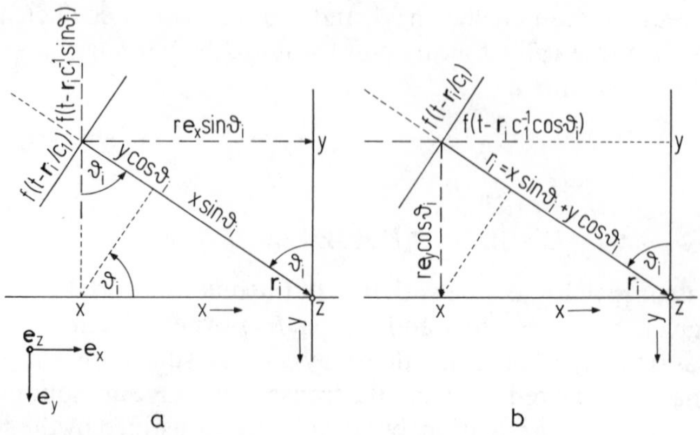

FIG. 4.1-2. Decomposition of a wavefront propagating in the direction $\mathbf{r}_i$ into a component propagating (a) in the direction $\mathbf{e}_x$ and (b) in the direction $\mathbf{e}_y$. $\mathbf{r}_i = r(\mathbf{e}_x \sin \vartheta_i + \mathbf{e}_y \cos \vartheta_i)$, $r_i = x \sin \vartheta_i + y \cos \vartheta_i$.

$f(t - \mathbf{r}_i/c_1)$ that moves with velocity c_1 in the direction of $\mathbf{r}_i$. This wavefront can be decomposed into a wavefront $f(t - \mathbf{r}_i c_1^{-1} \sin \vartheta_i) = f(t - r\mathbf{e}_x c_1^{-1} \sin \vartheta_i)$ propagating in the direction of the x axis, as shown in Fig. 4.1-2a, and a wavefront $f(t - \mathbf{r}_i c_1^{-1} \cos \vartheta_i) = f(t - r\mathbf{e}_y c_1^{-1} \cos \vartheta_i)$ propagating in the direction of the y axis, as shown in Fig. 4.1-2b. Hence, the tangential component of the function $f(t - \mathbf{r}_i/c_1)$ in Eq. (4) is found by dropping the term containing $\mathbf{e}_y$. By implication, the same applies to the reflected and the transmitted field strength.

We may thus write condition (a) by means of Eqs. (4), (10), and (12):

$$f(t - rc_1^{-1}\mathbf{e}_x \sin \vartheta_i) + \rho_s f(t - rc_1^{-1}\mathbf{e}_x \sin \vartheta_r) = \tau_s f(t - rc_2^{-1}\mathbf{e}_x \sin \vartheta_t) \quad (15)$$

Since ρ_s and τ_s were assumed to be constants, this equation can be satisfied for arbitrary values of t and r if and only if the three arguments of the function f are equal. This condition yields two equations:

$$t - rc_1^{-1}\mathbf{e}_x \sin \vartheta_i = t - rc_1^{-1}\mathbf{e}_x \sin \vartheta_r \quad (16)$$

$$t - rc_1^{-1}\mathbf{e}_x \sin \vartheta_i = t - rc_2^{-1}\mathbf{e}_x \sin \vartheta_t \quad (17)$$

Equation (16) yields Snell's[2] law of reflection for waves with arbitrary time variation,

$$\vartheta_i = \vartheta_r \quad (18)$$

[2] Willebrord Snell or Snellius, mathematician, 1591–1626; born in Leiden, The Netherlands.

while Eq. (17) yields Snell's law of refraction[3]:

$$\sin \vartheta_i/\sin \vartheta_t = c_1/c_2 = n \tag{19}$$

From Eq. (15) follows still a third relation if the three arguments of the function f are equal:

$$1 + \rho_s = \tau_s \tag{20}$$

We turn to the boundary condition (b), which yields with the help of Eqs. (5), (11), and (13) for the component propagating in the direction $\mathbf{e}_x$ the following relation:

$$Z_1^{-1}(\cos \vartheta_i) f(t - rc_1^{-1}\mathbf{e}_x \sin \vartheta_i) - \rho_s Z_1^{-1}(\cos \vartheta_r) f(t - rc_1^{-1}\mathbf{e}_x \sin \vartheta_r)$$
$$= \tau_s Z_2^{-1}(\cos \vartheta_t) f(t - rc_2^{-1}\mathbf{e}_x \sin \vartheta_t) \tag{21}$$

Again, this equation can be satisfied for arbitrary values of t and r if and only if the three arguments of the function f are equal. One obtains Eqs. (16) and (17) once more. But instead of Eq. (20) one now obtains:

$$\cos \vartheta_i - \rho_s \cos \vartheta_r = \tau_s(Z_1/Z_2)\cos \vartheta_t \tag{22}$$

With the help of Eqs. (18), (20), and (22) follows:

$$\rho_s = (Z_2 \cos \vartheta_i - Z_1 \cos \vartheta_t)/(Z_2 \cos \vartheta_i + Z_1 \cos \vartheta_t) \tag{23}$$

The term $\cos \vartheta_t$ can be eliminated by means of Eq. (19)

$$\cos^2 \vartheta_t = 1 - \sin^2 \vartheta_t = 1 - n^{-2} \sin^2 \vartheta_i \tag{24}$$

and one obtains with $\vartheta_i = \vartheta$:

$$\rho_s = \frac{Z_2 \cos \vartheta - Z_1(1 - n^{-2}\sin^2 \vartheta)^{1/2}}{Z_2 \cos \vartheta + Z_1(1 - n^{-2}\sin^2 \vartheta)^{1/2}}$$
$$= \frac{n^2 \cos \vartheta - (\epsilon_2/\epsilon_1)(n^2 - \sin^2 \vartheta)^{1/2}}{n^2 \cos \vartheta + (\epsilon_2/\epsilon_1)(n^2 - \sin^2 \vartheta)^{1/2}}$$
$$n = c_1/c_2 = (\mu_2\epsilon_2/\mu_1\epsilon_1)^{1/2}, \qquad nZ_1/Z_2 = \epsilon_2/\epsilon_1 \tag{25}$$

The constants Z_1, Z_2, c_1, and c_2 are defined in Eqs. (6) and (14).

---

[3] Neither Snell's laws nor Christian Huygens' (mathematician, 1629–1695; born in The Hague, The Netherlands) wave theory of light (Huygens, 1690) assumed sinusoidal time variation of light waves. Indeed one could hardly claim that light *not* produced by a laser has a sinusoidal time variation, and the experimentally verified laws of optics *not* based on laser light must thus generally hold for arbitrary time variation even if they are derived for sinusoidal time variation. The dominant role of the sinusoidal time variation in today's wave theory evolved after the introduction of partial differential equations and the discovery that particular solutions of these partial differential equations could easily be found by means of separation of variables using Bernoulli's product method.

This equation yields $\rho_2 \doteq -1$ for angles of incidence ϑ close to 90°. For normal incidence, $\vartheta = 0$, one obtains from Eq. (25)

$$\rho_s = (Z_2 - Z_1)/(Z_2 + Z_1) \tag{26}$$

and from Eq. (20):

$$\tau_s = 2Z_2/(Z_2 + Z_1) \tag{27}$$

If the impedances Z_1 and Z_2 are equal—which implies only the equality of the ratios $\mu_1/\epsilon_1 = \mu_2/\epsilon_2$ but not the equality of permittivity and permeability or velocity in the two media of Fig. 4.1-1—one obtains

$$\rho_s = 0, \qquad \tau_s = 1 \tag{28}$$

All these relations are well known from the theory of sinusoidal waves. The important points here are the method of derivation for nonsinusoidal waves, and the fact that these relations hold for waves with arbitrary time variation as long as the conductivity of medium 1 and 2 is zero. We will have to extend the theory at least to the case where the conductivity of one medium is not zero, if we want to study the distortion of radar signals produced in the air and reflected or scattered by the ground, seawater, fresh water, and a variety of objects made of metal, plastic, ferrites, etc.

We turn to the reflection of a plane wave polarized parallel to the plane of incidence as shown in Fig. 4.1-3. A comparison with Fig. 4.1-1 shows that the new vectors $\mathbf{H}_i$, $\mathbf{E}_r$, and $\mathbf{H}_t$ have the same direction as the vectors $\mathbf{E}_i$, $\mathbf{H}_r$, and $\mathbf{E}_t$ in Fig. 4.1-1, but the new vectors $\mathbf{E}_i$, $\mathbf{H}_r$, and $\mathbf{E}_t$ have the opposite direction of the vectors $\mathbf{H}_i$, $\mathbf{E}_r$, and $\mathbf{H}_t$ in Fig. 4.1-1; the direction of the vectors is chosen so that Figs. 4.1-1 and 4.1-3 yield the same directions of the vectors for $\vartheta_i = 0$ as noted in the caption of Fig. 4.1-3. We obtain from Eqs. (4), (5), (10), (11), (12), and (13):

$$\begin{aligned}\mathbf{E}_i &= -\mathbf{e}_i E_0 f(t - \mathbf{r}_i/c_1)\\ &= -(\mathbf{e}_x \cos\vartheta_i - \mathbf{e}_y \sin\vartheta_i) E_0 f[t - (\mathbf{e}_x \sin\vartheta_i + \mathbf{e}_y \cos\vartheta_i)r/c_1]\end{aligned} \tag{29}$$

$$\begin{aligned}\mathbf{H}_i &= \mathbf{e}_z E_0 Z_1^{-1} f(t - \mathbf{r}_i/c_1)\\ &= \mathbf{e}_z E_0 Z_1^{-1} f[t - (\mathbf{e}_x \sin\vartheta_i + \mathbf{e}_y \cos\vartheta_i)r/c_1]\end{aligned} \tag{30}$$

$$\begin{aligned}\mathbf{E}_r &= \mathbf{e}_r \rho_p E_0 f(t - \mathbf{r}_r/c_1)\\ &= (-\mathbf{e}_x \cos\vartheta_r - \mathbf{e}_y \sin\vartheta_r)\rho_p E_0 f[t - (\mathbf{e}_x \sin\vartheta_r - \mathbf{e}_y \cos\vartheta_r)r/c_1]\end{aligned} \tag{31}$$

$$\begin{aligned}\mathbf{H}_r &= -\mathbf{e}_z \rho_p E_0 Z_1^{-1} f(t - \mathbf{r}_r/c_1)\\ &= -\mathbf{e}_z \rho_p E_0 Z_1^{-1} f[t - (\mathbf{e}_x \sin\vartheta_r - \mathbf{e}_y \cos\vartheta_r)r/c_1]\end{aligned} \tag{32}$$

$$\begin{aligned}\mathbf{E}_t &= -\mathbf{e}_t \tau_p E_0 f(t - \mathbf{r}_t/c_2)\\ &= -(\mathbf{e}_x \cos\vartheta_t - \mathbf{e}_y \sin\vartheta_t)\tau_p E_0 f[t - (\mathbf{e}_x \sin\vartheta_t + \mathbf{e}_y \cos\vartheta_t)r/c_2]\end{aligned} \tag{33}$$

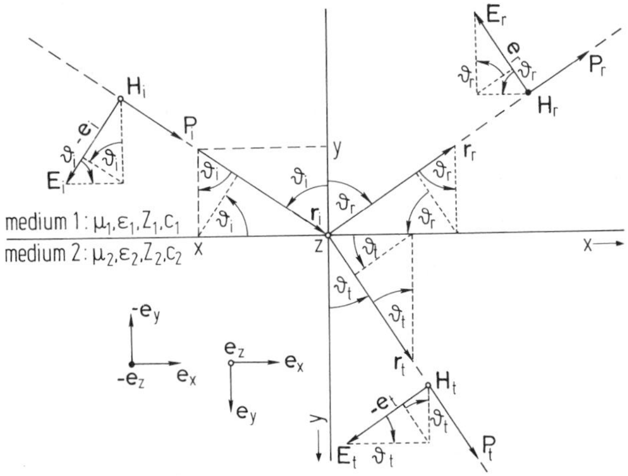

FIG. 4.1-3. Reflection and transmission of a wave that is polarized parallel to the plane of incidence. Note that the directions of $\mathbf{E}_i$, $\mathbf{E}_r$, $\mathbf{E}_t$, $\mathbf{H}_i$, $\mathbf{H}_r$, and $\mathbf{H}_t$ are chosen so that the same directions as in Fig. 4.1-1 are obtained for $\vartheta_i = 0$, which means equal direction for $\mathbf{E}_i$, $\mathbf{E}_r$, and $\mathbf{E}_t$ as well as $\mathbf{H}_i$ and $\mathbf{H}_t$ but opposite direction for $\mathbf{H}_i$ and $\mathbf{H}_r$.

$$\begin{aligned}\mathbf{H}_t &= \mathbf{e}_z \tau_p E_0 Z_2^{-1} t(t - \mathbf{r}_t/c_2) \\ &= \mathbf{e}_z \tau_p E_0 Z_2^{-1} f[t - (\mathbf{e}_x \sin \vartheta_t + \mathbf{e}_y \cos \vartheta_t) r/c_2]\end{aligned} \tag{34}$$

$$\rho_p = E_r/E_i, \qquad \tau_p = E_t/E_i \tag{35}$$

This time the following boundary conditions must be satisfied:

(c) The tangential components ($\mathbf{e}_z$) of the magnetic field strengths in both media must be equal at $y = 0$.

(d) The tangential components ($\mathbf{e}_x$) of the electric field strength in both media must be equal at $y = 0$.

Condition (c) yields the following relation if the components containing $\mathbf{e}_y$ are dropped as in Eq. (15):

$$\begin{aligned}Z_1^{-1} f(t - r c_1^{-1} \mathbf{e}_x \sin \vartheta_i) - \rho_p Z_1^{-1} f(t - r c_1^{-1} \mathbf{e}_x \sin \vartheta_r) \\ = \tau_p Z_2^{-1} f(t - r c_2^{-1} \mathbf{e}_x \sin \vartheta_t)\end{aligned} \tag{36}$$

Again, this equation can be satisfied for arbitrary values of r and t if and only if the conditions

$$t - r c_1^{-1} \mathbf{e}_x \sin \vartheta_i = t - r c_1^{-1} \mathbf{e}_x \sin \vartheta_r \tag{37}$$

$$t - r c_1^{-1} \mathbf{e}_x \sin \vartheta_i = t - r c_2^{-1} \mathbf{e}_x \sin \vartheta_t \tag{38}$$

and

$$1 - \rho_p = \tau_p Z_1/Z_2 \tag{39}$$

are satisfied. Equations (37) and (38) are the same as Eqs. (16) and (17). Hence, Snell's laws of reflection and refraction hold both for polarization perpendicular and parallel to the plane of incidence. However, Eq. (39) differs from Eq. (20).

Boundary condition (d) yields, with the help of Eqs. (29), (31), and (33) with the component $\mathbf{e}_y$ dropped:

$$-\cos\vartheta_{\rm i} f(t - rc_1^{-1}\mathbf{e}_x \sin\vartheta_{\rm i}) - \rho_p \cos\vartheta_{\rm r} f(t - rc_1^{-1}\mathbf{e}_x \sin\vartheta_{\rm r})$$
$$= -\tau_p \cos\vartheta_{\rm t} f(t - rc_2^{-1}\mathbf{e}_x \sin\vartheta_{\rm t}) \tag{40}$$

Equations (37) and (38) are again obtained, but in addition we find

$$\cos\vartheta_{\rm i} + \rho_p \cos\vartheta_{\rm i} = \tau_p \cos\vartheta_{\rm t} \tag{41}$$

Substitution of Eq. (39) yields:

$$\rho_p = (Z_2 \cos\vartheta_{\rm t} - Z_1 \cos\vartheta_{\rm i})/(Z_2 \cos\vartheta_{\rm t} + Z_1 \cos\vartheta_{\rm i}) \tag{42}$$

In analogy to the transition from Eq. (23) to Eq. (25) we obtain with $\vartheta_{\rm i} = \vartheta$:

$$\rho_p = \frac{Z_2(1 - n^{-2}\sin^2\vartheta)^{1/2} - Z_1\cos\vartheta}{Z_2(1 - n^{-2}\sin^2\vartheta)^{1/2} + Z_1\cos\vartheta}$$
$$= -\frac{n^2\cos\vartheta - (\mu_2/\mu_1)(n^2 - \sin^2\vartheta)^{1/2}}{n^2\cos\vartheta + (\mu_2/\mu_1)(n^2 - \sin^2\vartheta)^{1/2}}$$
$$n = c_1/c_2 = (\mu_2\epsilon_2/\mu_1\epsilon_1)^{1/2}, \qquad nZ_2/Z_1 = \mu_2/\mu_1 \tag{43}$$

A comparison between Eqs. (25) and (43) shows that perpendicular and parallel polarization yield substantially different results. Consider the case $\mu_2 = \mu_1$. In Eq. (25) the numerator is zero for $n = 1$ or $\epsilon_1 = \epsilon_2$, and it is negative for $n = (\epsilon_2/\epsilon_1)^{1/2} > 1$. On the other hand, the numerator of Eq. (43) vanishes for $n > 1$ and a certain value of the angle of incidence called the Brewster[4] angle $\vartheta_{\rm B}$:

$$n^2\cos\vartheta_{\rm B} = (n^2 - \sin^2\vartheta_{\rm B})^{1/2}, \qquad \cos\vartheta_{\rm B} = [(n^2-1)^{1/2}/(n^4-1)] \tag{44}$$

If medium 1 in Fig. 4.1-1 is air while medium 2 is a ferrite material with $\mu_2 > \mu_1$, one obtains $\rho_s = 0$ in Eq. (25) for a certain angle ϑ_μ:

$$n^2\cos\vartheta_\mu = (\epsilon_2/\epsilon_1)(n^2 - \sin^2\vartheta_\mu)^{1/2}$$
$$\cos\vartheta_\mu = \left(\frac{n^2-1}{n^4(\epsilon_1/\epsilon_2)^2 - 1}\right)^{1/2} \tag{45}$$

A value of $\cos\vartheta_\mu$ in the range $-1 < \cos\vartheta_\mu < +1$ is obtained if the condition

$$\epsilon_2/\epsilon_1 \leqq \mu_2/\mu_1 \tag{46}$$

---

[4] David Brewster, physicist, 1781–1868; born in Jedburg, Scotland.

is satisfied, as one may readily verify by inserting $n^2 = (c_1/c_2)^2 = \mu_2\epsilon_2/\mu_1\epsilon_1$ into Eq. (45). Hence, an angle ϑ_μ exists for which the reflection coefficient ρ_s is zero, provided the ratio ϵ_2/ϵ_1 of the absolute or relative permittivity is smaller than the ratio μ_2/μ_1 of the absolute or relative permeability.

From Eq. (43) follows for $\mu_2 \neq \mu_1$ a generalization of the Brewster angle:

$$n^2 \cos \vartheta_B = (\mu_2/\mu_1)(n^2 - \sin^2 \vartheta_B)^{1/2}$$

$$\cos \vartheta_B = \left(\frac{n^2 - 1}{n^4(\mu_1/\mu_2)^2 - 1}\right)^{1/2} \tag{47}$$

The condition for $\cos \vartheta_B$ being in the range $-1 < \cos \vartheta_B < 1$ becomes now

$$\epsilon_2/\epsilon_1 \geqq \mu_2/\mu_1 \tag{48}$$

Hence, there can be no material for which both reflection coefficients ρ_s and ρ_p vanish, except for $\epsilon_1 = \epsilon_2$ and $\mu_1 = \mu_2$.

4.2 Sinusoidal Wave Entering Conducting Medium

Air has almost no conductance while the conductivity of the surface of the Earth ranges from a typical value of 10^{-2} S/m for dry ground to 4 S/m for seawater. Hence, the investigation of the reflection of a wave propagating through a nonconducting medium and hitting the boundary with a conducting medium is of great interest in radar. Before we tackle this problem for nonsinusoidal electromagnetic waves, we follow Fresnel and study it for sinusoidal waves. The solution of this simpler case will help with the solution of the much more complicated nonsinusoidal case, and in addition show the difference between sinusoidal and nonsinusoidal waves.

In Section 2.2 we derived a solution of Maxwell's equations with sinusoidal time variation for a conducting medium. For a nonconducting medium 1 we obtain the simpler equations

$$k_1 = j\gamma_1'' = -\beta_1, \qquad \beta_1 = \omega/c_1, \qquad c_1 = (\mu_1\epsilon_1)^{-1/2} \tag{1}$$

$$Z_1 = (\mu_1/\epsilon_1)^{1/2} \tag{2}$$

while a conducting medium 2 with $\sigma_2 \neq 0$ yields:

$$k_2 = j\gamma_2'' = j[j\omega\mu_2(j\omega\epsilon_2 + \sigma_2)]^{1/2} = -(\beta_2 - j\alpha_2) \tag{3}$$

$$Z_{s2} = Z_2 = \left(\frac{j\omega\mu_2}{j\omega\epsilon_2 + \sigma_2}\right)^{1/2} \tag{4}$$

Next we rewrite $\mathbf{E}_i$, $\mathbf{H}_i$, $\mathbf{E}_r$, and $\mathbf{H}_r$ of Eqs. (4.1-4), (4.1-5), (4.1-10), and (4.1-11) for the particular functions

$$f(t - \mathbf{r}_i/c_1) = \exp[j(\omega t - \beta_1\mathbf{r}_i)], \qquad f(t - \mathbf{r}_r/c_1) = \exp[j(\omega t - \beta_1\mathbf{r}_r)] \tag{5}$$

and obtain:

$$\mathbf{E}_i = \mathbf{e}_z E_0 e^{j\omega t} \exp[-j\beta_1 r(\mathbf{e}_x \sin \vartheta_i + \mathbf{e}_y \cos \vartheta_i)] \tag{6}$$

$$\mathbf{H}_i = (\mathbf{e}_x \cos \vartheta_i - \mathbf{e}_y \sin \vartheta_i) E_0 Z_1^{-1} e^{j\omega t} \times \exp[-j\beta_1 r(\mathbf{e}_x \sin \vartheta_i + \mathbf{e}_y \cos \vartheta_i)] \tag{7}$$

$$\mathbf{E}_r = \mathbf{e}_z \rho_s E_0 e^{j\omega t} \exp[-j\beta_1 r(\mathbf{e}_x \sin \vartheta_r - \mathbf{e}_y \cos \vartheta_r)] \tag{8}$$

$$\mathbf{H}_r = (-\mathbf{e}_x \cos \vartheta_r - \mathbf{e}_y \sin \vartheta_r) \rho_s E_0 Z_1^{-1} e^{j\omega t} \times \exp[-j\beta_1 r(\mathbf{e}_x \sin \vartheta_r - \mathbf{e}_y \cos \vartheta_r)] \tag{9}$$

For $\mathbf{E}_t$ and $\mathbf{H}_t$ we must use k_2 and Z_2 of Eqs. (3) and (4),

$$f(t - \mathbf{r}_t/c_2) = \exp\{j[\omega t - (\beta_2 - j\alpha_2)\mathbf{r}_t]\} = e^{j\omega t} \exp[-(\alpha_2 + j\beta_2)\mathbf{r}_t] \tag{10}$$

to obtain from Eqs. (4.1-12) and (4.1-13):

$$\mathbf{E}_t = \mathbf{e}_z \tau_s E_0 e^{j\omega t} \exp[-(\alpha_2 + j\beta_2) r(\mathbf{e}_x \sin \vartheta_t + \mathbf{e}_y \cos \vartheta_t)] \tag{11}$$

$$\mathbf{H}_t = (\mathbf{e}_x \cos \vartheta_t - \mathbf{e}_y \sin \vartheta_t) \tau_s E_0 Z_2^{-1} e^{j\omega t} \times \exp[-(\alpha_2 + j\beta_2) r(\mathbf{e}_x \sin \vartheta_t + \mathbf{e}_y \cos \vartheta_t)] \tag{12}$$

The requirement for equality of the tangential components of the electric field strengths at the boundary $y = 0$ yields in analogy to Eq. (4.1-15) the relation:

$$\exp(-j\beta_1 r\mathbf{e}_x \sin \vartheta_i) + \rho_s \exp(-j\beta_1 r\mathbf{e}_x \sin \vartheta_r) = \tau_s \exp[-(\alpha_2 + j\beta_2) r\mathbf{e}_x \sin \vartheta_t] \tag{13}$$

For arbitrary values of r this equation can be satisfied if and only if the arguments of the exponential functions are equal. This requirement yields two equations:

$$\sin \vartheta_i = \sin \vartheta_r \tag{14}$$

$$j\beta_1 \sin \vartheta_i = (\alpha_2 + j\beta_2) \sin \vartheta_t \tag{15}$$

The exponential functions in Eq. (13) may then be cancelled, and a third equation is obtained:

$$1 + \rho_s = \tau_s \tag{16}$$

Equation (14) yields again Snell's law of reflection, Eq. (4.1-18), as in the case of two nonconducting media. Snell's law of refraction derived from Eq. (15) yields, however, a complex refraction coefficient n whose physical significance will have to be investigated:

$$(\sin \vartheta_i)/(\sin \vartheta_t) = (\beta_2 - j\alpha_2)/\beta_1 = k_2/k_1 = n \tag{17}$$

The requirement that the tangential components of the magnetic field strengths in both media must be equal at the boundary $y=0$ yields, in analogy to Eq. (4.1-21), the relation:

$$Z_1^{-1}(\cos\vartheta_i)\exp(-j\beta_1 r\mathbf{e}_x\sin\vartheta_i)-\rho_s Z_1^{-1}(\cos\vartheta_r)\exp(-j\beta_1 r\mathbf{e}_x\sin\vartheta_i)$$
$$=\tau_s Z_2^{-1}\cos\vartheta_t\exp[-(\alpha_2+j\beta_2)r\mathbf{e}_x\sin\vartheta_t] \qquad (18)$$

Again, this equation can exist for arbitrary values of r if and only if the arguments of the exponential functions are equal. This condition yields again Eqs. (14) and (15). Instead of Eq. (16) one obtains now

$$\cos\vartheta_i-\rho_s\cos\vartheta_i=\tau_s(Z_1/Z_2)\cos\vartheta_t,\qquad \vartheta_i=\vartheta_r \qquad (19)$$

From Eqs. (16) and (19) follows again Eq. (4.1-25) that held for the reflection of a wave with general time variation at the boundary of two nonconducting media. The difference is that Z_2 and n are now complex quantities:

$$\rho_s=\frac{Z_2\cos\vartheta-Z_1(1-n^{-2}\sin^2\vartheta)^{1/2}}{Z_2\cos\vartheta+Z_1(1-n^{-2}\sin^2\vartheta)^{1/2}}$$
$$=\frac{n^2\cos\vartheta-n(Z_1/Z_2)(n^2-\sin^2\vartheta)^{1/2}}{n^2\cos\vartheta+n(Z_1/Z_2)(n^2-\sin^2\vartheta)^{1/2}} \qquad (20)$$
$$\vartheta=\vartheta_i,\qquad Z_1=(\mu_1/\epsilon_1)^{1/2},\qquad Z_2=[j\omega\mu_2/(j\omega\epsilon_2+\sigma_2)]^{1/2},$$
$$n=(\beta_2-j\alpha_2)/\beta_1,\qquad \beta_1=\omega(\mu_1\epsilon_1)^{1/2}=2\pi/\lambda_1=2\pi\kappa_1,$$
$$\beta_2=\omega\{2^{-1}\mu_2\epsilon_2[(1+\sigma_2^2/\omega^2\epsilon_2^2)^{1/2}+1]\}^{1/2},$$
$$\alpha_2=\omega\{2^{-1}\mu_2\epsilon_2[(1+\sigma_2^2/\omega^2\epsilon_2^2)^{1/2}-1]\}^{1/2}$$

For $\sigma_2/\omega\epsilon_2 \ll 1$ the coefficient α_2/β_1 approaches zero, and we can use the results of Section 4.1 for the boundary between two nonconducting media. To see when this is so we list in Table 4.2-1 the conductivity and relative

TABLE 4.2-1

THE RATIO $\sigma_2/2\pi f\epsilon_2$ FOR SEAWATER, TWO TYPES OF GROUND, AND FRESH WATER[a]

	σ_2 [S/m]	ϵ_2/ϵ_1	ϵ_2 [F/m]	$\sigma_2/2\pi f\epsilon_2$
Seawater	4	80	7.1×10^{-10}	$900\times10^6/f$
High-conductivity ground	10^{-2}	16	1.4×10^{-10}	$110\times10^6/f$
Low-conductivity ground	10^{-3}	4	3.5×10^{-11}	$4.5\times10^6/f$
Fresh water	10^{-3}	80	7.1×10^{-10}	$0.22\times10^6/f$

[a] The permittivity $\epsilon_1=8.8543\times10^{-12}$ [F/m] is used for free space and air.

permittivity for seawater, high-conductivity ground, and low-conductivity ground. If the frequency f is large compared with 4.5 MHz we can treat low-conductivity ground like a nonconducting medium. In carrier-free high-resolution radar one is primarily interested in pulses with a duration of 1 to 0.1 ns, occupying the frequency bands $0 < f < 1$ GHz to $0 < f < 10$ GHz. For such short pulses most of the energy is contained in the Fourier components with a frequency well above 4.5 MHz, and the results of Section 4.1 are applicable. The situation is not as favorable for high-conductivity ground, while seawater can be treated neither as nonconductor nor as good conductor for pulses with a duration in the order of 1 to 0.1 ns. One needs the theory in full generality in this case.[1]

Since the angle ϑ_t in Fig. 4.1-1 must be real, the illustration cannot hold for the case of a conducting medium 2. However, our Eqs. (6) to (12) are more generally applicable than implied by Fig. 4.1-1, since they permit complex values for ρ_s, τ_s, and ϑ_t. This point is discussed in more detail in the literature (Reitz *et al.*, 1979, Sec. 18-4; Schumann, 1948; Wagner, 1953).

To elaborate the meaning of a complex refraction index n, we rewrite Eq. (17)

$$\sin\vartheta_t = \beta_1(\beta_2 - j\alpha_2)^{-1}\sin\vartheta_i = \beta_1(\beta_2 + j\alpha_2)(\beta_2^2 + \alpha_2^2)^{-1}\sin\vartheta_i$$
$$= (A + jB)\sin\vartheta_i,$$
$$A = \beta_1\beta_2/(\beta_2^2 + \alpha_2^2), \qquad B = \beta_1\alpha_2/(\beta_2^2 + \alpha_2^2) \tag{21}$$

and obtain

$$\cos\vartheta_t = (1 - \sin^2\vartheta_t)^{1/2} = (1 - A^2 + B^2 - 2jAB)^{1/2} = C - jD \tag{22}$$

The quantities C and D are obtained by squaring Eq. (22). This yields one equation for the real part and one for the imaginary part

$$1 - A^2 + B^2 = C^2 - D^2, \qquad AB = CD \tag{23}$$

with the solutions:

$$\begin{aligned} C &= \pm 2^{-1}\sqrt{2}\{[(1 - A^2 + B^2)^2 + 4A^2B^2]^{1/2} + 1 - A^2 + B^2\}^{1/2} \\ D &= \pm 2^{-1}\sqrt{2}\{[(1 - A^2 + B^2)^2 + 4A^2B^2]^{1/2} - (1 - A^2 + B^2)\}^{1/2} \end{aligned} \tag{24}$$

We insert $\sin\vartheta_t$ and $\cos\vartheta_t$ of Eqs. (21) and (22) into the argument of the exponential function of $\mathbf{E}_t$ in Eq. (11) and separate real and imaginary terms;

[1] Note that σ, ϵ, and μ are treated as constants. This assumption is correct for empty space but not for materials, when the atomic structure of matter becomes significant at sufficiently high rates of change of the field strengths.

the factor $A\alpha_2 - B\beta_2$ equals zero according to Eq. (21):

$$-(\alpha_2 + j\beta_2)r[\mathbf{e}_x(A + jB) - \mathbf{e}_y(C - jD)]$$
$$= -r\{-\mathbf{e}_y(C\alpha_2 + D\beta_2) + j[\mathbf{e}_x(A\beta_2 + B\alpha_2) - \mathbf{e}_y(C\beta_2 - D\alpha_2)]\}$$
$$= -\alpha_0 r\mathbf{e}_y - j\beta_0 r(\mathbf{e}_x \sin\vartheta_0 - \mathbf{e}_y \cos\vartheta_0)$$
$$\alpha_0 = -(C\alpha_2 + D\beta_2), \qquad \beta_0 \sin\vartheta_0 = A\beta_2 + B\alpha_2,$$
$$\beta_0 \cos\vartheta_0 = C\beta_2 - D\alpha_2 \tag{25}$$

One may solve these equations for ϑ_0 and β_0:

$$\tan\vartheta_0 = (A\beta_2 + B\alpha_2)/(C\beta_2 - D\alpha_2) \tag{26}$$
$$\beta_0 = [(A\beta_2 + B\alpha_2)^2 + (C\beta_2 - D\alpha_2)^2]^{1/2} \tag{27}$$

The negative values of C and D in Eq. (24) must be used for α_0 to assure an attenuated rather than an amplified wave in medium 2.

Figure 4.2-1 shows the resulting wave in medium 2. Its planes of constant phase propagate in the direction of the angle ϑ_0. However, the planes of constant attenuation due to the term $-\alpha_0 r\mathbf{e}_y$ in Eq. (25) which causes an attenuation factor $\exp(-\alpha_0 r\mathbf{e}_y)$ for $\mathbf{E}_t$ and $\mathbf{H}_t$, are parallel to the boundary between medium 1 and 2. A short reflection convinces one that in the steady-state case an unattenuated planar wave in medium 1 must make the attenuation in medium 2 a function of $y = |r\mathbf{e}_y|$ only, since the power density at the boundary $y = 0$ in medium 1 is the same for any value of x or z, and the power density in medium 2 can thus not vary with x or z either.

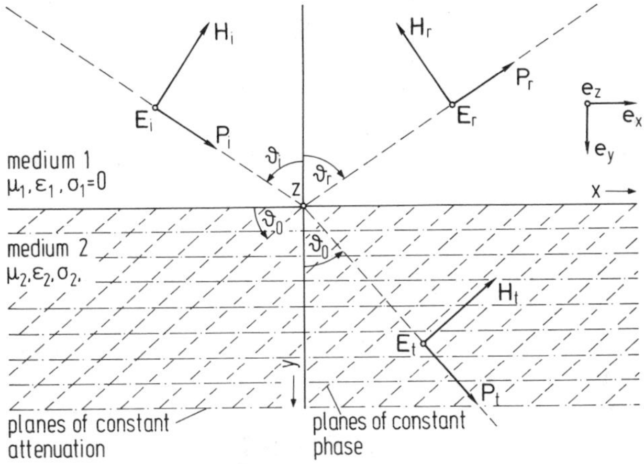

FIG. 4.2-1. Perpendicularly polarized sinusoidal wave propagating from a nonconducting medium 1 into a conducting medium 2.

For the evaluation of the complex reflection coefficient of Eq. (20) for $\mu_1 = \mu_2$ we write first n^2,

$$n^2 = n_r - jn_i, \qquad n_r = (\beta_2^2 - \alpha_2^2)/\beta_1^2, \qquad n_i = 2\alpha_2\beta_2/\beta_1^2 \tag{28}$$

and substitute into Eq. (20):

$$\rho_s = \frac{\cos\vartheta - (n_r - jn_i - \sin^2\vartheta)^{1/2}}{\cos\vartheta + (n_r - jn_i - \sin^2\vartheta)^{1/2}} \tag{29}$$

For the separation of the denominator into real and imaginary parts we put

$$(n_r - jn_i - \sin^2\vartheta)^{1/2} = X - jY \tag{30}$$

and solve for X and Y by squaring:

$$\begin{aligned} X &= \pm 2^{-1}\sqrt{2}\{[(n_r - \sin^2\vartheta)^2 + n_i^2]^{1/2} + (n_r - \sin^2\vartheta)\}^{1/2} \\ Y &= \pm 2^{-1}\sqrt{2}\{[(n_r - \sin^2\vartheta)^2 + n_i^2]^{1/2} - (n_r - \sin^2\vartheta)\}^{1/2} \end{aligned} \tag{31}$$

Equation (29) is then rationalized:

$$\rho_s = \frac{\cos\vartheta - X + jY}{\cos\vartheta + X - jY} = \frac{\cos^2\vartheta - X^2 - Y^2 + 2jY\cos\vartheta}{(\cos\vartheta + X)^2 + Y^2} = |\rho_s|e^{j\varphi},$$

$$|\rho_s| = \frac{[(\cos^2\vartheta - X^2 - Y^2)^2 + 4Y^2\cos^2\vartheta]^{1/2}}{(\cos\vartheta + X)^2 + Y^2},$$

$$\operatorname{tg}\varphi = \frac{2Y\cos\vartheta}{\cos^2\vartheta - X^2 - Y^2} \tag{32}$$

The amplitudes of the electric and magnetic field strengths $\mathbf{E}_r$ and $\mathbf{H}_r$ of the reflected wave equal $|\rho_s|\,E_0$ and $|\rho_s|\,E_0/Z_1$, where E_0 and E_0/Z_1 are the amplitudes of the electric and magnetic field strengths of the incident wave. In addition, the phase of the reflected wave jumps by φ at the boundary of the two media.

The transmission coefficient τ_s follows from Eqs. (16) and (32):

$$\tau_s = 1 + \rho_s = |\tau_s|e^{j\psi},$$

$$|\tau_s| = \frac{2\cos\vartheta}{[(\cos\vartheta + X)^2 + Y^2]^{1/2}}, \qquad \operatorname{tg}\psi = \frac{Y}{\cos\vartheta + X} \tag{33}$$

Again we have a change in amplitude and a phase jump at the boundary between the two media for the transmitted wave.

It is quite evident that expressing ϑ_0, α_0, $|\rho_s|$, $|\tau_s|$, φ, and ψ in terms of ω, ϵ_1, ϵ_2, and σ_2 would be a formidable task. However, the use of multiple substitutions—first β_1, β_2, α_2, then A, B, then C, D, and so on—is ideally suited for numerical evaluation by computer.

The calculation for a parallel polarized wave is very similar. We do not repeat it here but return to nonsinusoidal waves.

4.3 Step Function Wave Entering Conducting Medium

Let a plane wave with perpendicular incidence have the time variation $f_i(t)$. The time variation $f_r(t)$ of the reflected wave is written with the help of a reflection function $\rho_s(t)$:

$$f_r(t) = \rho_s(t) f_i(t) \tag{1}$$

From Eqs. (4.1-4), (4.1-5), (4.1-10), and (4.1-11) we obtain for the angle of incidence $\vartheta_i = 0$:

$$\mathbf{E}_i = \mathbf{e}_z E_0 f_i(t - \mathbf{r}\mathbf{e}_y/c_1) \tag{2}$$

$$\mathbf{H}_i = \mathbf{e}_x E_0 Z_1^{-1} f_i(t - \mathbf{r}\mathbf{e}_y/c_1) \tag{3}$$

$$\mathbf{E}_r = \mathbf{e}_z E_0 \rho_s(t) f_i(t + \mathbf{r}\mathbf{e}_y/c_1) \tag{4}$$

$$\mathbf{H}_r = -\mathbf{e}_x E_0 Z_1^{-1} \rho_s(t) f_i(t + \mathbf{r}\mathbf{e}_y/c_1) \tag{5}$$

We assume that the time variation $f_i(t)$ of the field strengths of the incident wave is that of a step function $E_0 S(t)$ that jumps from 0 to E_0 at the time $t = 0$. The field strengths at the boundary $y = 0$ are then:

$$\begin{aligned} \mathbf{E}_i + \mathbf{E}_r &= 0 && \text{for} \quad t < 0 \\ &= \mathbf{e}_z E_0 [1 + \rho_s(t)] && \text{for} \quad t \geqq 0 \end{aligned} \tag{6}$$

$$\begin{aligned} \mathbf{H}_i + \mathbf{H}_r &= 0 && \text{for} \quad t < 0 \\ &= \mathbf{e}_x E_0 Z_1^{-1} [1 - \rho_s(t)] && \text{for} \quad t \geqq 0 \end{aligned} \tag{7}$$

The electric field strength $\mathbf{E}_i + \mathbf{E}_r$ and the magnetic field strength $\mathbf{H}_i + \mathbf{H}_r$ at the boundary produce an electromagnetic wave in medium 2. The electric excitation force of Eq. (6) consists of the step function $E_0 S(t)$ plus the function $E_0 \rho(t) S(t)$. The magnitudes of the electric and magnetic field strengths due to the step function $E_0 S(t)$ are given by Eqs. (2.1-68) and (2.4-41) for $H_t(y) = 0$, while the field strengths due to $E_0 \rho_s(t) S(t)$ are given by Eqs. (2.8-7) and (2.8-8) with $E(0, t'') = E_0 \rho_s(t'')$. The sum of these field strengths yields:

$$\begin{aligned} E_E(y, t) = E_0 \bigg(1 + w(y, t) + \int_0^t \frac{d\rho_s(t'')}{dt''} \\ \times S(t - t'')[1 + w(y, t - t'')] dt'' \bigg) \end{aligned} \tag{8}$$

$$H_E(y, t) = E_0 Z_2^{-1} \Big(-Z_2\sigma_2 y + I'_{E1}(y, t) - I_{E2}(y, t)$$

$$+ \int_0^t \frac{d\rho_s(t'')}{dt''} S(t - t'')[-Z_2\sigma_2 y + I'_{E1}(y, t - t'')$$

$$- I_{E2}(y, t - t'')]dt'' \Big) \qquad (9)$$

The magnetic excitation force of Eq. (7) consists of the step function $E_0 Z_1^{-1} S(t)$ plus the function $-E_0 Z_1^{-1} \rho_s(t) S(t)$. The electric and magnetic field strengths due to the step function $E_0 Z_1^{-1} S(t)$ are given by Eqs.(2.6-19) and (2.6-3) for $E_t(y) = 0$, $H_0 = E_0/Z_1$, and $L \to \infty$, while the field strengths due to $-E_0 Z_1^{-1} \rho_s(t) S(t)$ are given by Eqs. (2.8-11) and (2.8-12) with $H(0, t'') = -E_0 Z_1^{-1} \rho_s(t'')$. The sum of these field strengths yields:

$$H_H(y, t) = E_0 Z_1^{-1} \Big(1 + w(y, t) - \int_0^t \frac{d\rho_s(t'')}{dt''}$$

$$\times S(t - t'')[1 + w(y, t - t'')]dt'' \Big) \qquad (10)$$

$$E_H(y, t) = E_0 Z_2 Z_1^{-1} \Big(I_{H1}(y, t) + I_{H2}(y, t)$$

$$- \int_0^t \frac{d\rho_s(t'')}{dt''} S(t - t'')[I_{H1}(y, t - t'') + I_{H2}(y, t - t'')]dt'' \Big) \qquad (11)$$

The field strengths E_t and H_t of the transmitted wave are the sums of Eqs. (8) and (11) as well as (9) and (10):

$$E_t(y, t) = E_0 \Big(1 + w(y, t) + Z_2 Z_1^{-1} [I_{H1}(y, t) + I_{H2}(y, t)]$$

$$+ \int_0^t \frac{d\rho_s(t'')}{dt''} S(t - t'')\{1 + w(y, t - t'')$$

$$- Z_2 Z_1^{-1} [I_{H1}(y, t - t'') + I_{H2}(y, t - t'')]\}dt'' \Big) \qquad (12)$$

$$
\begin{aligned}
H_t(y,t) = E_0 Z_1^{-1}\Bigg[& 1 + w(y,t) + Z_1 Z_2^{-1} \\
& \times [-Z_2\sigma_2 y + I'_{E1}(y,t) - I_{E2}(y,t)] \\
& - \int_0^t \frac{dp_s(t'')}{dt''} S(t-t'')\Big(1 + w(y,t-t'') \\
& - \frac{Z_1}{Z_2}[-Z_2\sigma_2 y + I'_{E1}(y,t-t'') - I_{E2}(y,t-t'')]\Big)dt''\Bigg] \qquad (13)
\end{aligned}
$$

Note that w, I_{H1}, I_{H2}, I'_{E1}, and I_{E2} refer to medium 2. Hence, the subscript 2 must be used,

$$\alpha = \alpha_2 = Z_2\sigma_2 c_2/2, \qquad Z_2 = (\mu_2/\epsilon_2)^{1/2}, \qquad c = c_2 = (\mu_2\epsilon_2)^{-1/2} \quad (14)$$

when these functions are computed from Eqs. (2.1-67), (2.6-16), (2.6-18), (2.4-39), and (2.4-40).

For the boundary $y = 0$ we write Eqs. (12) and (13) as follows:

$$E_t(0,t) = E_0\tau_{Es}(t) \qquad (15)$$

$$H_t(0,t) = E_0 Z_2^{-1}\tau_{Hs}(t) \qquad (16)$$

A vector diagram for $\mathbf{E}_i$, $\mathbf{H}_i$, $\mathbf{E}_r$, $\mathbf{H}_r$, $\mathbf{e}_z E_t(0,t)$, and $\mathbf{e}_x H_t(0,t)$ is shown in Fig. 4.3-1. Using Eqs. (2.8-15)–(2.8-19) we obtain for $\tau_{Es}(t)$ and $\tau_{Hs}(t)$ with the help of Eq. (1.3-7):

$$
\begin{aligned}
\tau_{Es}(t) = {} & 1 + Z_2 Z_1^{-1}[I_{H1}(0,t) + I_{H2}(0,t)] + p_s(t) \\
& - \frac{Z_2}{Z_1}\int_0^t \frac{dp_s(t'')}{dt''} S(t-t'')[I_{H1}(0,t-t'') + I_{H2}(0,t-t'')]dt'' \qquad (17)
\end{aligned}
$$

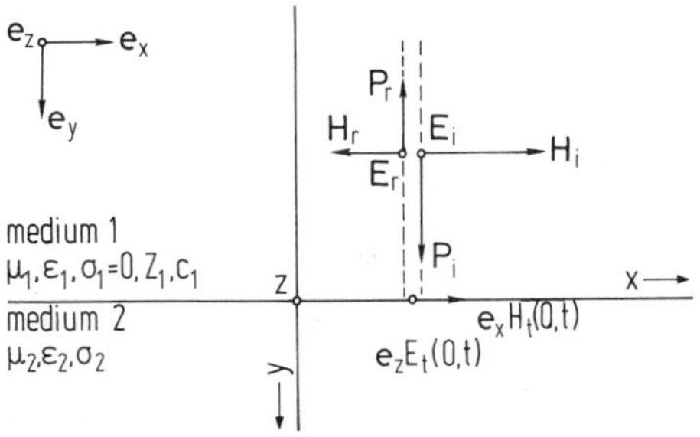

FIG. 4.3-1. Relation between the incident wave $\mathbf{E}_i$, $\mathbf{H}_i$, the reflected wave $\mathbf{E}_r$, $\mathbf{H}_r$, and the transmitted wave $\mathbf{e}_z E_t(0,t)$, $\mathbf{e}_x H_t(0,t)$ in the boundary plane $y = 0$.

$$\tau_{Hs}(t) = Z_2 Z_1^{-1} + I'_{E1}(0, t) - I_{E2}(0, t) - Z_2 Z_1^{-1} \rho_s(t) + \int_0^t \frac{d\rho_s(t'')}{dt''} S(t - t'')[I'_{E1}(0, t - t'') - I_{E2}(0, t - t'')]dt'' \tag{18}$$

The tangential components of the electric field strengths in medium 1 and 2 must be equal at the boundary $y = 0$. We obtain from Eqs. (6) and (15):

$$1 + \rho_s(t) = \tau_{Es}(t) \tag{19}$$

Similarly, continuity of the tangential component of the magnetic field strength requires, according to Eqs. (7) and (16):

$$1 - \rho_s(t) = Z_1 Z_2^{-1} \tau_{Hs}(t) \tag{20}$$

Insertion of Eq. (19) into Eq. (17) yields:

$$I_{H1}(0, t) + I_{H2}(0, t) = \int_0^t \frac{d\rho_s(t'')}{dt''} S(t - t'')[I_{H1}(0, t - t'') + I_{H2}(0, t - t'')]dt'' \tag{21}$$

We further insert Eq. (20) into Eq. (18):

$$I_{E2}(0, t) - I'_{E1}(0, t) = \int_0^t \frac{d\rho_s(t'')}{dt''} S(t - t'')[I'_{E1}(0, t - t'') - I_{E2}(0, t - t'')]dt'' \tag{22}$$

A comparison of Eqs. (21) and (22) with Eq. (1.4-1) shows that they can be written with the convolution symbol * as follows:

$$I_{H1}(0, t) + I_{H2}(0, t) = \frac{d\rho_s(t'')}{dt''} * S(t - t'')[I_{H1}(0, t - t'') + I_{H2}(0, t - t'')] \tag{23}$$

$$I_{E2}(0, t) - I'_{E1}(0, t) = \frac{d\rho_s(t'')}{dt''} * S(t - t'')[I'_{E1}(0, t - t'') - I_{E2}(0, t - t'')] \tag{24}$$

The formal deconvolution according to Eq. (1.4-3) yields

$$\rho_s(t) = \int_0^t [I_{H1}(0, t) + I_{H2}(0, t)] \mathbin{\overset{*}{*}} S(t - t'')[I_{H1}(0, t - t'') + I_{H2}(0, t - t'')]dt'' \tag{25}$$

and

$$\rho_s(t) = \int_0^t [I_{E2}(0, t) - I'_{E1}(0, t)] \overset{*}{*} S(t - t'')[I'_{E1}(0, t - t'') - I_{E2}(0, t - t'')]dt'' \tag{26}$$

The actual deconvolution requires Eq. (1.4-13) with

$$\rho_s(t) = F(k\,\Delta T) \tag{27}$$

and

$$I_{H1}(0, t) + I_{H2}(0, t) = F_c(k\,\Delta T) = g(k\,\Delta T) \tag{28}$$

or

$$I_{E2}(0, t) - I'_{E1}(0, t) = F_c(k\,\Delta T) = -g(k\,\Delta T) \tag{29}$$

Equations (25) and (26) must yield the same values for $\rho_s(t)$. One may use this fact to obtain a check on the accuracy of the computation.

The knowledge of the reflection function $\rho_s(t)$ for an incident wave with the time variation of a step function leads immediately to the solution of the general case. An incident wave with the time variation $F(t)S(t)$ yields the time variation $F_r(t)$ of the reflected wave according to Eq. (1.3-8):

$$F_r(t) = \int_0^t \frac{dF(t'')}{dt''} S(t - t'')\rho_s(t - t'')dt'' \tag{30}$$

The transmission functions $\tau_{Es}(t)$ and $\tau_{Hs}(t)$ follow with the reflection function $\rho_s(t)$ from Eqs. (19) and (20). These transmission functions give us in turn the magnitudes of the field strengths $\mathbf{E}_t(0, t)$ and $\mathbf{H}_t(0, t)$ according to Eqs. (15) and (16) that excite the transmitted wave in medium 2. The magnitudes of the field strengths $\mathbf{E}(y, t)$ and $\mathbf{H}(y, t)$ in medium 2 for any value of y and t are obtained by inserting $E_t(0, t'')$ and $H_t(0, t'')$ for $E(0, t'')$ and $H(0, t'')$ in Eqs. (2.8-13) and (2.8-14). These are the magnitudes of the field strengths of the transmitted wave due to an incident wave with the time variation of a step function for electric and magnetic field strength. The general time variation $F(t)$ will produce the magnitude of the transmitted field strengths

$$E_t(y, t) = \int_0^t \frac{dF(t'')}{dt''} S(t - t'')E(y, t - t'')dt'' \tag{31}$$

and

$$H_t(y, t) = \int_0^t \frac{dF(t'')}{dt''} S(t - t'')H(y, t - t'')dt'' \tag{32}$$

4.4 Perpendicularly Polarized Step Function Wave

We now generalize the results of the previous section from the angle of incidence $\vartheta_i = 0$ to an arbitrary value of ϑ_i. The case of perpendicular polarization is considered first. We assume again the time variation $f_i(t)$ for the incident wave and $f_r(t)$ for the reflected wave, and introduce a reflection function $\rho_s(t, \vartheta_i)$ that is a function of time as well as of the angle of incidence ϑ_i:

$$f_r(t) = \rho_s(t, \vartheta_i) f_i(t) \tag{1}$$

The electric and magnetic field strengths of incident and reflected wave are given by Eqs. (4.1-4), (4.1-5), (4.1-10), and (4.1-11):

$$\mathbf{E}_i = \mathbf{e}_z E_0 f_i[t - (\mathbf{e}_x \sin\vartheta_i + \mathbf{e}_y \cos\vartheta_i)\mathbf{r}/c_1] \tag{2}$$

$$\mathbf{H}_i = (\mathbf{e}_x \cos\vartheta_i - \mathbf{e}_y \sin\vartheta_i) E_0 Z_1^{-1} f_i[t - (\mathbf{e}_x \sin\vartheta_i + \mathbf{e}_y \cos\vartheta_i)\mathbf{r}/c_1] \tag{3}$$

$$\mathbf{E}_r = \mathbf{e}_z E_0 \rho_s(t, \vartheta_i) f_i[t - (\mathbf{e}_x \sin\vartheta_r - \mathbf{e}_y \cos\vartheta_r)\mathbf{r}/c_1] \tag{4}$$

$$\begin{aligned}\mathbf{H}_r = {} & -(\mathbf{e}_x \cos\vartheta_r + \mathbf{e}_y \sin\vartheta_r) E_0 Z_1^{-1} \rho_s(t, \vartheta_i) \\ & \times f_i[t - (\mathbf{e}_x \sin\vartheta_r - \mathbf{e}_y \cos\vartheta_r)\mathbf{r}/c_1]\end{aligned} \tag{5}$$

We assume that the time variation of the field strength of the incident wave is that of a step function $E_0 S(t)$. The field strengths at the boundary $y = 0$ have then the following values according to the results of Section 3.1 and Fig. 4.4-1:

$$\begin{aligned}\mathbf{E}_i + \mathbf{E}_r &= 0 && \text{for} \quad t' = t - x c_1^{-1} \sin\vartheta_i < 0 \\ &= \mathbf{e}_z E_0 [1 + \rho_s(t', \vartheta_i)] && \text{for} \quad t' \geqq 0\end{aligned} \tag{6}$$

$$\begin{aligned}\mathbf{H}_i + \mathbf{H}_r &= 0 && \text{for} \quad t' < 0 \\ &= E_0 Z_1^{-1} [\mathbf{e}_x \cos\vartheta_i - \mathbf{e}_y \sin\vartheta_i \\ &\quad - \rho_s(t', \vartheta_i)(\mathbf{e}_x \cos\vartheta_r + \mathbf{e}_y \sin\vartheta_r)] && \text{for} \quad t' \geqq 0\end{aligned} \tag{7}$$

The electric and magnetic field strengths $\mathbf{E}_i + \mathbf{E}_r$ and $\mathbf{H}_i + \mathbf{H}_r$ at the boundary produce an electromagnetic wave in medium 2. The electric excitation force defined by Eq. (6) consists of the step function $\mathbf{E}_0 S(t')$ reaching the line $y = 0$, x at the time $t' = 0$, plus the function $E_0 \rho_s(t', \vartheta_i) S(t')$ starting at the time $t' = 0$ at the line $y = 0$, x. The magnitude of the electric and magnetic field strengths due to the step function $E_0 S(t')$ are given by Eqs. (3.2-46) and (3.4-6) for $H_t(y \cos\vartheta) = 0$, while the magnitude of the field strengths due to $E_0 \rho_s(t', \vartheta_i) S(t')$ are given by Eqs. (3.6-7) and (3.6-8) with

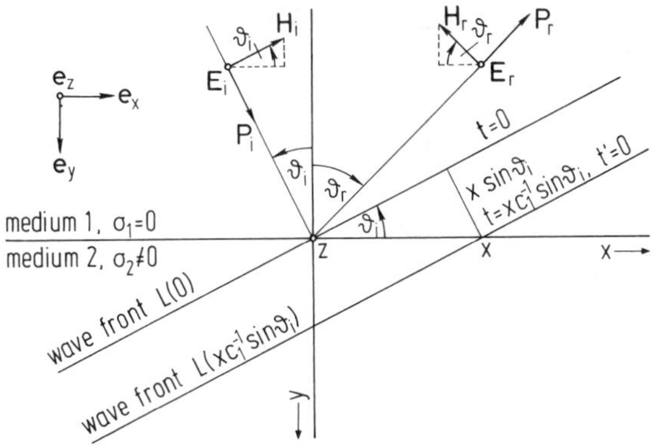

FIG. 4.4-1. Reflection of a perpendicularly polarized wave at the boundary between a non-conducting and a conducting medium.

$E_z(0, 0, t'') = E_0\rho_s(t'', \vartheta_i)$:

$$E_{zEs}(x, y, t) = E_0\Bigg(1 + w(y\cos\vartheta_i, t') + \int_0^{t'} \frac{d\rho_s(t'', \vartheta_i)}{dt''} S(t' - t'') \times [1 + w(y\cos\vartheta_i, t' - t'')]dt''\Bigg) \tag{8}$$

$$H_{xEs}(x, y, t) = E_0 Z_2^{-1}\cos\vartheta_i\Bigg(-Z_2\sigma_2 y\cos\vartheta_i + I'_{E1}(y\cos\vartheta_i, t') - I_{E2}(y\cos\vartheta_i, t') + \int_0^{t'} \frac{d\rho_s(t'', \vartheta_i)}{dt''} \times S(t' - t'')[-Z_2\sigma_2 y\cos\vartheta_i + I'_{E1}(y\cos\vartheta_i, t' - t'') - I_{E2}(y\cos\vartheta_i, t' - t'')]dt''\Bigg) \tag{9}$$

The magnetic excitation force of Eq. (7) has a tangential component $\mathbf{e}_x$ in the plane $y = 0$ in Fig. 4.4-1, and a component $\mathbf{e}_y$ perpendicular to this plane. The two components are connected by Eq. (3.1-16) for $\vartheta = \vartheta_i$. We are here primarily interested in the tangential component $\mathbf{e}_x$. It consists of the step function $E_0 Z_1^{-1} S(t')\cos\vartheta_i$ plus the function $-E_0 Z_1^{-1}\rho_s(t', \vartheta_i)S(t')\cos\vartheta_r$. The magnitude of the magnetic and electric field strengths due to the step function $E_0 Z_1^{-1} S(t')\cos\vartheta_i$ are given by Eqs. (3.5-4) and (3.5-6) for $H_0 = E_0 Z_1^{-1}\cos\vartheta_i$, while the magnitudes of the field strengths due to $-E_0 Z_1^{-1}\rho_s$

$(t', \vartheta_i)S(t')\cos\vartheta_r$ are given by Eqs. (3.6-11) and (3.6-12) with $H_x(0, 0, t'') = -E_0 Z_1^{-1} \rho_s(t'', \vartheta_i)\cos\vartheta_r$:

$$H_{xHs}(x, y, t') = E_0 Z_1^{-1}\Bigg(\cos\vartheta_i[1 + w(y\cos\vartheta_i, t')] - \cos\vartheta_r \times \int_0^{t'} \frac{d\rho_s(t'', \vartheta_i)}{dt''} S(t' - t'') \times [1 + w(y\cos\vartheta_i, t' - t'')]dt''\Bigg) \tag{10}$$

$$E_{zHs}(x, y, t') = E_0 Z_2 Z_1^{-1}\Bigg(I_{H1}(y\cos\vartheta_i, t') + I_{H2}(y\cos\vartheta_i, t') - \frac{\cos\vartheta_r}{\cos\vartheta_i}\int_0^{t'} \frac{d\rho_s(t'', \vartheta_i)}{dt''} \times S(t' - t'')[I_{H1}(y\cos\vartheta_i, t' - t'') + I_{H2}(y\cos\vartheta_i, t' - t'')]dt''\Bigg) \tag{11}$$

The magnitudes of the field strengths $\mathbf{E}_{zt}$ and $\mathbf{H}_{xt}$ of the transmitted wave are the sums of Eqs. (8) and (11) as well as (9) and (10):

$$E_{zt}(x, y, t) = E_0\Bigg[1 + w(y\cos\vartheta_i, t') + Z_2 Z_1^{-1}[I_{H1}(y\cos\vartheta_i, t') + I_{H2}(y\cos\vartheta_i, t')] + \int_0^{t'} \frac{d\rho_s(t'', \vartheta_i)}{dt''} S(t' - t'') \times \Bigg(1 + w(y\cos\vartheta_i, t' - t'') - \frac{Z_2\cos\vartheta_r}{Z_1\cos\vartheta_i} \times [I_{H1}(y\cos\vartheta_i, t' - t'') + I_{H2}(y\cos\vartheta_i, t' - t'')]\Bigg)dt''\Bigg] \tag{12}$$

$$H_{xt}(x, y, t) = E_0 Z_1^{-1}\Bigg(\{1 + w(y\cos\vartheta_i, t') + Z_1 Z_2^{-1}[-Z_2\sigma_2 y\cos\vartheta_i + I'_{E1}(y\cos\vartheta_i, t') + I_{E2}(y\cos\vartheta_i, t')]\}\cos\vartheta_i - \int_0^{t'} \frac{d\rho_s(t'', \vartheta_i)}{dt''} S(t' - t'')\{[1 + w(y\cos\vartheta_i, t' - t'')] \times \cos\vartheta_r - Z_1 Z_2^{-1}[-Z_2\sigma_2 y\cos\vartheta_i + I'_{E1}(y\cos\vartheta_i, t' - t'') - I_{E2}(y\cos\vartheta_i, t' - t'')]\cos\vartheta_i\}dt''\Bigg) \tag{13}$$

The functions w, I_{H1}, I_{H2}, I'_{E1}, and I_{E2} refer to medium 2; the subscript 2 must be used for σ, μ, and ϵ when these functions are computed from Eqs. (3.2-46), (3.5-6), (3.4-7), and (3.4-8).

For the boundary $y=0$ we write Eqs. (12) and (13) as follows:

$$E_{zt}(x, 0, t) = E_0 \tau_{Es}(t', \vartheta_i) \tag{14}$$

$$H_{xt}(x, 0, t) = E_0 Z_2^{-1} \tau_{Hs}(t', \vartheta_i) \tag{15}$$

Using Eqs. (3.6-15)–(3.6-19) we obtain for $\tau_{Es}(t', \vartheta_i)$ and $\tau_{Hs}(t', \vartheta_i)$ with the help of Eq. (1.3-7):

$$\begin{aligned}\tau_{Es}(t', \vartheta_i) &= 1 + Z_2 Z_1^{-1}[I_{H1}(0, t') + I_{H2}(0, t')] + \rho_s(t', \vartheta_i) \\ &\quad - \frac{Z_2 \cos \vartheta_r}{Z_1 \cos \vartheta_i} \int_0^{t'} \frac{d\rho_s(t'', \vartheta_i)}{dt''} S(t'-t'')[I_{H1}(0, t'-t'') \\ &\quad + I_{H2}(0, t'-t'')]dt''\end{aligned} \tag{16}$$

$$\begin{aligned}\tau_{Hs}(t', \vartheta_i) &= [Z_2 Z_1^{-1} + I'_{E1}(0, t') - I_{E2}(0, t')]\cos \vartheta_i \\ &\quad - Z_2 Z_1^{-1} \rho_s(t', \vartheta_i) \cos \vartheta_r \\ &\quad + \cos \vartheta_i \int_0^{t'} \frac{d\rho_s(t'', \vartheta_i)}{dt''} S(t'-t'') \\ &\quad \times [I'_{E1}(0, t'-t'') - I_{E2}(0, t'-t'')]dt''\end{aligned} \tag{17}$$

The tangential components of the electric field strengths in mediums 1 and 2 must be equal at the boundary $y=0$. We obtain from Eqs. (6) and (14):

$$1 + \rho_s(t', \vartheta_i) = \tau_{Es}(t', \vartheta_i) \tag{18}$$

The continuity of the tangential component of the magnetic field strength requires according to Eqs. (7) and (15):

$$\cos \vartheta_i - \rho_s(t', \vartheta_i)\cos \vartheta_r = (Z_1/Z_2)\tau_{Hs}(t', \vartheta_i) \tag{19}$$

Equation (19) is inserted into Eq. (17). Four terms cancel and one obtains:

$$\begin{aligned}I_{E2}(0, t') - I'_{E1}(0, t') &= \int_0^{t'} \frac{d\rho_s(t'', \vartheta_i)}{dt''} S(t'-t'')[I'_{E1}(0, t'-t'') \\ &\quad - I_{E2}(0, t'-t'')]dt''\end{aligned} \tag{20}$$

This is the same equation as Eq. (4.3-22) except for the replacement of t by t'. Its deconvolution yields:

$$\begin{aligned}\rho_s(t', \vartheta_i) &= \int_0^{t'} [I_{E2}(0, t') - I'_{E1}(0, t')] \circledast S(t'-t'') \\ &\quad \times [I'_{E1}(0, t'-t'') - I_{E2}(0, t'-t'')]dt'', \\ t' &= t - xc_1^{-1} \sin \vartheta_i\end{aligned} \tag{21}$$

Except for the replacement of t by t', we have the same equation as Eq. (4.3-26). The actual deconvolution may thus be performed as discussed for Eq. (4.3-26).

We insert Eq. (18) into Eq. (16):

$$\cos\vartheta_i = R_s(t', \vartheta_i)\cos\vartheta_r \tag{22}$$

$$R_s(t', \vartheta_i) = [I_{H1}(0, t') + I_{H2}(0, t')]^{-1}\int_0^{t'} \frac{d\rho_s(t'', \vartheta_i)}{dt''}[I_{H1}(0, t'-t'') + I_{H2}(0, t'-t'')]dt'' \tag{23}$$

In order to evaluate $R_s(t', \vartheta_i)$ we change the notation t and $\rho_s(t'')$ in Eq. (4.3-21) to t' and $\rho_s(t'', \vartheta_i)$:

$$I_{H1}(0, t') + I_{H2}(0, t') = \int_0^{t'} \frac{d\rho_s(t'', \vartheta_i)}{dt''} S(t'-t'')[I_{H1}(0, t'-t'') + I_{H2}(0, t'-t'')]dt'' \tag{24}$$

Insertion of Eq. (24) into Eq. (23) yields

$$R_s(t', \vartheta_i) = 1 \tag{25}$$

for any time t' and any angle ϑ_i, and Eq. (22) yields Snell's law of reflection

$$\vartheta_i = \vartheta_r \tag{26}$$

The transmission function $\tau_{Es}(t', \vartheta_i)$ follows by insertion of $\rho_s(t')$ into Eq. (18), while $\tau_{Hs}(t', \vartheta_i)$ follows from Eqs. (19) and (26):

$$\tau_{Hs}(t', \vartheta_i) = Z_2 Z_1^{-1}[1 - \rho_s(t')]\cos\vartheta_i \tag{27}$$

From τ_{Es} and τ_{Hs} follow the magnitudes of the electric and magnetic field strengths, $E_{zt}(x, 0, t)$ and $H_{xt}(x, 0, t)$ of Eqs. (14) and (15), that excite a wave in medium 2. The magnitudes of the field strengths of this wave are obtained by inserting $E_{zt}(x, 0, t) = E_{zt}(0, 0, t')$, $H_{xt}(x, 0, t) = H_{xt}(0, 0, t')$ and ϑ_i for $E_z(0, 0, t'')$, $H_x(0, 0, t'')$, and ϑ in Eqs. (3.6-13) and (3.6-14).

We have so far discussed the reflected and transmitted wave excited by an incident perpendicularly polarized wave with the time variation of a step function. The generalization to an incident wave with general time variation $F(t)S(t)$ follows closely the procedure discussed in Section 4.3 in connection with Eqs. (4.3-30)–(4.3-32).

4.5 Parallel Polarized Step Function Wave

The calculation for parallel polarization follows closely the calculation for perpendicular polarization in Section 4.4. We connect the time variation $f_i(t)$

of the incident wave with the time variation $f_r(t)$ of the reflected wave by a reflection function $\rho_p(t, \vartheta_i)$:

$$f_r(t) = \rho_p(t, \vartheta_i) f_i(t) \tag{1}$$

The electric and magnetic field strengths of the parallel polarized incident and reflected wave are given by Eqs. (4.1-29)–(4.1-32):

$$\mathbf{E}_i = -(\mathbf{e}_x \cos\vartheta_i - \mathbf{e}_y \sin\vartheta_i) E_0 f_i[t - (\mathbf{e}_x \sin\vartheta_i + \mathbf{e}_y \cos\vartheta_i) r/c_1] \tag{2}$$

$$\mathbf{H}_i = \mathbf{e}_z E_0 Z_1^{-1} f_i[t - (\mathbf{e}_x \sin\vartheta_i + \mathbf{e}_y \cos\vartheta_i) r/c_1] \tag{3}$$

$$\begin{aligned}\mathbf{E}_r = &-(\mathbf{e}_x \cos\vartheta_r + \mathbf{e}_y \sin\vartheta_r) E_0 \rho_p(t, \vartheta_i)\\ &\times f_i[t - (\mathbf{e}_x \sin\vartheta_r - \mathbf{e}_y \cos\vartheta_r) r/c_1]\end{aligned} \tag{4}$$

$$\mathbf{H}_r = -\mathbf{e}_z E_0 Z_1^{-1} \rho_p(t, \vartheta_i) f_i[t - (\mathbf{e}_x \sin\vartheta_r - \mathbf{e}_y \cos\vartheta_r) r/c_1] \tag{5}$$

Let the field strengths of the incident wave have the time variation of the step function $E_0 S(t)$. According to Fig. 4.5-1 and the results of Section 3.1, the field strengths at the boundary $y = 0$ between the nonconducting medium 1 and the conducting medium 2 have the following values:

$$\begin{aligned}\mathbf{E}_i + \mathbf{E}_r &= 0 && \text{for} \quad t' < 0\\ &= E_0[-\mathbf{e}_x \cos\vartheta_i + \mathbf{e}_y \sin\vartheta_i\\ &\qquad - \rho_p(t', \vartheta_i)(\mathbf{e}_x \cos\vartheta_r + \mathbf{e}_y \sin\vartheta_r)] && \text{for} \quad t' \geqq 0\end{aligned} \tag{6}$$

$$\begin{aligned}\mathbf{H}_i + \mathbf{H}_r &= 0 && \text{for} \quad t' < 0\\ &= \mathbf{e}_z E_0 Z_1^{-1}[1 - \rho_p(t', \vartheta_i)] && \text{for} \quad t' \geqq 0\end{aligned} \tag{7}$$

The electric and magnetic field strengths $\mathbf{E}_i + \mathbf{E}_r$ and $\mathbf{H}_i + \mathbf{H}_r$ at the boundary $y = 0$ produce an electromagnetic wave in medium 2. The electric excitation force of Eq. (6) has a tangential component characterized by $\mathbf{e}_x$ in the plane $y = 0$ in Fig. 4.5-1, and a component characterized by $\mathbf{e}_y$ perpendicular to this plane. The two components are connected by Eq. (3.1-26). We are here primarily interested in the tangential component characterized by $\mathbf{e}_x$. It consists of the step function $-E_0 S(t')\cos\vartheta_i$ reaching the line $y = 0, x$ at the time $t' = 0$, plus the function $-E_0 \rho_p(t', \vartheta_i) S(t') \cos\vartheta_r$ starting at the time $t' = 0$ at the line $y = 0, x$. The magnitude of the electric field strength due to the step function is given by Eq. (3.3-15) with E_0 replaced by $-E_0 \cos\vartheta_i$, while the magnitude of the field strength due to $-E_0 \rho_p(t', \vartheta_i) S(t') \cos\vartheta_r$ is given by Eq. (3.6-23) with $E_x(0, 0, t'')$ replaced by $-E_0 \rho_p(t'', \vartheta_i) \cos\vartheta_r$:

$$\begin{aligned}E_{xEp}(x, y, t) = -E_0\Bigg(&1 + w(y\cos\vartheta_i, t')]\cos\vartheta_i + \cos\vartheta_r \int_0^{t'} \frac{d\rho_p(t'', \vartheta_i)}{dt''}\\ &\times S(t' - t'')[1 + w(y\cos\vartheta_i, t' - t'')]dt''\Bigg)\end{aligned} \tag{8}$$

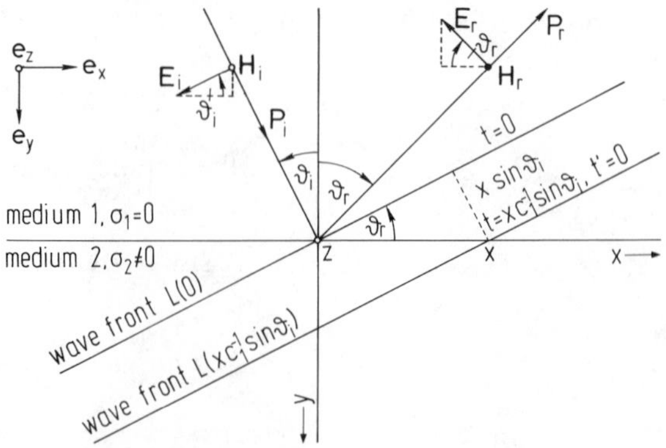

FIG. 4.5-1. Reflection of a parallel polarized wave at the boundary between a nonconducting and a conducting medium. The directions of $\mathbf{E}_i$, $\mathbf{H}_i$, $\mathbf{E}_r$, and $\mathbf{H}_r$ are chosen so that one obtains Fig. 4.3-1 for $\sin \vartheta_i = 0$, showing equal direction for $\mathbf{E}_i$ and $\mathbf{E}_r$ but opposite direction for $\mathbf{H}_i$ and $\mathbf{H}_r$.

The magnitude of the magnetic field strength produced by the electric excitation of Eq. (6) is given by Eqs. (3.4-20) with $H_{E0} = 0$ and (3.6-24):

$$\begin{aligned} H_{zEp}(x, y, t) = E_0 Z_2^{-1}\Big(&-Z_2\sigma_2 y \cos \vartheta_i + I'_{E1}(y \cos \vartheta_i, t') \\ &- I_{E2}(y \cos \vartheta_i, t') \\ &+ \frac{\cos \vartheta_r}{\cos \vartheta_i} \int_0^{t'} \frac{d\rho_p(t'', \vartheta_i)}{dt''} S(t' - t'')[-Z_2\sigma_2 y \cos \vartheta_i \\ &+ I'_{E1}(y \cos \vartheta_i, t' - t'') - I_{E2}(y \cos \vartheta_i, t' - t'')]dt''\Big) \end{aligned} \quad (9)$$

The magnetic excitation force of Eq. (7) consists of the step function $E_0 Z_1^{-1} S(t')$ reaching the line $y = 0$, x at the time $t' = 0$, plus the function $-E_0 Z_1^{-1} \rho_p(t', \vartheta_i) S(t')$ starting at the time $t' = 0$ at the line $y = 0$, x. The magnitude of the magnetic field strength due to the step function is given by Eq. (3.6-25) with H_0 replaced by $E_0 Z_1^{-1}$, while the magnitude of the field strength due to $-E_0 Z_1^{-1} \rho_p(t', \vartheta_i) S(t')$ is given by Eq. (3.6-27) with $H_z(0, 0, t'')$ replaced by $-E_0 Z_1^{-1} \rho_p(t'', \vartheta_i)$:

$$\begin{aligned} H_{zHp}(x, y, t) = E_0 Z_1^{-1}\Big(&1 + w(y \cos \vartheta_i, t') \\ &- \int_0^{t'} \frac{d\rho_p(t'', \vartheta_i)}{dt''} S(t' - t'') \\ &\times [1 + w(y \cos \vartheta_i, t' - t'')]dt''\Big) \end{aligned} \quad (10)$$

The magnitude of the electric field strength produced by the magnetic excitation of Eq. (7) is given by Eqs. (3.6-26) and (3.6-28):

$$E_{xHp}(x, y, t) = -E_0 Z_2 Z_1^{-1} \cos\vartheta_i \Big(I_{H1}(y\cos\vartheta_i, t') + I_{H2}(y\cos\vartheta_i, t')$$
$$- \int_0^{t'} \frac{d\rho_p(t'', \vartheta_i)}{dt''} S(t'-t'')[I_{H1}(y\cos\vartheta_i, t'-t'')$$
$$+ I_{H2}(y\cos\vartheta_i, t'-t'')]dt''\Big) \qquad (11)$$

The magnitude of the field strengths of the wave transmitted into medium 2 are the sums of Eqs. (8) and (11) as well as (9) and (10):

$$E_{xt}(x, y, t) = -E_0 \Big(\{1 + w(y\cos\vartheta_i, t') + Z_2 Z_1^{-1}[I_{H1}(y\cos\vartheta_i, t')$$
$$+ I_{H2}(y\cos\vartheta_i, t')]\}\cos\vartheta_i$$
$$+ \int_0^{t'} \frac{d\rho_p(t'', \vartheta_i)}{dt''} S(t'-t'')$$
$$\times \{[1 + w(y\cos\vartheta_i, t'-t'')]\cos\vartheta_r$$
$$- Z_2 Z_1^{-1}[I_{H1}(y\cos\vartheta_i, t'-t'')$$
$$+ I_{H2}(y\cos\vartheta_i, t'-t'')]\cos\vartheta_i\}dt''\Big) \qquad (12)$$

$$H_{zt}(x, y, t) = E_0 Z_1^{-1}\Big[1 + w(y\cos\vartheta_i, t') + Z_1 Z_2^{-1}$$
$$[-Z_2\sigma_2 y\cos\vartheta_i + I'_{E1}(y\cos\vartheta_i, t')$$
$$- I_{E2}(y\cos\vartheta_i, t')] - \int_0^{t'} \frac{d\rho_p(t'', \vartheta_i)}{dt''} S(t'-t'')$$
$$\times \Big(1 + w(y\cos\vartheta_i, t'-t'') - \frac{Z_1\cos\vartheta_r}{Z_2\cos\vartheta_i}$$
$$\times [-Z_2\sigma_2 y\cos\vartheta_i + I'_{E1}(y\cos\vartheta_i, t'-t'')$$
$$- I_{E2}(y\cos\vartheta_i, t'-t'')]\Big)dt''\Big] \qquad (13)$$

Since the functions w, I_{H1}, I_{H2}, I'_{E1}, and I_{E2} refer to medium 2 one must use the subscript 2 for σ, μ, and ϵ when these functions are computed from Eqs. (3.2-46), (3.5-6), (3.4-7), and (3.4-8).

At the boundary $y = 0$ we write E_{xt} and H_{zt} as follows:

$$E_{xt}(x, 0, t) = E_0 \tau_{Ep}(t', \vartheta_i) \tag{14}$$

$$H_{zt}(x, 0, t) = E_0 Z_2^{-1} \tau_{Hp}(t', \vartheta_i) \tag{15}$$

With the help of Eqs. (3.6-15)–(3.6-19) as well as Eq. (1.3-7) we obtain for $\tau_{Ep}(t', \vartheta_i)$ and $\tau_{Hp}(t', \vartheta_i)$:

$$\begin{aligned}\tau_{Ep}(t', \vartheta_i) = -\Bigg(&\{1 + Z_2 Z_1^{-1}[I_{H1}(0, t') + I_{H2}(0, t')]\}\cos\vartheta_i \\ &+ \rho_p(t', \vartheta_i)\cos\vartheta_r - \frac{Z_2\cos\vartheta_i}{Z_1}\int_0^{t'} \frac{d\rho_p(t'', \vartheta_i)}{dt''} \\ &\times S(t' - t'')[I_{H1}(0, t' - t'') + I_{H2}(0, t' - t'')]dt''\Bigg)\end{aligned} \tag{16}$$

$$\begin{aligned}\tau_{Hp}(t', \vartheta_i) = {}& Z_2 Z_1^{-1} + I'_{E1}(0, t') - I_{E2}(0, t') - Z_2 Z_1^{-1}\rho_p(t', \vartheta_i) \\ &+ \frac{\cos\vartheta_r}{\cos\vartheta_i}\int_0^{t'} \frac{d\rho_p(t'', \vartheta_i)}{dt''} S(t' - t'')[I'_{E1}(0, t' - t'') \\ &- I_{E2}(0, t' - t'')]dt''\end{aligned} \tag{17}$$

The tangential components of the magnetic field strengths in medium 1 and 2 must be equal at the boundary $y = 0$. We obtain from Eqs. (7) and (15):

$$Z_1^{-1}[1 - \rho_p(t', \vartheta_i)] = Z_2^{-1}\tau_{Hp}(t', \vartheta_i) \tag{18}$$

The continuity of the tangential components of the electric field strength requires according to Eqs. (6) and (14):

$$-\cos\vartheta_i - \rho_p(t', \vartheta_i)\cos\vartheta_r = \tau_{Ep}(t', \vartheta_i) \tag{19}$$

Equation (19) is inserted into Eq. (16):

$$\begin{aligned}I_{H1}(0, t') + I_{H2}(0, t') = \int_0^{t'} \frac{d\rho_p(t'', \vartheta_i)}{dt''} S(t' - t'')[I_{H1}(0, t' - t'') \\ + I_{H2}(0, t' - t'')]dt''\end{aligned} \tag{20}$$

This is the same equation as Eq. (4.3-21) except that t is replaced by t'. The formal deconvolution yields:

$$\begin{aligned}\rho_p(t', \vartheta_i) = \int_0^{t'} [I_{H1}(0, t') + I_{H2}(0, t')] \mathbin{\ddagger} S(t' - t'')[I_{H1}(0, t' - t'') \\ + I_{H2}(0, t' - t'')]dt'', \qquad t' = t - xc_1^{-1}\sin\vartheta_i\end{aligned} \tag{21}$$

The actual deconvolution can be performed as discussed for Eq. (4.3-25).

Insertion of Eq. (18) into Eq. (7) yields

$$\cos\vartheta_{\mathrm{i}} = R_p(t', \vartheta_{\mathrm{i}})\cos\vartheta_{\mathrm{r}} \tag{22}$$

$$R_p(t', \vartheta_{\mathrm{i}}) = [I_{E2}(0, t') - I'_{E1}(0, t')]^{-1} \int_0^{t'} \frac{d\rho_p(t'', \vartheta_{\mathrm{i}})}{dt''} S(t' - t'') \times [I'_{E1}(0, t' - t'') - I_{E2}(0, t' - t'')]dt'' \tag{23}$$

In order to evaluate $R_p(t', \vartheta_{\mathrm{i}})$ we consider Eq. (4.3-22) rewritten with a change of notation from t and $\rho_s(t'')$ to t' and $\rho_p(t'', \vartheta_{\mathrm{i}})$. Insertion of this rewritten equation into Eq. (23) yields

$$R_p(t', \vartheta_{\mathrm{i}}) = 1 \tag{24}$$

for any time t' and any angle ϑ_{i}. Equation (22) thus yields Snell's law of reflection

$$\vartheta_{\mathrm{i}} = \vartheta_{\mathrm{r}} \tag{25}$$

for parallel polarized waves.

4.6 Exponential Ramp Wave with Perpendicular Incidence

In Sections 4.3 to 4.5 we studied the reflection and transmission of planar waves at the boundary of a nonconducting and a conducting medium using waves with the time variation of a step function. We extend now the theory to waves with the time variation of the exponential ramp function $E_1(1 - e^{-t/\tau})$ of Fig. 1.2-4b and Eq. (2.3-3). In analogy to Eq. (4.3-1) we introduce a reflection function $\rho_s(t)(1 - e^{-t/\tau})^{-1}$:

$$f_{\mathrm{r}}(t) = \rho_s(1 - e^{-t/\tau})^{-1} f_{\mathrm{i}}(t) \tag{1}$$

Equations (4.3-2) and (4.3-3) remain unchanged, but Eqs. (4.3-4) and (4.3-5) assume the following form:

$$\mathbf{E}_{\mathrm{r}} = \mathbf{e}_z E_0 \rho_s(t)(1 - e^{-t/\tau})^{-1} f_{\mathrm{i}}(t - \mathbf{r}\mathbf{e}_y/c_1) \tag{2}$$

$$\mathbf{H}_{\mathrm{r}} = -\mathbf{e}_x E_0 Z_1^{-1} \rho_s(t)(1 - e^{-t/\tau})^{-1} f_{\mathrm{i}}(t - \mathbf{r}\mathbf{e}_y/c_1) \tag{3}$$

We assume that the time variation $f_{\mathrm{i}}(t)$ of the field strengths of the incident wave is that of an exponential ramp function $E_0 Q(1 - e^{-t/\tau})S(t)$ that is zero for $t < 0$ and varies like $1 - e^{-t/\tau}$ for $t \geqq 0$. The field strengths at the boundary $y = 0$ equal:

$$\begin{aligned} \mathbf{E}_i + \mathbf{E}_r &= 0 && \text{for} \quad t < 0 \\ &= \mathbf{e}_z E_1[1 - e^{-t/\tau} + \rho_s(t)) && \text{for} \quad t \geqq 0 \end{aligned} \tag{4}$$

$$\begin{aligned} \mathbf{H}_i + \mathbf{H}_r &= 0 && \text{for} \quad t < 0 \\ &= \mathbf{e}_x E_1 Z_1^{-1}[1 - e^{-t/\tau} - \rho_s(t)] && \text{for} \quad t \geqq 0 \end{aligned} \tag{5}$$

$$E_1 = E_0 Q = E_0(1 - e^{-2\alpha \Delta T})^{-1}$$

The electric and magnetic field strengths $\mathbf{E}_i + \mathbf{E}_r$ and $\mathbf{H}_i + \mathbf{H}_r$ at the boundary $y = 0$ produce an electromagnetic wave in medium 2. The electric excitation force of Eq. (4) consists of the exponential ramp function $E_1(1 - e^{-t/\tau})S(t)$ plus the function $E_1\rho_s(t)S(t)$. The magnitude of the electric and magnetic field strengths due to the ramp function $E_1(1 - e^{-t/\tau})$ are given by Eqs. (2.3-36) and (2.9-20) with $H_t(y) = 0$, while the magnitude of the field strengths due to $E_1\rho(t)$ are given by Eqs. (2.11-10) and (2.11-12) with $E(0, t'') = E_1\rho(t'')$. The sum of the magnitude of these field strengths yields:

$$E_E(y, t) = E_1\bigg(1 - e^{-t/\tau} + u(y, t) + \tau \int_0^t \frac{d^2\rho_s(t'')}{dt''^2} S(t - t'')[1 - e^{-(t-t'')/\tau} + u(y, t - t'')]dt''\bigg) \tag{6}$$

$$H_E(y, t) = E_1 Z_2^{-1}\bigg(-Z_2\sigma_2 y + I'_{E3}(y, t) - I_{E4}(y, t) + \tau \int_0^t \frac{d^2\rho_s(t'')}{dt''^2} S(t - t'')[-Z_2\sigma_2 y + I'_{E3}(y, t - t'') - I_{E4}(y, t - t'')]dt''\bigg) \tag{7}$$

$$\tau = 1/2\alpha = \epsilon/\sigma$$

The magnetic excitation force of Eq. (5) consists of the exponential ramp function $E_1 Z_1^{-1}(1 - e^{-t/\tau})S(t)$ plus the function $-E_1 Z_1^{-1}\rho_s(t)S(t)$. The magnitude of the electric and magnetic field strengths due to the ramp function are given by Eqs. (2.10-2) for $L \to \infty$ and (2.10-21) with $E_t(y) = 0$, while the magnitudes of the field strengths due to $-E_1 Z_1^{-1}\rho_s(t)S(t)$ are given by Eqs. (2.11-15) and (2.11-16). The sum of the magnitude of these field strengths yields with $E_1 = H_1 Z_1$:

$$H_H(y, t) = E_1 Z_1^{-1}\Bigg(1 - e^{-t/\tau} + u(y, t) - \tau \int_0^t \frac{d^2 \rho_s(t'')}{dt''^2} S(t - t'')[1 - e^{-(t-t'')/\tau} + u(y, t - t'')]dt''\Bigg) \quad (8)$$

$$E_H(y, t) = E_1 Z_2 Z_1^{-1}\Bigg(-Z_2\sigma_2 y e^{-t/\tau} + I'_{H3}(y, t) - I_{H4}(y, t) - \tau \int_0^t \frac{d^2 \rho_s(t'')}{dt''^2} S(t - t'') \times [-Z_2\sigma_2 y e^{-(t-t'')/\tau} + I'_{H3}(y, t - t'') - I_{H4}(y, t - t'')]dt''\Bigg) \quad (9)$$

The magnitudes of the field strengths $\mathbf{E}_t$ and $\mathbf{H}_t$ of the transmitted wave are the sums of Eqs. (6) and (9) as well as (7) and (8):

$$E_t(y, t) = E_1\Bigg(1 - e^{-t/\tau} + u(y, t) + Z_2 Z_1^{-1}[-Z_2\sigma_2 y e^{-t/\tau} + I'_{H3}(y, t) - I_{H4}(y, t) + \tau \int_0^t \frac{d^2 \rho_s(t'')}{dt''^2} S(t - t'')\{1 - e^{-(t-t'')/\tau} + u(y, t) - Z_2 Z_1^{-1}[-Z_2\sigma_2 y e^{-(t-t'')/\tau} + I'_{H3}(y, t - t'') - I_{H4}(y, t - t'')]\}dt''\Bigg) \quad (10)$$

$$H_t(y, t) = E_1 Z_1^{-1}\Bigg(1 - e^{-t/\tau} + u(y, t) + Z_1 Z_2^{-1}[-Z_2\sigma_2 y + I'_{E3}(y, t) - I_{E4}(y, t)] - \tau \int_0^t \frac{d^2 \rho_s(t'')}{dt''^2} S(T - t'')\{1 - e^{-(t-t'')/\tau} + u(y, t - t'') - Z_1 Z_2^{-1}[-Z_2\sigma_2 y + I'_{E3}(y, t - t'') - I_{E4}(y, t - t'')]\}dt''\Bigg) \quad (11)$$

The functions u, I'_{H3}, I_{H4}, I'_{E3}, and I_{E4} refer to medium 2. Hence, the subscript 2 must be used,

$$\tau = \epsilon_2/\sigma_2, \qquad Z_2 = (\mu_2/\epsilon_2)^{1/2}, \qquad c = (\mu_2\epsilon_2)^{-1/2} \quad (12)$$

when these function are computed from Eqs. (2.3-33), (2.10-19), (2.10-20), (2.9-18), and (2.9-19).

For the boundary $y=0$ we write Eqs. (10) and (11) as follows:

$$E_t(0,t) = E_1 \tau_{Es}(t) \tag{13}$$

$$H_t(0,t) = E_1 Z_2^{-1} \tau_{Hs}(t) \tag{14}$$

The vector diagram of Fig. 4.3-1 applies here as well as in the case of the step function analyzed in Section 4.3. Using Eqs. (2.11-21)–(2.11-25) we obtain for $\tau_{Es}(t)$ and $\tau_{Hs}(t)$ with the help of Eq. (1.3-33):

$$\begin{aligned}\tau_{Es}(t) = {} & 1 - e^{-t/\tau} + Z_2 Z_1^{-1}[I'_{H3}(0,t) - I_{H4}(0,t)] + \rho_s(t) \\ & - Z_2 Z_1^{-1}\,\tau \int_0^t \frac{d^2\rho_s(t'')}{dt''^2} S(t-t'')[I'_{H3}(0,t-t'') \\ & - I_{H4}(0,t-t'')]dt''\end{aligned} \tag{15}$$

$$\begin{aligned}\tau_{Hs}(t) = {} & Z_2 Z_1^{-1}(1 - e^{-t/\tau}) + I'_{E3}(0,t) - I_{E4}(0,t) - Z_2 Z_1^{-1}\rho_s(t) \\ & + \tau \int_0^t \frac{d^2\rho_s(t'')}{dt''^2} S(t-t'')[I'_{E3}(0,t-t'') - I_{E4}(0,t-t'')]dt''\end{aligned} \tag{16}$$

The tangential components of the electric field strengths in medium 1 and 2 must be equal at the boundary $y=0$. We obtain from Eqs. (4) and (13):

$$1 - e^{-t/\tau} + \rho_s(t) = \tau_{Es}(t) \tag{17}$$

Similarly, continuity of the tangential component of the magnetic field strength requires according to Eqs. (5) and (14):

$$Z_1^{-1}[1 - e^{-t/\tau} - \rho_s(t)] = Z_2^{-1}\tau_{Hs}(t) \tag{18}$$

Insertion of Eq. (17) into Eq. (15) yields:

$$\begin{aligned}I'_{H3}(0,t) - I_{H4}(0,t) = \tau \int_0^t \frac{d^2\rho_s(t'')}{dt''^2} S(t-t'')[I'_{H3}(0,t-t'') \\ - I_{H4}(0,t-t'')]dt''\end{aligned} \tag{19}$$

Furthermore, we insert Eq. (18) into Eq. (16):

$$\begin{aligned}I_{E4}(0,t) - I'_{E3}(0,t) = \tau \int_0^t \frac{d^2\rho_s(t'')}{dt''^2} S(t-t'')[I'_{E3}(0,t-t'') \\ - I_{E4}(0,t-t'')]dt''\end{aligned} \tag{20}$$

A comparison of Eqs. (19) and (20) with Eqs. (1.4-14) to (1.4-16) shows that

$\rho_s(t)$ is formally defined by the following two equations:

$$\rho_s(t) = \int_0^t \left(\int_0^{t\dagger} [I'_{H3}(0, t) - I_{H4}(0, t)] \ddagger S(t - t'')\tau[I'_{H3}(0, t - t'') - I_{H4}(0, t - t'')]dt'' \right) dt^\dagger \tag{21}$$

$$\rho_s(t) = \int_0^t \left(\int_0^{t\dagger} [I_{E4}(0, t) - I'_{E3}(0, t)] \ddagger S(t - t'')\tau[I'_{E3}(0, t - t'') - I_{E4}(0, t - \iota'')]dt'' \right) dt^\dagger \tag{22}$$

The actual deconvolution for sampled functions can be carried out with Eq. (1.4-27). Since Eqs. (21) and (22) must yield the same result for $\rho_s(t)$, one may use this fact as a check on the accuracy of the computation.

Let us note that $\rho_s(t)$ is the *reflected function* due to an incident function $1 - e^{-t/\tau}$, since the *reflection function* was defined as $\rho_s(t)(1 - e^{-t/\tau})^{-1}$ in Eq. (1).

The transmission functions $\tau_{Es}(t)$ and $\tau_{Hs}(t)$ follow from $\rho_s(t)$ with the help of Eqs. (15) and (16). These transmission functions give us in turn the magnitude of the field strengths $\mathbf{E}_t(0, t)$ and $\mathbf{H}_t(0, t)$ according to Eqs. (13) and (14) that excite the transmitted wave in medium 2. The magnitudes of the field strengths $\mathbf{E}_t(y, t)$ and $\mathbf{H}_t(y, t)$ in medium 2 for any value of y and t are obtained by inserting $E_t(0, t'')$ and $H_t(0, t'')$ into Eqs. (2.11-17) and (2.11-18), or Eqs. (2.11-19) and (2.11-20) for sampled functions. These are the magnitudes of the field strengths of the transmitted wave due to an incident wave with the time variation of an exponential ramp function for electric and magnetic field strength.

The knowledge of the reflected function $\rho_s(t)$ for an incident wave with the time variation of an exponential ramp function leads to the solution of the general case. An incident wave with the time variation $F(t)S(t)$ yields the time variation $F_r(t)$ of the reflected wave according to Eq. (1.3-34)

$$F_r(k\,\Delta T) = \sum_{i=0}^{k} \{F[i + 1)\Delta T] - (1 + e^{-\Delta T/\tau})F(i\,\Delta T) + e^{-\Delta T/\tau}F[(i - 1)\Delta T]\}S(t - i\,\Delta T)\rho_s(t/\Delta T - i) \tag{23}$$

where $t = k\,\Delta T$. The magnitudes of the transmitted field strengths due to $F(t)S(t)$ become:

$$E_{Ft}(y, k\,\Delta T) = \sum_{i=0}^{k} \{F[(i + 1)\Delta T] - (1 + e^{-\Delta T/\tau})F(i\,\Delta T) + e^{-\Delta T/\tau}F[(i - 1)\Delta T]\}S(t - i\,\Delta T)E_t(y, t/\Delta T - i) \tag{24}$$

$$H_{\mathrm{Ft}}(y, k\,\Delta T) = \sum_{i=0}^{k} \{F[(i+1)\Delta T] - (1 + e^{-\Delta T/\tau})F(i\,\Delta T) + e^{-\Delta T/\tau}F[(i-1)\Delta T]\}S(t - i\,\Delta T)H_{\mathrm{t}}(y, t/\Delta T - i) \quad (25)$$

4.7 Perpendicularly Polarized Exponential Ramp Wave

The results of Section 4.6 must be generalized for angles of incidence $\vartheta_{\mathrm{i}} \neq 0$, which introduces the distinction between perpendicular and parallel polarization. We consider perpendicular polarization first. Let $f_{\mathrm{i}}(t)$ and $f_{\mathrm{r}}(t)$ stand for the time variation of the incident and the reflected wave. We define the reflection function $\rho_s(t, \vartheta_{\mathrm{i}})(1 - e^{-t/\tau})^{-1}$:

$$f_{\mathrm{r}}(t) = \rho_s(t, \vartheta_{\mathrm{i}})(1 - e^{-t/\tau})^{-1} f_{\mathrm{i}}(t) \quad (1)$$

The electric and magnetic field strengths of the incident wave are given by Eqs. (4.4-2) and (4.4-3), but the field strengths of the reflected wave assume the following form:

$$\mathbf{E}_{\mathrm{r}} = \mathbf{e}_z E_0 \rho_s(t, \vartheta_{\mathrm{i}})(1 - e^{-t/\tau})^{-1} f_{\mathrm{i}}[t - (\mathbf{e}_x \sin\vartheta_{\mathrm{r}} - \mathbf{e}_y \cos\vartheta_{\mathrm{r}})r/c_1] \quad (2)$$

$$\mathbf{H}_{\mathrm{r}} = -(\mathbf{e}_x \cos\vartheta_{\mathrm{r}} + \mathbf{e}_y \sin\vartheta_{\mathrm{i}})E_0 Z_1^{-1} \rho_s(t, \vartheta_{\mathrm{i}})(1 - e^{-t/\tau})^{-1} \times f_{\mathrm{i}}[t - (\mathbf{e}_x \sin\vartheta_{\mathrm{r}} - \mathbf{e}_y \cos\vartheta_{\mathrm{r}})r/c_1] \quad (3)$$

The time variation of $f_{\mathrm{i}}(t)$ shall be that of an exponential ramp function $Q(1 - e^{-t/\tau})S(t)$. The field strengths at the boundary have then the following values according to Fig. 4.4-1:

$$\begin{aligned} \mathbf{E}_{\mathrm{i}} + \mathbf{E}_{\mathrm{r}} &= 0 && \text{for} \quad t' < 0 \\ &= \mathbf{e}_z E_1[1 - e^{-t'/\tau} + \rho_s(t', \vartheta_{\mathrm{i}})] && \text{for} \quad t' \geqq 0 \end{aligned} \quad (4)$$

$$\begin{aligned} \mathbf{H}_{\mathrm{i}} + \mathbf{H}_{\mathrm{r}} &= 0 \qquad \text{for} \quad t' < 0 \\ &= E_1 Z_1^{-1}[(\mathbf{e}_x \cos\vartheta_{\mathrm{i}} - \mathbf{e}_y \sin\vartheta_{\mathrm{i}})(1 - e^{-t'/\tau}) \\ &\quad - \rho_s(t', \vartheta_{\mathrm{i}})(\mathbf{e}_x \cos\vartheta_{\mathrm{r}} + \mathbf{e}_y \sin\vartheta_{\mathrm{r}})] \qquad \text{for} \quad t' \geqq 0 \end{aligned} \quad (5)$$

$$E_1 = E_0 Q = E_0(1 - e^{-2\alpha\Delta T})^{-1}, \qquad 2\alpha = 1/\tau$$

The electric and magnetic field strengths $\mathbf{E}_{\mathrm{i}} + \mathbf{E}_{\mathrm{r}}$ and $\mathbf{H}_{\mathrm{i}} + \mathbf{H}_{\mathrm{r}}$ at the boundary produce an electromagnetic wave in medium 2. The electric excitation force defined by Eq. (4) consists of the exponential ramp function $E_1(1 - e^{-t'/\tau})S(t')$ reaching the line $y = 0$, x at the time $t' = 0$, plus the function $E_1\rho_s(t', \vartheta_{\mathrm{i}})S(t')$ starting at the time $t' = 0$ at the line $y = 0$, x. The magnitudes of the electric and magnetic field stengths due to the exponential ramp function are given by Eqs. (3.7-30) and (3.7-39), while the magnitudes of the field strengths due to $E_1\rho_s(t', \vartheta_{\mathrm{i}})S(t')$ are given by Eqs. (3.10-6) and

(3.10-7) with $E_z(0, 0, t'') = E_1\rho_s(t'', \vartheta_i)$:

$$E_{zEs}(x, y, t) = E_1\bigg(1 - e^{-t'/\tau} + u(y\cos\vartheta_i, t') + \tau\int_0^{t'} \frac{d^2\rho_s(t'', \vartheta_i)}{dt''^2} S(t' - t'')[1 - e^{-(t'-t'')/\tau} + u(y\cos\vartheta_i, t' - t'')]dt''\bigg) \tag{6}$$

$$H_{xEs}(x, y, t) = E_1 Z_2^{-1}\cos\vartheta_i\bigg(-Z_2\sigma_2 y\cos\vartheta_i + I'_{E3}(y\cos\vartheta_i, t') - I_{E4}(y\cos\vartheta_i, t') + \tau\int_0^{t'} \frac{d^2\rho_s(t'', \vartheta_i)}{dt''^2} S(t' - t'')[-Z_2\sigma_2 y\cos\vartheta_i + I'_{E3}(y\cos\vartheta_i, t' - t'') - I_{E4}(y\cos\vartheta_i, t' - t'')]dt''\bigg) \tag{7}$$

The magnetic excitation force of Eq. (5) has a tangential component $\mathbf{e}_x$ in the plane $y = 0$ and a component $\mathbf{e}_y$ perpendicular to this plane. We are here interested in the tangential component $\mathbf{e}_x$. It consists of the exponential ramp function $E_1 Z_1^{-1}\cos\vartheta_i(1 - e^{-t'/\tau})S(t')$ plus the function $-E_1 Z_1^{-1}\rho_s(t', \vartheta_i)S(t')\cos\vartheta_i$. The magnitudes of the magnetic and electric field strengths due to the exponential ramp function are given by Eqs. (3.9-4) and (3.9-7) with $H_1 = E_1 Z_1^{-1}$, while the magnitudes of the field strengths due to $\rho_s(t', \vartheta_i)$ are given by Eqs. (3.10-10) and (3.10-11):

$$H_{xHs}(x, y, t) = E_1 Z_1^{-1}\bigg(\cos\vartheta_i[1 - e^{-t'/\tau} + u(y\cos\vartheta_i, t')] - \tau\cos\vartheta_r + \int_0^{t'} \frac{d^2\rho_s(t'', \vartheta_i)}{dt''^2} S(t' - t'') \times [1 - e^{-(t'-t'')/\tau} + u(y\cos\vartheta_i, t' - t'')]dt''\bigg) \tag{8}$$

$$E_{zHs}(x, y, t) = E_1 Z_2 Z_1^{-1}\bigg(-Z_2\sigma_2 e^{-t'/\tau}\cos\vartheta_i + I'_{H3}(y\cos\vartheta_i, t') - I_{H4}(y\cos\vartheta_i, t') - \tau\frac{\cos\vartheta_r}{\cos\vartheta_i}\int_0^{t'} \frac{d^2\rho_s(t'', \vartheta_i)}{dt''^2} S(t' - t'') \times [-Z_2\sigma_2 e^{-(t'-t'')/\tau} y\cos\vartheta_i + I'_{H3}(y\cos\vartheta_i, t' - t'') - I_{H4}(y\cos\vartheta_i, t' - t'')]dt''\bigg) \tag{9}$$

The magnitudes of the field strengths $\mathbf{E}_{zt}$ and $\mathbf{H}_{xt}$ of the transmitted wave are the sums of Eqs. (6) and (9) as well as (7) and (8):

$$
\begin{aligned}
E_{zt}(x, y, t) = E_1 \Bigg[& 1 - e^{-t'/\tau} + u(y \cos \vartheta_i, t') \\
& + Z_2 Z_1^{-1}[-Z_2\sigma_2 e^{-t'/\tau} y \cos \vartheta_i \\
& + I'_{H3}(y \cos \vartheta_i, t') - I_{H4}(y \cos \vartheta_i, t')] \\
& + \tau \int_0^{t'} \frac{d^2 p_s(t'', \vartheta_i)}{dt''^2} S(t' - t'') \\
& \times \Bigg(1 - e^{-(t'-t'')/\tau} + u(y \cos \vartheta_i, t' - t'') \\
& - \frac{Z_2 \cos \vartheta_r}{Z_1 \cos \vartheta_i} [-Z_2\sigma_2 e^{-(t'-t'')/\tau} y \cos \vartheta_i \\
& + I'_{H3}(y \cos \vartheta_i, t' - t'') - I_{H4}(y \cos \vartheta_i, t' - t'')] \Bigg) dt'' \Bigg]
\end{aligned} \tag{10}
$$

$$
\begin{aligned}
H_{xt}(x, y, t) = E_1 Z_1^{-1} \Bigg(& \{1 - e^{-t'/\tau} + u(y \cos \vartheta_i, t') \\
& + Z_1 Z_2^{-1}[-Z_2\sigma_2 y \cos \vartheta_i \\
& + I'_{E3}(y \cos \vartheta_i, t') - I_{E4}(y \cos \vartheta_i, t')]\} \cos \vartheta_i \\
& - \tau \int_0^{t'} \frac{d^2 p_s(t'', \vartheta_i)}{dt''^2} S(t' - t'') \{[1 - e^{-(t'-t'')/\tau} \\
& + u(y \cos \vartheta_i, t' - t'')] \cos \vartheta_r - Z_1 Z_2^{-1}[-Z_2\sigma_2 y \cos \vartheta_i \\
& + I'_{E3}(y \cos \vartheta_i, t' - t'') \\
& - I_{E4}(y \cos \vartheta_i, t' - t'')] \cos \vartheta_i\} dt'' \Bigg)
\end{aligned} \tag{11}
$$

The functions u, I'_{H3}, I_{H4}, I'_{E3}, and I_{E4} refer to medium 2; the subscript 2 must be used for σ, μ, and ϵ when these functions are computed from Eqs. (3.7-29), (3.9-8), (3.9-9), (3.7-40), and (3.7-41).

For the boundary $y = 0$ we write Eqs. (10) and (11) as follows:

$$E_{zt}(x, 0, t) = E_1 \tau_{Es}(t', \vartheta_i) \tag{12}$$

$$H_{xt}(x, 0, t) = E_1 Z_2^{-1} \tau_{Hs}(t', \vartheta_i) \tag{13}$$

Using Eqs. (2.11-21)–(2.11-25) for $t = t'$ we obtain for $\tau_{Es}(t', \vartheta_i)$ and

$\tau_{Hs}(t', \vartheta_i)$ with the help of Eq. (1.3-33):

$$\tau_{Es}(t', \vartheta_i) = 1 - e^{-t'/\tau} + Z_2 Z_1^{-1}[I'_{H3}(0, t') - I_{H4}(0, t')] + \rho_s(t', \vartheta_i)$$
$$- \tau \frac{Z_2 \cos \vartheta_r}{Z_1 \cos \vartheta_i} \int_0^{t'} \frac{d^2\rho_s(t'', \vartheta_i)}{dt''^2} S(t' - t'')$$
$$\times [I'_{H3}(0, t' - t'') - I_{H4}(0, t' - t'')]dt'' \qquad (14)$$

$$\tau_{Hs}(t', \vartheta_i) = [Z_2 Z_1^{-1}(1 - e^{-t'/\tau}) + I'_{E3}(0, t') - I_{E4}(0, t')]\cos \vartheta_i$$
$$- Z_2 Z_1^{-1}\rho_s(t', \vartheta_i)\cos \vartheta_r$$
$$+ \tau \cos \vartheta_i \int_0^{t'} \frac{d^2\rho_s(t'', \vartheta_i)}{dt''^2} S(t' - t'')$$
$$\times [I'_{E3}(0, t' - t'') - I_{E4}(0, t' - t'')]dt'' \qquad (15)$$

The tangential components of the electric field strengths in medium 1 and 2 must be equal at the boundary $y = 0$. We obtain from Eqs. (4) and (12):

$$1 - e^{-t'/\tau} + \rho_s(t', \vartheta_i) = \tau_{Es}(t', \vartheta_i) \qquad (16)$$

The continuity of the tangential component of the magnetic field strength requires according to Eqs. (5) and (13):

$$(1 - e^{-t'/\tau})\cos \vartheta_i - \rho_s(t', \vartheta_i)\cos \vartheta_r = (Z_1/Z_2)\tau_{Hs}(t', \vartheta_i) \qquad (17)$$

Equation (17) is inserted into Eq. (15):

$$I_{E4}(0, t') - I'_{E3}(0, t') = \tau \int_0^{t'} \frac{d^2\rho_s(t'', \vartheta_i)}{dt''^2} S(t' - t'')$$
$$\times [I'_{E3}(0, t' - t'') - I_{E4}(0, t' - t'')]dt'' \qquad (18)$$

This is again Eq. (4.6-20), except for the replacement of t by t'. Its formal deconvolution yields:

$$\rho_s(t', \vartheta_i) = \int_0^{t'} \left(\int_0^{t\dagger} [I_{E4}(0, t') - I'_{E3}(0, t')] \ast S(t' - t'')\tau \right.$$
$$\left. \times [I'_{E3}(0, t' - t'') - I_{E4}(0, t' - t'')]dt'' \right) dt^\dagger \qquad (19)$$

We insert Eq. (16) into Eq. (14):

$$\cos \vartheta_i = R_{es}(t', \vartheta_i)\cos \vartheta_r \qquad (20)$$

$$R_{es}(t', \vartheta_i) = [I'_{H3}(0, t') - I_{H4}(0, t')]^{-1}\tau \int_0^{t'} \frac{d^2\rho_s(t'', \vartheta_i)}{dt''^2}$$
$$\times S(t' - t'')[I'_{H3}(0, t' - t'') - I_{H4}(0, t' - t'')]dt'' \qquad (21)$$

Consider Eq. (4.6-19) rewritten with the notation t and $\rho_s(t'')$ replaced by t' and $\rho_s(t'', \vartheta_i)$. Insertion of the rewritten equation into Eq. (21) yields

$$R_{es}(t', \vartheta_i) = 1 \tag{22}$$

for any value of t' and ϑ_i. From Eq. (20) thus follows Snell's law of reflection.

4.8 Parallel Polarized Exponential Ramp Wave

For the calculation of the parallel polarization case we connect the time variation $f_i(t)$ of the incident wave with the time variation $f_r(t)$ of the reflected wave by a reflection function $\rho_p(t, \vartheta_i)(1 - e^{-t/\tau})^{-1}$:

$$f_r(t) = \rho_p(t, \vartheta_i)(1 - e^{-t/\tau})^{-1} f_i(t) \tag{1}$$

The electric and magnetic field strengths of the parallel polarized incident wave are given by Eqs. (4.5-2) and (4.5-3). The field strengths of the reflected wave follow with the help of Eq. (1) from Eqs. (4.5-4) and (4.5-5):

$$\mathbf{E}_r = -(\mathbf{e}_x \cos \vartheta_r + \mathbf{e}_y \sin \vartheta_r) E_0 \rho_p(t, \vartheta_i)(1 - e^{-t/\tau})^{-1} \times f_i[t - (\mathbf{e}_x \sin \vartheta_r - \mathbf{e}_y \cos \vartheta_r) r/c_1] \tag{2}$$

$$\mathbf{H}_r = -\mathbf{e}_z E_0 Z_1^{-1} \rho_p(t, \vartheta_i)(1 - e^{-t/\tau})^{-1} f_i[t - (\mathbf{e}_x \sin \vartheta_r - \mathbf{e}_y \cos \vartheta_r) r/c_1] \tag{3}$$

Let the field strength of the incident wave have the time variation of an exponential ramp function $Q(1 - e^{-t/\tau})S(t)$. The field strengths at the boundary $y = 0$ between the nonconducting medium 1 and the conducting medium 2 have the following values:

$$\begin{aligned} \mathbf{E}_i + \mathbf{E}_r &= 0 \qquad \text{for} \quad t' < 0 \\ &= E_1[(-\mathbf{e}_x \cos \vartheta_i + \mathbf{e}_y \sin \vartheta_i)(1 - e^{-t'/\tau}) \\ &\quad - \rho_p(t', \vartheta_i)(\mathbf{e}_x \cos \vartheta_r + \mathbf{e}_y \sin \vartheta_r)] \qquad \text{for} \quad t' \geqq 0 \end{aligned} \tag{4}$$

$$\begin{aligned} \mathbf{H}_i + \mathbf{H}_r &= 0 \qquad \text{for} \quad t' < 0 \\ &= \mathbf{e}_z E_1 Z_1^{-1}[1 - e^{-t'/\tau} - \rho_p(t', \vartheta_i)] \qquad \text{for} \quad t' \geqq 0 \end{aligned} \tag{5}$$

The electric and magnetic field strengths $\mathbf{E}_i + \mathbf{E}_r$ and $\mathbf{H}_i + \mathbf{H}_r$ at the boundary $y = 0$ produce an electromagnetic wave in medium 2. The electric excitation force of Eq. (4) has a tangential component characterized by $\mathbf{e}_x$ in the plane $y = 0$ in Fig. 4.5-1, and a component characterized by $\mathbf{e}_y$ perpendicular to this plane. The two components are connected by Eq. (3.1-26). We are here interested in the tangential component. It consists of the exponential ramp function $-E_1(1 - e^{-t'/\tau})S(t')\cos \vartheta_i$ reaching the line $y = 0, x$ at the time $t' = 0$, plus the function $-E_1\rho_p(t', \vartheta_i)S(t')\cos \vartheta_r$ starting at the time $t' = 0$ at the line $y = 0, x$. The magnitude of the electric field strength due to the ramp function is given by Eq. (3.8-7) with E_1 replaced by $-E_1 \cos \vartheta_i$, while the magnitude of the field strength due to $-E_1 \rho_p(t', \vartheta_i) S(t') \cos \vartheta_r$ is given by Eq. (3.10-18) with $E_x(0, 0, t'')$ replaced by $-E_1 \rho_p(t', \vartheta_i) \cos \vartheta_r$:

$$E_{xEp}(x, y, t) = -E_1\Bigg([1 - e^{-t'/\tau} + u(y \cos \vartheta_i, t')]\cos \vartheta_i$$
$$+ \tau \cos \vartheta_r \int_0^{t'} \frac{d^2 \rho_p(t'', \vartheta_i)}{dt''^2} S(t' - t'')[1 - e^{-(t'-t'')/\tau}$$
$$+ u(y \cos \vartheta_i, t' - t'')]dt''\Bigg) \tag{6}$$

The magnitude of the magnetic field strength produced by the electric excitation of Eq. (4) is given by Eqs. (3.8-11) and (3.10-19):

$$H_{zEp}(x, y, t) = E_1 Z_2^{-1}\Bigg(-Z_2 \sigma_2 y \cos \vartheta_i + I'_{E3}(y \cos \vartheta_i, t')$$
$$- I_{E4}(y \cos \vartheta_i, t')$$
$$+ \tau \frac{\cos \vartheta_r}{\cos \vartheta_i} \int_0^{t'} \frac{d^2 \rho_p(t'', \vartheta_i)}{dt''^2} S(t' - t'')[-Z_2 \sigma_2 y \cos \vartheta_i$$
$$+ I'_{E3}(y \cos \vartheta_i, t' - t'') - I_{E4}(y \cos \vartheta_i, t' - t'')]dt''\Bigg) \tag{7}$$

The magnetic excitation force of Eq. (5) consists of the exponential ramp function $E_1 Z_1^{-1}(1 - e^{-t'/\tau})S(t')$ reaching the line $y = 0$, x at the time $t' = 0$, plus the function $-E_1 Z_1^{-1} \rho_p(t', \vartheta_i) S(t')$ starting at the time $t' = 0$ at the line $y = 0$, x. The magnitude of the magnetic field strength due to the exponential ramp function is given by Eq. (3.9-13) with H_1 replaced by $E_1 Z_1^{-1}$, while the magnitude of the field strength due to $-E_1 Z_1^{-1} \rho_p(t', \vartheta_i) S(t')$ is given by Eq. (3.10-20) with $H_z(0, 0, t'')$ replaced by $-E_1 Z_1^{-1} \rho_p(t'', \vartheta_i)$:

$$H_{zHp}(x, y, t) = E_1 Z_1^{-1}\Bigg(1 - e^{-t'/\tau} + u(y \cos \vartheta_i, t')$$
$$- \tau \int_0^{t'} \frac{d^2 \rho_p(t'', \vartheta_i)}{dt''^2} S(t' - t'')[1 - e^{-(t'-t'')/\tau}$$
$$+ u(y \cos \vartheta_i, t' - t'')]dt''\Bigg) \tag{8}$$

The magnitude of the electric field strength produced by the magnetic excitation of Eq. (5) is given by Eqs. (3.9-15) and (3.10-21):

$$
\begin{aligned}
E_{xHp}(x, y, t) = -E_1 Z_2 Z_1^{-1} \cos \vartheta_i \Big(& -Z_2 \sigma_2 e^{-t'/\tau} y \cos \vartheta_i \\
& + I'_{H3}(y \cos \vartheta_i, t') \\
& - I_{H4}(y \cos \vartheta_i, t') - \tau \int_0^{t'} \frac{d^2 p_p(t'', \vartheta_i)}{dt''^2} S(t' - t'') \\
& \times [-Z_2 \sigma_2 e^{-(t'-t'')/\tau} y \cos \vartheta_i \\
& + I'_{H3}(y \cos \vartheta_i, t' - t'') - I_{H4}(y \cos \vartheta_i, t' - t'')] dt'' \Big) \qquad (9)
\end{aligned}
$$

The magnitudes of the field strengths of the wave transmitted into medium 2 are the sums of Eqs. (6) and (9) as well as (7) and (8):

$$
\begin{aligned}
E_{xt}(x, y, t) = -E_1 \Big(& \{1 - e^{-t'/\tau} + u(y \cos \vartheta_i, t') \\
& + Z_2 Z_1^{-1}[-Z_2 \sigma_2 e^{-t'/\tau} y \cos \vartheta_i \\
& + I'_{H3}(y \cos \vartheta_i, t') - I_{H4}(y \cos \vartheta_i, t')]\} \cos \vartheta_i \\
& + \tau \int_0^{t'} \frac{d^2 p_p(t'', \vartheta_i)}{dt''^2} S(t' - t'') \\
& \times \{[1 - e^{-(t'-t'')/\tau} + u(y \cos \vartheta_i, t' - t'')] \cos \vartheta_r \\
& - Z_2 Z_1^{-1}[-Z_2 \sigma_2 e^{-(t'-t'')/\tau} y \cos \vartheta_i \\
& + I'_{H3}(y \cos \vartheta_i, t' - t'') \\
& - I_{H4}(y \cos \vartheta_i, t' - t'')] \cos \vartheta_i\} dt'' \Big) \qquad (10)
\end{aligned}
$$

$$
\begin{aligned}
H_{zt}(x, y, t) = E_1 Z_1^{-1} \Big[& 1 - e^{-t'/\tau} + u(y \cos \vartheta_i, t') \\
& + Z_1 Z_2^{-1}[-Z_2 \sigma_2 y \cos \vartheta_i + I'_{E3}(y \cos \vartheta_i, t') \\
& - I_{E4}(y \cos \vartheta_i, t')] - \tau \int_0^{t'} \frac{d^2 p_p(t'', \vartheta_i)}{dt''^2} S(t' - t'') \\
& \times \Big(1 - e^{-(t'-t'')/\tau} + u(y \cos \vartheta_i, t' - t'')
\end{aligned}
$$

$$- \frac{Z_1 \cos \vartheta_r}{Z_2 \cos \vartheta_i} [-Z_2 \sigma_2 y \cos \vartheta_i$$

$$+ I'_{E3}(y \cos \vartheta_i, t' - t'') - I_{E4}(y \cos \vartheta_i, t' - t'')] \Big) dt'' \Big] \quad (11)$$

One must use the subscript 2 for σ, μ, and ϵ when the functions u, I'_{H3}, I_{H4}, I'_{E3}, and I_{E4} are computed from Eqs. (3.7-29), (3.9-8), (3.9-9), (3.7-40), and (3.7-41), since these function refer to medium 2.

At the boundary $y = 0$ we write E_{xt} and H_{xt} as follows:

$$E_{xt}(x, 0, t) = E_1 \tau_{Ep}(t', \vartheta_i) \quad (12)$$

$$H_{zt}(x, 0, t) = E_1 Z_2^{-1} \tau_{Hp}(t', \vartheta_i) \quad (13)$$

With the help of Eqs. (2.11-21) to (2.11-25) for $t = t'$ we obtain for $\tau_{Ep}(t', \vartheta_i)$ and $\tau_{Hp}(t', \vartheta_i)$ with the help of Eq. (1.3-33):

$$\tau_{Ep}(t', \vartheta_i) = -\Big(\{1 - e^{-t'/\tau} + Z_2 Z_1^{-1}[I'_{H3}(0, t') - I_{H4}(0, t')]\} \cos \vartheta_i$$

$$+ \rho_p(t', \vartheta_i) \cos \vartheta_r - \tau Z_2 Z_1^{-1} \cos \vartheta_i \int_0^{t'} \frac{d^2 \rho_p(t'', \vartheta_i)}{dt''^2}$$

$$\times S(t' - t'')[I'_{H3}(0, t' - t'') - I_{H4}(0, t' - t'')] dt'' \Big) \quad (14)$$

$$\tau_{Hp}(t', \vartheta_i) = Z_2 Z_1^{-1}(1 - e^{-t'/\tau}) + I'_{E3}(0, t')$$

$$- I_{E4}(0, t') - Z_2 Z_1 \rho_p(t', \vartheta_i)$$

$$+ \tau \frac{\cos \vartheta_r}{\cos \vartheta_i} \int_0^{t'} \frac{d^2 \rho_p(t'', \vartheta_i)}{dt''^2} S(t' - t'')[I'_{E3}(0, t' - t'')$$

$$- I_{E4}(0, t' - t'')] dt'' \quad (15)$$

The tangential components of the magnetic field strength in medium 1 and 2 must be equal at the boundary $y = 0$. We obtain from Eqs. (5) and (13):

$$Z_1^{-1}[1 - e^{-t'/\tau} - \rho_p(t', \vartheta_i)] = Z_2^{-1} \tau_{Hp}(t', \vartheta_i) \quad (16)$$

The continuity of the tangential components of the electric field strengths requires according to Eqs. (4) and (12):

$$-\cos \vartheta_i (1 - e^{-t'/\tau}) - \rho_p(t', \vartheta_i) \cos \vartheta_r = \tau_{Ep}(t', \vartheta_i) \quad (17)$$

Equation (17) is inserted into Eq. (14):

$$I'_{H3}(0, t') - I_{H4}(0, t') = \int_0^{t'} \frac{d^2 \rho_s(t'', \vartheta_i)}{dt''^2} S(t' - t'')$$

$$\times [I'_{H3}(0, t' - t'') - I_{H4}(0, t' - t'')] dt'' \quad (18)$$

This is the same equation as Eq. (4.6-19) except that t is replaced by t'. The formal deconvolution yields:

$$\rho_p(t', \vartheta_i) = \int_0^t \left(\int_0^{t\dagger} [I'_{H3}(0, t') - I_{H4}(0, t')] \circledast S(t' - t'')\tau \right.$$
$$\left. \times [I'_{H3}(0, t' - t'') - I_{H4}(0, t' - t'')]dt'' \right) dt^\dagger \qquad (19)$$

The actual deconvolution may be performed with Eq. (1.4-27).

Insertion of Eq. (16) into Eq. (15) yields:

$$\cos \vartheta_i = R_{ep}(t', \vartheta_i) \cos \vartheta_r \qquad (20)$$

$$R_{ep}(t', \vartheta_i) = [I_{E4}(0, t') - I_{E3}(0, t')]^{-1} \int_0^{t'} \frac{d^2\rho_p(t'', \vartheta_i)}{dt''^2} S(t' - t'')$$
$$\times [I'_{E3}(0, t' - t'') - I_{E4}(0, t' - t'')]dt'' \qquad (21)$$

To obtain R_{ep} we consider Eq. (4.6-20) rewritten with a change of notation from t and $\rho_s(t'')$ to t' and $\rho_p(t'', \vartheta_i)$. The resulting equation inserted into Eq. (21) yields

$$R_{ep}(t', \vartheta_i) = 1 \qquad (22)$$

for any value of t' and ϑ_i. From Eq. (20) thus follows Snell's law of reflection.

4.9 Reflection-Modifying Layers on Metal Surfaces

Consider a metal plate covered with a layer of a nonconducting material. We denote with $Z_0 = (\mu_0/\epsilon_0)^{1/2}$ the wave impedance of free space, and with $Z_1 = (\mu_1/\epsilon_1)^{1/2}$ the wave impedance of the nonconducting material. In analogy we have $c_0 = (\mu_0\epsilon_0)^{-1/2}$ and $c_1 = (\mu_1\epsilon_1)^{-1/2}$.

A wave with perpendicular incidence is reflected by a metal plate with essentially infinite conductivity so that the electric field strength is amplitude reversed and the magnetic field strength remains unchanged. A nonconducting material will reflect a wave with reversed amplitude of the electric field strength for $Z_2 < Z_0$, and with a reversed amplitude of the magnetic field strength for $Z_2 > Z_0$ according to Eq. (4.1-26) with $Z_1 = Z_0$. Since the metal plate reverses the amplitude of the electric field strength, we want a nonreversing material for its cover. This implies $Z_2 > Z_0$ and $\mu_2/\mu_0 = \mu_r > 1$ due to the relation $Z_2 = (\mu_2/\epsilon_2)^{1/2}$ and the impossibility of making ϵ_2 smaller than ϵ_0. The typical materials with $\mu_r > 1$ in wave propagation are ferrites. These are ceramics with the general formula $M\mathrm{Fe_2O_3}$, where M stands for a bivalent metal such as Mn, Ni, Cu, Zn, Mg, Co, Li, or Cd.

Figure 4.9-1 shows a metal plate covered by a layer of ferrite material. A

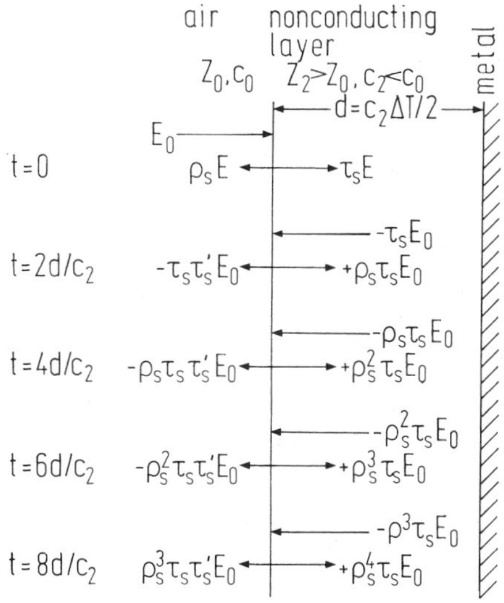

FIG. 4.9-1. Magnitude of the electric field strengths produced by a wave with time variation $E_0 S(t)$ incident on a metal plate covered with a loss-free layer having the impedance $Z_2 > Z_0$.

wave with vertical incidence and the time variation $E_0 S(t)$ hits the ferrite material, as shown on top of the illustration. The amplitude of the reflected and the transmitted electric field strengths follow from Eqs. (4.1-26) and (4.1-27):

$$\rho_s E_0 = E_0 (Z_2 - Z_0)/(Z_2 + Z_0) \tag{1}$$

$$\tau_s E_0 = 2 E_0 Z_2/(Z_2 + Z_0) \tag{2}$$

The transmitted field strength reaches the metal plate with a delay d/c_2, and becomes the reflected field strength with amplitude $-\tau_s E_0$, which reaches the boundary between air and the nonconducting layer with the delay $2d/c_2$. The wave is now going from the medium with impedance Z_2 to the medium with impedance Z_0. Reflected and transmitted electric field strengths become, according to Eqs. (4.1-26) and (4.1-27) with $Z_1 = Z_2$:

$$\rho_s'(-\tau_s E_0) = -\tau_s E_0 (Z_0 - Z_2)/(Z_0 + Z_2) = \rho_s \tau_s E_0 \tag{3}$$

$$\tau_s'(-\tau_s E_0) = -2\tau_s E_0 Z_0/(Z_0 + Z_2) = -2\tau_s (1 - \rho_s) E_0 \tag{4}$$

$$\tau_s \tau_s' = (1 + \rho_s)(1 - \rho_s) = 1 - \rho_s^2, \qquad \tau_s' = 2Z_0/(Z_0 + Z_2) \tag{5}$$

The amplitude of the reflected and transmitted waves with delays $4d/c_2$, $6d/c_2$, and $8d/c_2$ follow readily from this discussion, and they are shown for successive times t in Fig. 4.9-1.

The amplitude of the reflected electric field strength at the time $t \to \infty$ is given by the sum of the amplitudes of the field strengths at the boundary of air and ferrite in Fig. 4.9-1 that propagate to the left:

$$E_0[\rho_s - \tau_s\tau_s'(1 + \rho_s + \rho_s^2 + \rho_s^3 + \cdots)] = \left(\rho_s - \frac{\tau_s\tau_s'}{1 - \rho_s}\right)E_0 = -E_0 \quad (6)$$

Hence, for a step function the amplitude of the reflected electric field strength will eventually approach $-E_0$ regardless of the value of the reflection coefficient ρ_s. Figure 4.9-2a shows the normalized incident electric field strength $E_0S(t)/E(0) = S(t)$, while Figure 4.9-2b shows the normalized reflected field strength for $\rho_s = 0.618$. This choice of ρ_s makes the waves reflected at the time $t = 0$ and $t = 2d/c_2$ in Fig. 4.9-1 cancel:

$$\rho_s - \tau_s\tau_s' = \rho_s - (1 - \rho_s^2) = 0 \quad (7)$$

Consider next an incident electric field strength with the time variation of

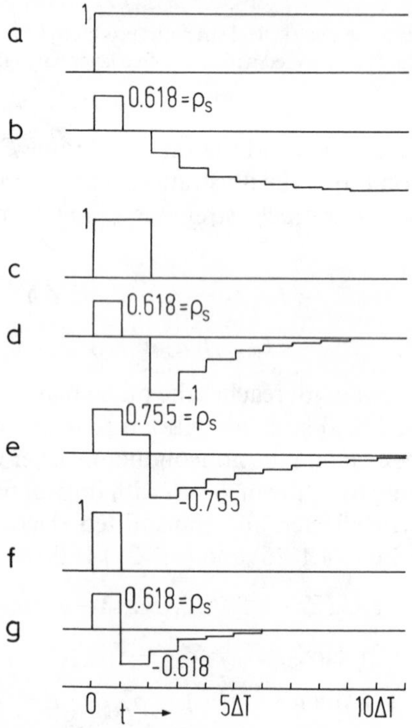

FIG. 4.9-2. Time variation of the amplitude of (a), (c), (f) incident and (b), (d), (e), (g) reflected electric field strengths in Fig. 4.9-1.

a rectangular pulse of duration $2\,\Delta T = 4d/c_2$ as shown in Fig. 4.9-2c. For $\rho_s = 0.618$ we get the time variation of the electric field strength shown in Fig. 4.9-2d. The magnitude of the reflected pulse in the interval $0 \leqq t \leqq \Delta T$ is reduced to 0.618, but the one in the interval $2\,\Delta T \leqq t \leqq 3\,\Delta T$ has the same value 1 as the incident pulse. We may reduce the negative peak amplitude by choosing ρ_s so that the pulses in the intervals $0 \leqq t \leqq \Delta T$ and $2\,\Delta T \leqq t \leqq 3\,\Delta T$ have the same magnitude. Since the rectangular pulse in Fig. 4.9-2c may be written $S(t) - S(t - 2\,\Delta T)$ we get readily from Fig. 4.9-1

$$\rho_s = \rho_s + \tau_s\tau_s' + \rho_s\tau_s\tau_s' - \rho_s \tag{8}$$

or with the help of Eq. (5)

$$\rho_s^3 + \rho_s^2 - 1 = 0 \tag{9}$$

One of the roots of this equation is $\rho_s = 0.755$. Figure 4.9-2e shows the resulting time variation of the reflected electric field strength caused by the pulse of Fig. 4.9-2c.

Consider now a rectangular pulse of duration $\Delta T = 2d/c_2$ as shown in Fig. 4.9-2f. The time variation of the reflected electric field strength is shown for $\rho_s = 0.618$ in Fig. 4.9-2g. The peak amplitude is reduced from 1 to 0.618 or by $10 \log(0.618)^2 = -4.18$ dB. Of course, the energy is conserved since we have assumed no losses in the nonconducting layer; the peak amplitude is reduced but the total duration of the reflected pulse is increased. This is a partial explanation of the experimental observation mentioned at the end of Section 1.7.

The plots in Fig. 4.9-2 represent interference phenomena of the step function and rectangular pulses. Interference is usually associated with sinusoidal waves, but this is strictly due to the almost universal neglect of waves with other than (almost) sinusoidal time variation. Interference phenomena of nonsinusoidal waves can be qualitatively different from those of sinusoidal waves. For instance, the peak amplitude of the electric field strength of a wave with sinusoidal time variation reflected by a metal plate cannot be reduced by a layer of loss-free material, contrary to the results of Fig. 4.9-2e and g.

Let us return to the reflection of the magnetic field strength. According to Figs.4.1-1 and 4.1-3 the incident and the reflected magnetic field strengths have opposite amplitudes for perpendicular incidence. The ratio of reflected to incident magnetic field strength follows from Eqs. (4.1-5) and (4.1-11) or (4.1-30) and (4.1-32)

$$\mathbf{H}_r/\mathbf{H}_i = -\rho_s \tag{10}$$

while the ratio of transmitted to incident field strength follows from Eqs.

(4.1-13) and (4.1-5) or (4.1-30) and (4.1-34):

$$\mathbf{H}_t/\mathbf{H}_i = \tau_s Z_1 Z_2 = \tau_s Z_0/Z_2 = 2Z_0/(Z_2 + Z_0) = \tau_s' \qquad (11)$$

Figure 4.9-3 shows a diagram for the reflection and transmission of the magnetic field strengths in analogy to Fig. 4.9-1 holding for the electric field strengths. There are differences in signs and τ_s is replaced by τ_s'. The sum of the magnetic field strengths at the boundary of air and ferrite that propagate to the left are given by the infinite sum

$$H_0[-\rho_s + \tau_s\tau_s'(1 + \rho_s + \rho_s^2 + \cdots)] = -\left(\rho_s - \frac{\tau_s\tau_s'}{1-\rho_s}\right)H_0 = H_0 \qquad (12)$$

Hence, for $t \to \infty$ the amplitude of the reflected magnetic field strength equals the amplitude of the incident field strength if the time variation is that of a step function.

4.10 Step Function Wave Leaving Conducting Medium

Let the nonconducting layer in Fig. 4.9-1 be replaced by a conducting one as shown in Fig. 4.10-1. The reflected and transmitted electric and magnetic field strengths at the boundary between air and the conducting layer can be calculated according to Section 4.3 for a wave entering the conducting layer from the air. At the metal plate, the magnitude of the transmitted electric and

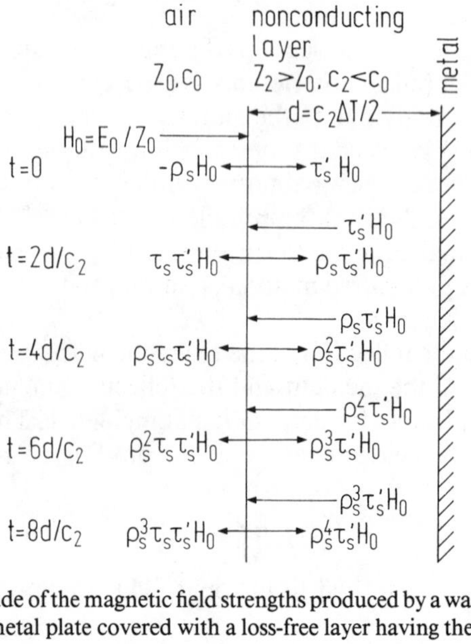

FIG. 4.9-3. Magnitude of the magnetic field strengths produced by a wave with time variation $H_0S(t)$ incident on a metal plate covered with a loss-free layer having the impedance $Z_2 > Z_0$.

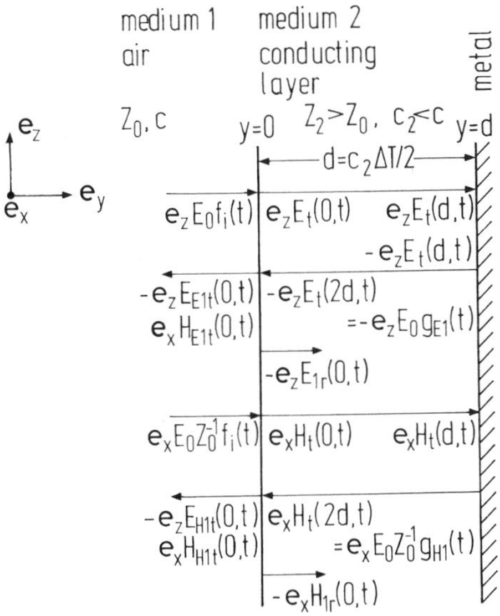

FIG. 4.10-1. Electric and magnetic field strengths produced by a wave with time variation $f_i(t)$ incident on a metal plate covered with a lossy layer having the impedance $Z_2 > Z_0$.

magnetic field strengths are given by Eqs. (4.3-12) and (4.3-13) as $E_t(d, t)$ and $H_t(d, t)$. The direction of the electric field strengh is reversed at the metal plate from $\mathbf{e}_z$ to $-\mathbf{e}_z$, while the direction of the magnetic field strength remains e_x. The magnitudes of the field strengths of the wave returned to the boundary between the conducting layer and air are given by Eqs. (4.3-12) and (4.3-13) for $y = 2d$:

$$E_t(2d, t) = E_0 g_{E1}(t) \tag{1}$$

$$H_t(2d, t) = E_0 Z_0^{-1} g_{H1}(t) \tag{2}$$

Let the wave reflected to the right at the boundary $y = 0$ in Fig. 4.10-1 have the field strengths $-\mathbf{e}_z E_{1r}(0, t)$ and $-\mathbf{e}_x H_{1r}(0, t)$; this wave propagates again toward the metal. The wave transmitted at the boundary $y = 0$ to the left into the air has the field strengths $-\mathbf{e}_z E_{E1t}(0, t)$ and $\mathbf{e}_x H_{E1t}(0, t)$ caused by $-\mathbf{e}_z E_t(2d, t)$, plus the field strengths $-\mathbf{e}_z E_{H1t}(0, t)$ and $\mathbf{e}_z H_{H1t}(0, t)$ caused by $\mathbf{e}_x H_t(2d, t)$. We may thus write the field strengths of incident and reflected wave at the boundary $y = 0$ as follows:

$$\mathbf{E}_{1i}(0, t) = -\mathbf{e}_z E_0 g_{E1}(t) \tag{3}$$

$$\mathbf{H}_{1i}(0, t) = \mathbf{e}_x E_0 Z_0^{-1} g_{H1}(t) \tag{4}$$

$$\mathbf{E}_{1r}(0, t) = -\mathbf{e}_z E_0 \rho_E(t) \tag{5}$$

$$\mathbf{H}_{1r}(0, t) = -\mathbf{e}_x E_0 Z_0^{-1} \rho_H(t) \tag{6}$$

The sums of these field strengths are:

$$\begin{aligned}\mathbf{E}_{1i}(0, t) + \mathbf{E}_{1r}(0, t) &= 0 & t < 2d/c_2 \\ &= -\mathbf{e}_z E_0[g_{E1}(t) + \rho_E(t)] & t \geqq 2d/c_2\end{aligned} \tag{7}$$

$$\begin{aligned}\mathbf{H}_{1i}(0, t) + \mathbf{H}_{1r}(0, t) &= 0 & t < 2d/c_2 \\ &= \mathbf{e}_x E_0 Z_0^{-1}[g_{H1}(t) - \rho_H(t)] & t \geqq 2d/c_2\end{aligned} \tag{8}$$

The electric field strength $\mathbf{E}_{1i} + \mathbf{E}_{1r}$ and the magnetic field strength $\mathbf{H}_{1i} + \mathbf{H}_{1r}$ at the boundary produce an electromagnetic wave in medium 1 in Fig. 4.10-1. The magnitudes of the electric and magnetic field strength due to the electric excitation force of Eq. (7) are given by Eqs. (2.8-7) and (2.8-8) with $E(0, t'') = E_0[g_{E1}(t'') + \rho_E(t'')]$:

$$E_{E1t}(y, t) = E_0 \int_0^t \frac{d}{dt''}[g_{E1}(t'') + \rho_E(t'')]S(t - t'')[1 + w(y, t - t'')]dt'' \tag{9}$$

$$\begin{aligned}H_{E1t}(y, t) = E_0 Z_0^{-1} &\int_0^t \frac{d}{dt''} + [g_{E1}(t'')\rho_E(t'')]S(t - t'')[I'_{E1}(y, t - t'') \\ &- I_{E2}(y, t - t'')]dt'', \qquad t \geqq 2d/c_2\end{aligned} \tag{10}$$

The magnitudes of the electric and magnetic field strength due to the magnetic excitation force of Eq. (8) are given by Eqs. (2.8-11) and (2.8-12); with

$$H(0, t'') = E_0 Z_0^{-1}[g_{H1}(t'') - \rho_H(t'')]$$

follows:

$$\begin{aligned}H_{H1t}(y, t) = E_0 Z_0^{-1} &\int_0^t \frac{d}{dt''}[g_{H1}(t'') \\ &- \rho_H(t'')]S(t - t'')[1 + w(y, t - t'')]dt''\end{aligned} \tag{11}$$

$$\begin{aligned}E_{H1t}(y, t) = E_0 &\int_0^t \frac{d}{dt''}[g_{H1}(t'') - \rho_H(t'')]S(t - t'') \\ &\times [I_{H1}(y, t - t'') \\ &+ I_{H2}(y, t - t'')]dt'', \qquad t \geqq 2d/c_2\end{aligned} \tag{12}$$

The magnitudes E_{1t} and H_{1t} of the field strengths of the transmitted wave are the sums of Eqs. (9) and (12) as well as (10) and (11):

$$\begin{aligned} E_{1t}(y,t) = E_0\bigg(&\int_0^t \frac{d}{dt''}[g_{E1}(t'') + \rho_E(t'')]S(t-t'') \\ &\times [1 + w(y, t-t'')]dt'' \\ &+ \int_0^t \frac{d}{dt''}[g_{H1}(t'') - \rho_H(t'')]S(t-t'')[I_{H1}(y, t-t'') \\ &+ I_{H2}(y, t-t'')]dt''\bigg) \end{aligned} \tag{13}$$

$$\begin{aligned} H_{1t}(y,t) = E_0 Z_0^{-1}\bigg(&\int_0^t \frac{d}{dt''}[g_{H1}(t'') - \rho_H(t'')]S(t-t'') \\ &\times [1 + w(y, t-t'')]dt'' \\ &+ \int_0^t \frac{d}{dt''}[g_{E1}(t'') + \rho_E(t'')]S(t-t'')[I'_{E1}(y, t-t'') \\ &- I_{E2}(y, t-t'')]dt''\bigg), \qquad t \geqq 2d/c_2 \end{aligned} \tag{14}$$

Note that w, I_{H1}, I_{H2}, I'_{E1}, and I_{E2} refer to air. Hence, we must use

$$\alpha = Z_0\sigma c/2 = 0, \qquad Z_0 = (\mu_0/\epsilon_0)^{1/2} = 377\ \Omega,$$
$$c = (\mu_0\epsilon_0)^{-1/2} = 3 \times 10^8\ \text{m/s} \tag{15}$$

when these functions are computed. From Eqs. (2.1-67), (2.6-16), and (2.6-18) we obtain readily:

$$\begin{aligned} w(y,t) = -\frac{2}{\pi}\int_0^\infty \frac{\cos\beta ct \sin\beta y}{\beta}\,d\beta &= 0 \qquad \text{for} \quad y \leqq 0, \quad ct > |y| \\ &= \tfrac{1}{2} \qquad \text{for} \quad y < 0, \quad ct = |y| \\ &= 1 \qquad \text{for} \quad y < 0, \quad ct < |y| \end{aligned} \tag{16}$$

$$I_{H1}(y,t) = 0 \tag{17}$$

$$\begin{aligned} I_{H2}(y,t) = \frac{2}{\pi}\int_0^\infty \frac{\sin\beta ct \cos\beta y}{\beta}\,d\beta &= 1 \qquad \text{for} \quad y \leqq 0, \quad ct > |y| \\ &= \tfrac{1}{2} \qquad \text{for} \quad y < 0, \quad ct = |y| \\ &= 0 \qquad \text{for} \quad y < 0, \quad ct < |y| \end{aligned} \tag{18}$$

$$t \geqq 2d/c_2$$

The computation of I'_{E1} and I_{E2} from Eqs. (2.4-39) and (2.4-40) is more difficult:

$$I'_{E1}(y, t) = I'_{E1}(0, t) = \lim_{\alpha \to 0} \frac{4\alpha}{\pi c} \left\{ \frac{1}{d} - \int_d^{\alpha/c} \left(\operatorname{ch}(\alpha^2 - \beta^2 c^2)^{1/2} t + \frac{\alpha \operatorname{sh}(\alpha^2 - \beta^2 c^2)^{1/2} t}{(\alpha^2 - \beta^2 c^2)^{1/2}} \right) \frac{d\beta}{\beta} - \int_0^d \left[\exp\left(\frac{-\beta^2 c^2 t}{2\alpha} \right) - 1 \right] \frac{d\beta}{\beta^2} \right\} \tag{19}$$

$$I_{E2}(y, t) = \lim_{\alpha \to 0} \frac{4\alpha}{\pi c} \int_{\alpha/c}^{\infty} \frac{\cos(\beta^2 c^2 - \alpha^2)^{1/2} t \cos \beta y}{\beta^2} d\beta - \frac{2}{\pi} \int_0^{\infty} \frac{\sin \beta c t \cos \beta y}{\beta} d\beta, \qquad t \geqq 2d/c_2 \tag{20}$$

For $y = 0$ one obtains

$$w(0, t) = 0 \tag{21}$$

$$I_{H2}(0, t) = \frac{2}{\pi} \int_0^{\infty} \frac{\sin \beta c t}{\beta} d\beta = 1 \tag{22}$$

$$I_{E2}(0, t) = \lim_{\alpha \to 0} \frac{4\alpha}{\pi c} \int_{\alpha/c}^{\infty} \frac{\cos(\beta^2 c^2 - \alpha^2)^{1/2} t}{\beta^2} d\beta - 1 \tag{23}$$

$$t \geqq 2d/c_2$$

For the boundary $y = 0$ we write Eqs. (13) and (14) as follows:

$$E_{1t}(0, t) = E_0 \tau_E(t) \tag{24}$$

$$H_{1t}(0, t) = E_0 Z_0^{-1} \tau_H(t) \tag{25}$$

Using Eqs. (17), (21), and (22) we obtain for $\tau_E(t)$ and $\tau_H(t)$ with the help of Eq. (1.3-7):

$$\tau_E(t) = g_{E1}(t) + \rho_E(t) + g_{H1}(t) - \rho_H(t) \tag{26}$$

$$\tau_H(t) = g_{H1}(t) - \rho_H(t) + \int_0^t \frac{dg_{E1}(t'')}{dt''} S(t - t'') \times [I'_{E1}(0, t - t'') - I_{E2}(0, t - t'')] dt'' + \int_0^t \frac{d\rho_E(t'')}{dt''} S(t - t'') [I'_{E1}(0, t - t'') - I_{E2}(0, t - t'')] dt'' \tag{27}$$

The magnitude of the tangential components of the electric field strengths in mediums 1 and 2 must be equal[1] at the boundary $y=0$. We obtain from Eqs. (7) and (24):

$$g_{E1}(t)+\rho_E(t)\tau_E(t), \qquad t \geqq 2d/c_2 \tag{28}$$

Similarly, continuity of the tangential component of the magnetic field strength requires according to Eqs. (8) and (25):

$$g_{H1}(t)-\rho_H(t)=\tau_H(t) \tag{29}$$

Insertion of Eq. (28) into Eq. (26) yields:

$$\rho_H(t)=g_{H1}(t), \qquad t \geqq 2d/c_2 \tag{30}$$

We further insert Eq. (29) into Eq. (27):

$$\begin{aligned}\int_0^t \frac{dg_{E1}(t'')}{dt''}\, S(t-t'')[I_{E2}(0,t-t'')-I'_{E1}(0,t-t'')]dt''\\ =\int_0^t \frac{d\rho_E(t'')}{dt''}\, S(t-t'')[I'_{E1}(0,t-t'')-I_{E2}(0,t-t'')]dt''\end{aligned} \tag{31}$$

Comparison of this equation with Eq. (4.3-22) shows that we may obtain $\rho_E(t)$ formally by deconvolution according to Eq. (4.3-26):

$$\begin{aligned}\rho_E(t)=\int_0^t\biggl(\int_0^t \frac{dg_{E1}(t'')}{dt''}\, S(t-t'')[I_{E2}(0,t-t'')\\ -I'_{E1}(0,t-t'')]dt''\biggr)\\ \overset{*}{*}\, S(t-t'')[I'_{E1}(0,t-t'')-I_{E2}(0,t-t'')]dt'', \qquad t \geqq 2d/c_2\end{aligned} \tag{32}$$

From Eqs. (29) and (30) follows

$$\tau_H(t)=0 \tag{33}$$

which is characteristic for a wave propagating from a medium with conductivity $\sigma>0$ to a medium with $\sigma=0$. According to Eqs. (11) and (12) the components H_{H1t} and E_{H1t} of the transmitted wave vanish, but the components E_{E1t} and H_{E1t} remain and assume the following form with the help of Eq. (28):

$$\begin{aligned}E_{1t}(y,t)=E_{E1t}(y,t)=E_0\int_0^t \frac{d\tau_E(t'')}{dt''}\, S(t-t'')\\ \times[1+w(y,t-t'')]dt''\end{aligned} \tag{34}$$

---

[1] Note that we are interested in the magnitude of the field strengths. The direction of $\mathbf{E}_{1i}+\mathbf{E}_{1r}$ as well as of $\mathbf{E}_{1t}$ is $-\mathbf{e}_z$ according to Eq. (7).

$$H_{1t}(y, t) = E_{E1t}(y, t) = E_0 \int_0^t \frac{d\tau_E(t'')}{dt''} S(t - t'') \times [I'_{E1}(y, t - t'') - I_{E2}(y, t - t'')]dt'', \quad t \geqq 2d/c_2, \quad y \leqq 0 \tag{35}$$

Since this wave propagates in air, the relation

$$E_{1t}(y, t) = Z_0 H_{1t}(y, t) \tag{36}$$

must hold. Hence, only one of Eqs. (34) or (35) must be evaluated; the other only provides a check on the accuracy of the calculation.

It is not difficult to extend the theory to the second reflection of the wave at the metal plate in Fig. 4.10-1, which arrives at the boundary $y = 0$ with a delay of $4d/c_2$. Furthermore, it is not difficult to assume that medium 1 in Fig. 4.10-1 is also a conducting layer, but with values $\sigma_1, \mu_1, \epsilon_1$ different from $\sigma_2, \mu_2, \epsilon_2$ holding in medium 2. However, it is also not difficult to see that the computational effort becomes severe. The analysis of the effect of layers or "stratified media" on radar pulses is clearly in a different class of mathematical and computational complexity than the analysis for periodic sinusoidal waves (Wait, 1970).

4.11 Exponential Ramp Wave Leaving Conducting Medium

We extend the theory of Section 4.10 to waves with the time variation of the exponential ramp function $E_1(1 - e^{-t/\tau})$ of Fig. 1.2-4b and Eq. (2.3-3). Equations (4.10-1)–(4.10-8) remain unchanged, since $g_{E1}(t)$ and $g_{H1}(t)$ may be caused by an exponential ramp function as well as by a step function. Instead of Eqs. (4.10-9) and (4.10-10) we obtain from Eqs. (2.11-10) and (2.11-12):

$$E_{E1t}(y, t) = \frac{E_0}{2\alpha} \int_0^t \frac{d^2}{dt''^2} [g_{E1}(t'') + \rho_E(t'')] S(t - t'') \times [1 - e^{-2\alpha(t - t'')} + u(y, t - t'')]dt'' \tag{1}$$

$$H_{E1t}(y, t) = \frac{E_0}{2\alpha Z_0} \int_0^t \frac{d^2}{dt''^2} [g_{E1}(t'') + \rho_E(t'')] S(t - t'') \times [-Z\sigma y + I'_{E3}(y, t - t'') - I_{E4}(y, t - t'')]dt'', \quad t \geqq 2d/c_2 \tag{2}$$

The magnitudes of the electric and magnetic field strength due to the magnetic excitation force of Eq. (4.10-8) are given by Eqs. (2.11-15) and (2.11-16):

$$H_{H1t}(y, t) = \frac{E_0}{2\alpha Z_0} \int_0^t \frac{d^2}{dt''^2} [g_{H1}(t'') - \rho_H(t'')] S(t - t'') \times [1 - e^{-2\alpha(t-t'')} + u(y, t - t'')] dt'' \tag{3}$$

$$E_{H1t}(y, t) = \frac{E_0}{2\alpha} \int_0^t \frac{d^2}{dt''^2} [g_{H1}(t'') - \rho_H(t'')] S(t - t'') \times [-Z\sigma y e^{-2\alpha(t-t'')} + I'_{H3}(y, t - t'') - I_{H4}(y, t - t'')] dt'' \tag{4}$$

The magnitudes E_{1t} and H_{1t} of the field strengths of the transmitted wave are the sums of Eqs. (1) and (4) as well as (2) and (3):

$$\begin{aligned} E_{1t}(y, t) = \frac{E_0}{2\alpha} \Bigg(\int_0^t \frac{d^2}{dt''^2} & [g_{E1}(t'') + \rho_E(t'')] S(t - t'') \\ & \times [1 - e^{-2\alpha(t-t'')} + u(y, t - t'')] dt'' \\ & + \int_0^t \frac{d^2}{dt''^2} [g_{H1}(t'') - \rho_H(t'')] S(t - t'') [-Z\sigma y e^{-2\alpha(t-t'')} \\ & + I'_{H3}(y, t - t'') - I_{H4}(y, t - t'')] dt'' \Bigg) \end{aligned} \tag{5}$$

$$\begin{aligned} H_{1t}(y, t) = \frac{E_0}{2\alpha Z_0} \int_0^t \frac{d^2}{dt''^2} & [g_{E1}(t'') + \rho_E(t'')] S(t - t'') \\ & \times [1 - e^{-2\alpha(t-t'')} + u(y, t - t'')] dt'' \\ & + \int_0^t \frac{d^2}{dt''^2} [g_{E1}(t'') + \rho_E(t'')] S(t - t'') \\ & \times [-Z\sigma y + I'_{E3}(y, t - t'') \\ & - I_{E4}(y, t - t'')] dt'', \qquad t \geqq 2d/c_2 \end{aligned} \tag{6}$$

The functions $u, I'_{H3}, I_{H4}, I'_{E3}$, and I_{E4} refer to air, and the conductivity $\sigma = 0$ must be used for the computation. From Eqs. (2.3-35), (2.10-19), (2.10-20), (2.9-18), and (2.9-19) we obtain:

$$\begin{aligned} (2\alpha)^{-1} u(x, y) &= -\frac{2}{\pi c} \int_0^\infty \frac{\sin \beta ct \sin \beta y}{\beta^2} d\beta \\ &= -y/c \qquad \text{for} \quad ct > y \geqq 0 \\ &= -t \qquad \text{for} \quad y \geqq ct > 0 \end{aligned} \tag{7}$$

$$(2\alpha)^{-1}I'_{H3}(y, t-t'') = (2\alpha)^{-1}I'_{H3}(0, t)$$

$$= \lim_{\alpha\to 0}\frac{2}{\pi c}\left[\frac{1}{d} - \int_d^{\alpha/c}[\mathrm{ch}(\alpha^2-\beta^2c^2)^{1/2}t - \frac{\alpha\,\mathrm{sh}(\alpha^2-\beta^2c^2)^{1/2}t}{(\alpha^2-\beta^2c^2)^{1/2}}]\frac{d\beta}{\beta^2} - 2\int_0^d \mathrm{sh}^2\left(\frac{\beta^2c^2t}{4\alpha}\right)\frac{d\beta}{\beta^2}\right] \tag{8}$$

$$\begin{aligned}(2\alpha)^{-1}I_{H4}(y, t) &= \frac{2}{\pi c}\int_0^\infty \frac{\cos\beta ct\cos\beta y}{\beta^2}\,d\beta\\ &= -t \qquad \text{for}\quad ct \geqq y \geqq 0\\ &= -y/c \qquad \text{for}\quad y > ct > 0\end{aligned} \tag{9}$$

$$(2\alpha)^{-1}I'_{E3}(y, t) = \frac{2}{\pi c}\left\{\frac{1}{d} - \int_d^{\alpha/c}\left(\mathrm{ch}(\alpha^2-\beta^2c^2)^{1/2}t + \frac{\alpha\,\mathrm{sh}(\alpha^2-\beta^2c^2)^{1/2}t}{(\alpha^2-\beta^2c^2)^{1/2}}\right)\frac{d\beta}{\beta^2} - \int_0^d\left[\exp\left(\frac{-\beta^2c^2t}{2\alpha}\right) - 1\right]\frac{d\beta}{\beta^2}\right\} \tag{10}$$

$$(2\alpha)^{-1}I_{E4}(y, t) = \frac{2}{\pi c}\int_0^\infty \frac{\cos\beta ct\cos\beta y}{\beta^2}\,d\beta = I_{H4}(y, t),\qquad t \geqq 2d/c_2 \tag{11}$$

For $y = 0$ one obtains:

$$(2\alpha)^{-1}u(0, t) = 0 \tag{12}$$

$$(2\alpha)^{-1}I_{H4}(0, t) = -t \tag{13}$$

$$(2\alpha)^{-1}I_{E4}(0, t) = -t \tag{14}$$

For the boundary $y = 0$ in Fig. 4.10-1 we write Eqs. (5) and (6) as follows:

$$E_{1t}(0, t)\,t = E_0\tau_E(t) \tag{15}$$

$$H_{1t}(0, t) = E_0Z_0^{-1}\tau_H(t) \tag{16}$$

Using Eqs. (8), (12), (13), and (14) we obtain for $\tau_E(t)$ and $\tau_H(t)$ with the help of Eq. (1.3-33) for $\tau = 1/2\alpha$:

$$\tau_E(t) = g_{E1}(t) + \rho_E(t) + \int_0^t \frac{d^2 g_{H1}(t'')}{dt''^2} S(t-t'') \times [(2\alpha)^{-1} I'_{H3}(0, t-t'') + (t+t'')]dt'' - \int_0^t \frac{d^2 \rho_H(t'')}{dt''^2} S(t-t'')[(2\alpha)^{-1} I'_{H3}(0, t-t'') + (t+t'')]dt'' \quad (17)$$

$$\tau_H(t) = g_{H1}(t) - \rho_H(t) + \int_0^t \frac{d^2 g_{E1}(t'')}{dt''^2} S(t-t'') \times [(2\alpha)^{-1} I'_{E3}(0, t-t'') + (t+t'')]dt'' + \int_0^t \frac{d^2 \rho_E(t'')}{dt''^2} S(t-t'')[(2\alpha)^{-1} I'_{E3}(0, t-t'') + (t+t'')]dt'' \quad (18)$$

The magnitude as well as the direction of the tangential components of the electric field strength in medium 1 and 2 must be equal at the boundary $y = 0$. We obtain from Eqs. (4.10-7) and (15):

$$g_{E1}(t) + \rho_E(t) = \tau_E(t), \qquad t \geqq 2d/c_2 \quad (19)$$

Similarly, continuity of the tangential component of the magnetic field strength requires according to Eqs. (4.10-8) and (16):

$$g_{H1}(t) - \rho_H(t) = \tau_H(t) \quad (20)$$

Insertion of Eq. (19) into Eq. (17) yields:

$$\int_0^t \frac{d^2 g_{H1}(t'')}{dt''^2} S(t-t'')[(2\alpha)^{-1} I'_{H3}(0, t-t'') + t + t'']dt'' = \int_0^t \frac{d^2 \rho_H(t'')}{dt''^2} S(t-t'')[(2\alpha)^{-1} I'_{H3}(0, t-t'') + t + t'']dt'' \quad (21)$$

Comparison of this equation with Eq. (4.6-19) shows that we may obtain $\rho_H(t)$ formally by deconvolution according to Eq. (4.6-21):

$$\rho_H(t) = \int_0^t \left[\int_0^{t\dagger} \left(\int_0^t \frac{d^2 g_{H1}(t'')}{dt''^2} S(t-t'') \times [(2\alpha)^{-1} I'_{H3}(0, t-t'') + t + t'']dt'' \right) \ast S(t-t'')[(2\alpha)^{-1} I'_{H3}(0, t-t'') + t + t'']dt'' \right] dt^{\dagger} \quad (22)$$

We further insert Eq. (20) into Eq. (18):

$$\int_0^t \frac{d^2 g_{E1}(t'')}{dt''^2} S(t-t'')[(2\alpha)^{-1} I'_{E3}(0, t-t'') + t + t'']dt''$$
$$= -\int_0^t \frac{d^2 \rho_E(t'')}{dt''^2} S(t-t'')[(2\alpha)^{-1} I'_{E3}(0, t-t'') + t + t'']dt'' \quad (23)$$

Formal deconvolution in analogy to Eq. (22) brings:

$$\rho_E(t) = \int_0^t \Bigg[\int_0^{t\dagger} \Bigg(\int_0^t \frac{d^2 g_{E1}(t'')}{dt''^2} S(t-t'') \times [(2\alpha)^{-1} I'_{E3}(0, t-t'') + t + t'']dt'' \Bigg) \circledast S(t-t'')[-(2\alpha)^{-1} I'_{E3}(0, t-t'') - (t+t'')]dt'' \Bigg] dt^{\dagger} \quad (24)$$

We have thus formally derived $\rho_E(t)$ and $\rho_H(t)$. From Eqs. (19) and (20) follow then $\tau_E(t)$ and $\tau_H(t)$. However, a good deal more work is needed before our formulas can be used for computer processing. For instance, one suspects that Eqs. (21) and (22) actually yield $\rho_H(t) = g_{H1}(t)$. This result would be obtained if Eq. (8) yields:

$$(2\alpha)^{-1} I'_{H3}(0, t) = (2\alpha)^{-1}(1 - e^{-2\alpha t}) - t \quad (25)$$

5 Propagation Velocity of Signals

5.1 Classical Phase and Group Velocity

The propagation velocity of electromagnetic signals has been a vexing problem for about a century. Initially the group velocity of sinusoidal waves with infinitesimal frequency difference was believed to represent the velocity of propagation of signals or of energy. A controversy ensued when the special theory of relativity demanded that neither signals nor energy could propagate faster than with the velocity of light in vacuum, since it was recognized that the group velocity could exceed the velocity of light. Papers on this topic published in the quarter century following the introduction of the special theory of relativity make fascinating reading today, since any serious attempt to connect the group velocity with the propagation velocity of signals is missing. In textbooks on radio transmission it is usual to assume a sinusoidal carrier cos ωt that is amplitude modulated by a sinusoidal voltage E cos $\Delta\omega t$ to yield the modulated carrier

$$e(t) = E(1 + M \cos \Delta\omega t)\cos \omega t \tag{1}$$

The envelope of $e(t)$ moves with the group velocity for $\Delta\omega \to d\omega$. At this point the following conclusion is drawn[1]: "The intelligence conveyed by the modulation moves with the velocity of the envelope, i.e., at the group velocity." This seems to be very much the basis—*the whole basis*—for the belief that the group velocity represents the propagation velocity of signals.

Let us note the fundamental theorem of communications, that a periodic sinusoidal wave or function transmits information at the rate zero, as we discussed in Section 1.1. The function $e(t)$ of Eq. (1) is the sum of three sinusoidal functions with the frequencies ω, $\omega - \Delta\omega$, and $\omega + \Delta\omega$. This function thus transmits information at the rate three times zero, which is still zero. Instead of saying that the transmission rate of information is zero, we may just as well say that the information is transmitted with the velocity zero. The connection of group velocity and propagation velocity of signals is

[1] Kraus (1984, p. 438) or Kraus and Carver (1973, 1981, p. 377), last sentence in third paragraph from end of page. We cite here from recent editions of a textbook by renowned authors used worldwide. Very much the same reasoning may be found in older books by no less renowned authors (Stratton, 1941, pp. 330–340; Brillouin, 1946, Ch. 5). Note the vague expression "intelligence conveyed" instead of "information transmitted."

thus based on the violation of one of the most basic concepts of information theory.

A serious discussion of the propagation velocity of signals was not possible as long as one only had solutions of Maxwell's equations that held for periodic sinusoidal time functions. An important advance was made by Sommerfeld (1914) and Brillouin (1914), who investigated[2] the propagation of a sinusoidal step function $S(t)\sin \omega t$ that started at the time $t = 0$. There are three major reasons why this work needs updating and why we obtain results that are qualitatively different:

(1) The concept of noise and signal-to-noise ratio did not exist in 1914.

(2) Functions defined by complicated integrals like $w(y, t)$ in Eq. (2.1-67) could not be plotted before the development of the electronic computer.

(3) The results of Sommerfeld and Brillouin applied to light, and this required assumptions about the atomistic structure of the medium in which the light propagated. In radio transmission, we are usually not concerned with the atomistic structure of the medium, and we should be able to obtain a signal propagation velocity within the framework of the continuum theory of Maxwell's equations, or rather the modified Maxwell equations. The investigation of Sommerfeld and Brillouin went beyond Maxwell's theory into what is now called quantum electrodynamics, but quantum electrodynamics did not yet exist in 1914.

The introduction of the signal-to-noise ratio implies that velocity is a statistical quantity and that we should talk about the *average velocity* and a *velocity distribution.* However, in addition to these anticipated results we will obtain an unexpected and rather surprising result. We give first a short historical review of the concept of propagation velocity before we investigate the statistical propagation velocity.

A sinusoidal wave may be written in the form

$$\sin \beta(x - \omega t/\beta) = \sin 2\pi\kappa(x - ft/\kappa) \tag{2}$$

where β is the phase constant—or circular wave number—κ the wave number, f the frequency, and ω the circular frequency. The phase velocity is defined as the factor of t:

$$v_p = \omega/\beta \tag{3}$$

The phase velocity gives the velocity with which a particular point of the sinusoidal wave—such as a zero crossing with positive slope—moves. Since the phase velocity is a feature of periodic sinusoidal waves, it cannot tell us anything about the velocity of propagation of signals or information.

---

[2] An English summary of these two papers may be found in Stratton (1941), pp. 330–340.

Consider next the sum of two sinusoidal waves with different frequencies and phase constants:

$$\sin \beta(x - \omega t/\beta) + \sin \beta_0(x - \omega_0 t/\beta_0)$$
$$= 2 \cos\{2^{-1}(\beta - \beta_0)[x - (\omega - \omega_0)t/(\beta - \beta_0)]\}$$
$$\times \sin\{2^{-1}(\beta + \beta_0)[x - (\omega + \omega_0)t/(\beta + \beta_0)]\} \tag{4}$$

If β and β_0 as well as ω and ω_0 have nearly the same value, this wave consists of a wave with high-frequency $\omega + \omega_0$ amplitude modulated by a wave with low-frequency $\omega - \omega_0$. The high-frequency wave has in the limit $\omega_0 = \omega + d\omega$ and $\beta_0 = \beta + d\beta$ the phase velocity

$$v_p = \lim_{\omega_0 \to \omega} (\omega + \omega_0)/(\beta + \beta_0) = \omega/\beta \tag{5}$$

while the low-frequency wave has the phase velocity

$$v_g = \lim_{\omega_0 \to \omega} (\omega - \omega_0)/(\beta - \beta_0) = d\omega/d\beta \tag{6}$$

which is called the group velocity.

One may connect phase and group velocity by writing Eq. (5) as $\beta = \omega/v_p$, differentiating

$$d\beta/d\omega = d(\omega/v_p)/d\omega = v_p^{-1}(1 - \omega v_p^{-1}\, dv_p/d\omega) \tag{7}$$

and inserting this expression into Eq. (6):

$$v_g = v_p(1 - \omega v_p^{-1}\, dv_p/d\omega)^{-1} \tag{8}$$

Since the sum of a finite number of sinusoidal waves must transmit information at the same rate zero as one sinusoidal wave, the group velocity cannot have anything to do with the velocity of propagation of signals or information. This statement can be extended to sums of denumerably many sinusoidal waves.

To show explicitly that the group velocity fails even in simple cases, we use the results of Section 2.2 for a sinusoidal wave in a medium with ohmic losses. The phase constant β of Eq. (2.2-6) is written in the form

$$\beta = 2^{-1}\sqrt{2}\sigma Z(\omega\epsilon/\sigma)[1 + (1 + \sigma^2/\omega^2\epsilon^2)^{1/2}]^{1/2} \tag{9}$$

where $Z = (\mu/\epsilon)^{1/2}$, and plotted in the normalized form $\beta/(\sigma Z/\sqrt{2})$ in Fig. 5.1-1 as function of the normalized frequency $\omega\epsilon/\sigma$. For large values of $\omega\epsilon/\sigma$ the function approaches the asymptote

$$\beta = Z\epsilon\omega = (\mu\epsilon)^{1/2}\omega = \omega/c \tag{10}$$

shown in Fig. 5.1-1 by the dashed line. The derivative $d\beta/d\omega = 1/c$ implies the group velocity according to Eq. (6). For any value of $\omega\epsilon/\sigma$ for which the

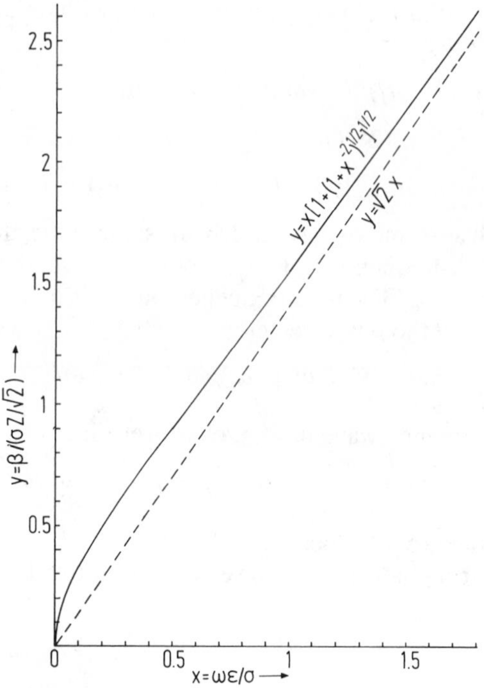

FIG. 5.1-1. Normalized phase constant $\sqrt{2}\,\beta/\sigma Z$ as a function of the normalized frequency $\omega\epsilon/\sigma$ for a conducting medium.

solid line in Fig. 5.1-1 has a larger derivative than the dashed line, the group velocity will be smaller than c, but wherever the derivative is smaller the group velocity will be larger than c. Since the solid line in Fig. 5.1-1 has a larger derivative than the dashed line for small values of $\omega\epsilon/\sigma$ and the same derivative for large values of $\omega\epsilon/\sigma$, there must be a region where the derivative dy/dx is smaller than $\sqrt{2}$ and the group velocity is thus larger than c.

Differentiation of Eq. (9) yields:

$$v_g = \frac{1}{d\beta/d\omega} = c\frac{2\sqrt{2}\{x[x+(x^2+1)^{1/2}](x^2+1)\}^{1/2}}{2[x+(x^2+1)^{1/2}](x^2+1)^{1/2}-1}, \qquad x = \omega\epsilon/\sigma \quad (11)$$

The normalized group velocity v_g/c is plotted in Fig. 5.1-2. The group velocity is larger than c when $\omega\epsilon/\sigma$ exceeds the approximate value 0.25, or for

$$f > 4.49 \times 10^9 \sigma/\epsilon_r, \qquad f \text{ in Hz}, \qquad \sigma \text{ in S/m}$$

where ϵ_r is the relative dielectric constant. From Table 4.2-1 we find for seawater $\sigma/\epsilon_r = 4/80$ and $f > 225$ MHz, for fresh water we get $f > 56$ kHz.

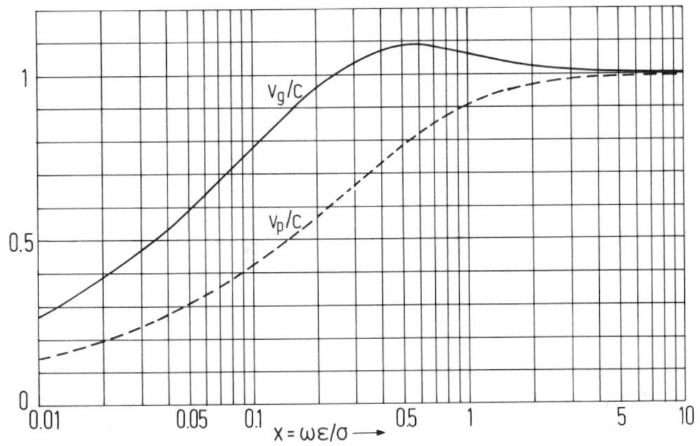

FIG. 5.1-2. Normalized phase velocity v_p/c and group velocity v_g/c as a function of the normalized frequency $\omega\epsilon/\sigma$ for a conducting medium.

Since the conductivity of the atmosphere is well below the conductivity of fresh water,[3] we deduce that the group velocity of an electromagnetic wave in air is larger than c at all frequencies of practical interest (the group velocity may be smaller at the molecular resonances of water vapor, oxygen, etc., but we avoid these frequency ranges due to their high absorption losses).

Figure 5.1-2 shows also the normalized phase velocity according to Eqs. (5) and (9):

$$v_p = \frac{\omega}{\beta} = c\left(\frac{2}{1 + (1 + 1/x^2)^{1/2}}\right)^{1/2}, \qquad x = \omega\epsilon/\sigma \tag{12}$$

The phase velocity of an electromagnetic wave in a conducting medium is thus always smaller than c, and it approaches c for $\sigma \to 0$.

5.2 Propagation Velocity of a Step Wave

In Section 2.1 we investigated the electromagnetic wave excited by an electric step function $E_0S(t)$ of Eq. (2.1-28). The magnitude of the electric field strength $\mathbf{E}_E(\xi, \theta)$ of such a step wave is given by Eq. (2.1-80) with the normalized variables

$$\theta = \alpha t = (\sigma/2\epsilon)t, \qquad \xi = \alpha y/c = (Z\sigma/2)y, \qquad Z = (\mu/\epsilon)^{1/2} \tag{1}$$

---

[3] The conductivity of the ionized layers in the upper atmosphere is of the order of 10^{-8} S/m. With $\epsilon_r = 1$ one obtains $f > 44.9$ Hz. The conductivity of the lower atmosphere is substantially less in dry weather.

Figure 5.2-1 shows plots of $E_E(\xi, \theta)/E_0$ for $\xi = 0, 1, 2, 3$ and θ in the range $0 \leqq \theta \leqq 26$. The unit step of $E_E(0, \theta)/E_0$ is reduced for $\xi = 1, 2, 3$, but the step remains for any finite value of ξ. From this step the function $E_E(\xi, \theta)/E_0$ rises very slowly to the value 1 reached for $\theta \to \infty$. The step occurs at $\theta = \xi$ for any value of ξ. Using Eq. (1) we obtain for the propagation velocity of the step:

$$\xi/\theta = y/ct = 1, \qquad y/t = c \tag{2}$$

Hence, the step always propagates with the velocity $c = (\mu\epsilon)^{-1/2}$ of light in the particular medium.

In order to observe the arrival time of a step wave at the location ξ or y we may use a threshold detector set to the threshold voltage V_{th} shown by the dashed-dotted line in Fig. 5.2-1. The step waves $E_E(1, \theta)$ and $E_E(2, \theta)$ cross the threshold voltage at the distances $\xi = 1$ and $\xi = 2$, or $y = c/\alpha$ and $y = 2c/\alpha$, at the times $\theta = 1$ and $\theta = 2$, or $t = 1/\alpha$ and $t = 2/\alpha$, which yields the propagation velocity $y/t = c$. However, the step wave $E_E(3, \theta)$ crosses the threshold voltage at the distance $\xi = 3$ or $y = 3c/\alpha$ at the time $\theta = 4$ or $t = 4/\alpha$, and we get the *observed propagation velocity* $y/t = 3c/4$.

It is evident from Fig. 5.2-1 that the observed propagation velocity of a step

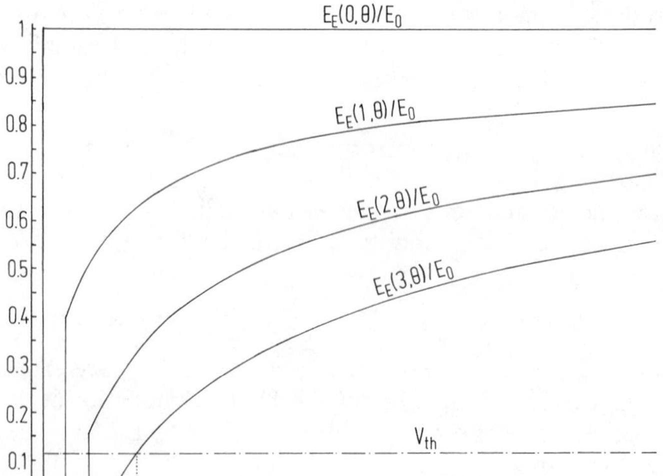

FIG. 5.2-1. The function $E_E(\xi, \theta)/E_0$ according to Eq. (2.1-80) for $\xi = 0, 1, 2, 3$ and θ in the range of $0 \leqq \theta \leqq 26$. The non-normalized variables are $t = \theta/\alpha = (2\epsilon/\sigma)\theta$ and $y = (c/\alpha)\xi = (2/\sigma)(\epsilon/\mu)^{1/2}\xi$. This illustration is based on computer plots by R. Boules, University of Alexandria, Egypt.

wave will be equal to c for a distance $\xi \leqq \xi_0(V_{th})$, but for $\xi > \xi_0(V_{th})$ it will decrease monotonously and eventually approach zero.

Assuming a technically perfect threshold detector and the absence of noise, one could choose the threshold V_{th} in Fig. 5.2-1 so low that one would observe the propagation velocity c for any finite distance ξ or y. In the presence of noise, one cannot set V_{th} arbitrarily low without increasing arbitrarily the false alarm probability.[1] The value of the threshold V_{th} thus depends on the false alarm probability one is willing to accept and on the signal-to-noise ratio.[2]

Some qualitative[3] results on the influence of noise may be seen from Fig. 5.2-1. The influence of noise on the propagation velocity will be small if the threshold V_{th} intersects the discontinuous section of the step wave $E_E(\xi, \theta)$. But the influence will be greater if V_{th} intersects $E_E(\xi, \theta)$ in a section where the derivative $dE_E(\xi, \theta)/d\theta$ is not infinite; the smaller $dE_E(\xi, \theta)/d\theta$, the more will the noise affect the observed propagation velocity.

Astrophysicists have speculated in the past about whether signals propagate with fixed velocity over any distance in the universe, since a variety of attractive theories becomes possible if the velocity of signals decreases with increasing distance and eventually approaches zero. In this case a finite universe would appear infinite to an observer surveying it by means of signals. Such theories have not found much acceptance since all our observations have shown that light inherently propagates with constant velocity; only secondary effects like propagation through matter or strong gravitational fields slow it down. We see from Fig. 5.2-1 that the observed propagation velocity of a signal may be substantially less than c and actually approach zero, even though the velocity of light has the constant value c.

The observed propagation velocity also depends on the amplitude of the excitation field strength. Figure 5.2-2 shows the two electric excitation step functions $2E_0(0, \theta)$ and $E_0(0, \theta)$ at the location $\xi = 0$. These two step functions produce the step waves $2E_E(2, \theta)$ and $E_E(2, \theta)$ at the location $\xi = 2$. If the arrival of the step waves is observed with a threshold detector with

---

[1] In radar terminology *false alarm* means that an observed signal is created by noise, while *error* means that an actual signal is made unobserved by noise.

[2] We tacitly assume that a meaningful signal-to-noise ratio can be determined. We do not yet discuss this point here, since signals have a beginning and end like a pulse and unlike a step function, which only has a beginning.

[3] For quantitative results we refer to Boules (1987). One does not need to use threshold detection in the form of Fig. 5.2-1, but one can use a sliding correlator and use threshold detection of the produced cross-correlation function. This creates a noise suppression effect and makes it easier to define a signal-to-noise ratio. Since the time variation of a step wave $E_E(\xi, \theta)$ depends on the distance ξ, the question arises which sliding correlator will give the "best" result, and it is not evident which result one should call "best," etc.

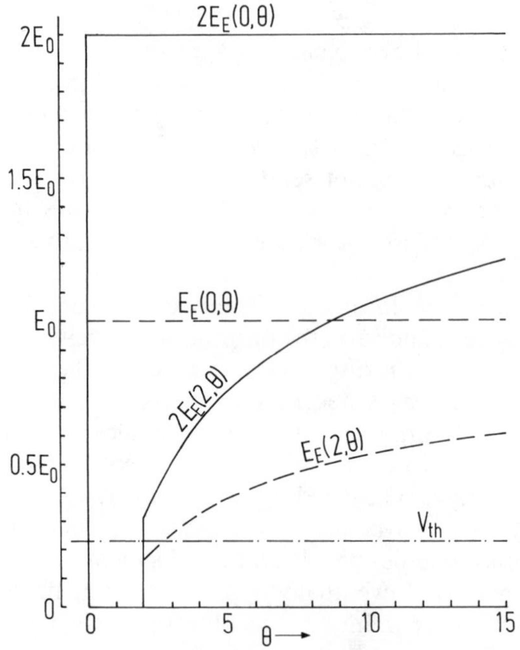

FIG. 5.2-2. Electric excitation step functions $2E_E(0, \theta)$ and $E_E(0, \theta)$ with the field strengths $2E_E(2, \theta)$ and $E_E(2, \theta)$ produced at the location $\xi = 2$. The step wave produced by $E_E(0, \theta)$ is observed to propagate with lesser velocity than the step wave produced by $2E_E(0, \theta)$, if a threshold detector is used for reception.

threshold voltage V_{th}, one observes the propagation velocity c for the step wave $2E_E(2, \theta)$ but a lesser velocity for the step wave $E_E(2, \theta)$.

One may object to Fig. 5.2-2 in that the step wave $E_E(2, \theta)$ can be amplified and thus made equal to $2E_E(2, \theta)$. This is correct in communications if noise is not a problem. In physics, one must usually rely on the amplitude, power, or energy of the received signal to trigger an effect, since amplifiers are manmade.

5.3 PROPAGATION OF SIGNALS

A step wave or step function has a beginning but no end. One must be careful in using it for the representation of a signal, since signals have both a beginning and an end. The simplest signal is a rectangular pulse that is obtained as the difference $S(t) - S(t - \Delta T)$ of two step functions. Figure 5.3-1 shows as an example the rectangular pulse $[E_E(0, \theta) - E_E(0, \theta - 10)]/E_0$ with a duration $\Delta\theta = 10$ or $\Delta T = 10/\alpha = 20\epsilon/\sigma$. The time variation of

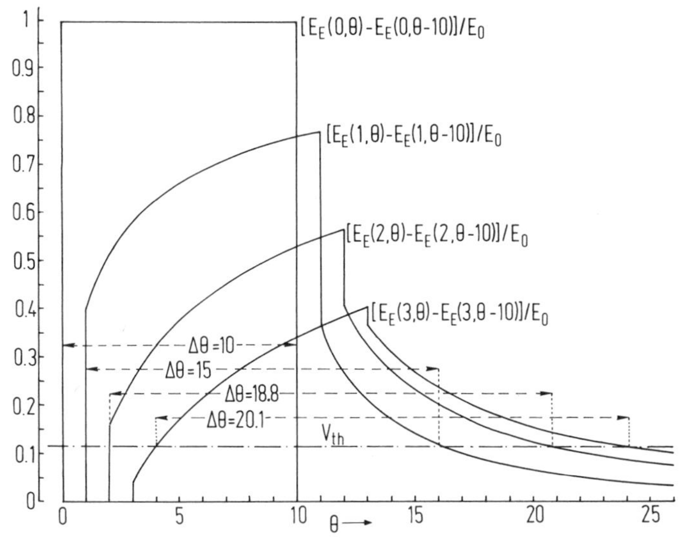

FIG. 5.3-1. Rectangular electric excitation pulse of duration $\Delta\theta = 10$ at the location $\xi = 0$ and the electric field strengths at the locations $\xi = 1, 2, 3$ as a function of time θ. The non-normalized variables are $t = (2\epsilon/\sigma)\theta$, $y = (2/\sigma)(\epsilon/\mu)^{\frac{1}{2}}\xi$, and $\Delta T = (2\epsilon/\sigma)\Delta\theta$.

the pulse at the location ξ is given by the difference $[E_E(\xi, \theta) - E_E(\xi, \theta - \Delta\theta)]/E_0$. The pulses of Fig. 5.3-1 follow thus readily from the step functions of Fig. 5.2-1.

Except for $\xi = 0$, all the pulses in Fig. 5.3-1 have infinitely long trailing edges. Hence, a signal traveling through a medium with ohmic losses not only propagates with slower and slower observed velocity as the distance increases, but it also gets longer and longer. We may use the intersection of the threshold voltage V_{th} with the rising and falling edge of the pulses as a measure of their widening and obtain for $\xi = 0, 1, 2, 3$ the durations $\Delta\theta = 10, 15, 18.8, 20.1$. However, the matter is more complicated than these numbers suggest.

We may see this from Fig. 5.3-2, which is a repetition of Fig. 5.3-1 except that the duration of the pulse at $\xi = 0$ is reduced from $\Delta\theta = 10$ to $\Delta\theta = 1.5$. The durations of the pulses at $\xi = 1, 2, 3$ is now $\Delta\theta = 2, 2, 0.5$. We recognize that the pulse duration increases with the distance ξ to a maximum, but then decreases again and eventually becomes zero. In Fig. 5.3-2 the duration $\Delta\theta = 0$ is reached—for the fixed threshold voltage V_{th}—at a location ξ between 3 and 4, while in Fig. 5.3-1 the respective value of ξ is substantially larger.

Let the pulses of Figs. 5.3-1 and 5.3-2 be received with a sliding correlator

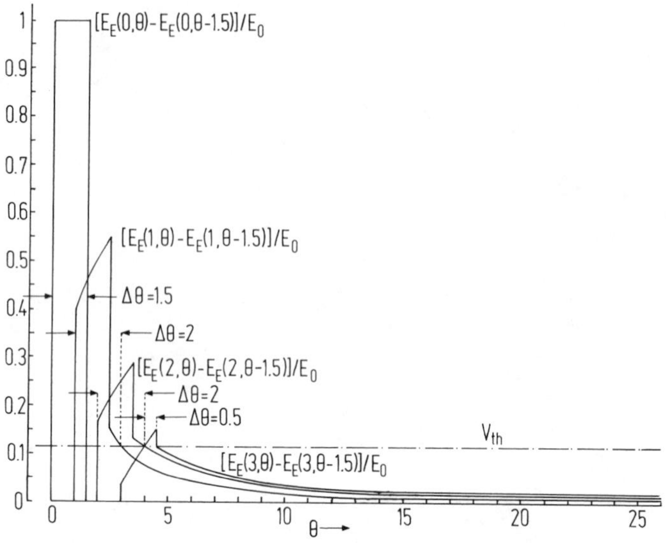

FIG. 5.3-2. Rectangular electric excitation pulse of duration $\Delta\theta = 1.5$ at the location $\xi = 0$ and the electric field strengths produced at the locations $\xi = 1, 2, 3$ as a function of time θ.

in order to suppress superimposed thermal noise. It is evident that the number of stages and the total delay of the sliding correlator must increase with ξ if one wants to make use of "most" of the energy in the signal. This increase of the delay implies a reduction of the observed propagation velocity of the received pulse.

For very large values of θ the amplitude of the trailing edges of the pulses in Figs. 5.3-1 and 5.3-2 becomes very small. The energy in a certain time interval where the amplitude is very small may be lower than the average noise energy in that interval. This effect limits the useful length of a sliding correlator used for reception. For a given signal energy at the location $\xi = 0$ and a certain noise background at the location ξ one thus arrives at an optimal delay time for the sliding correlator and a propagation velocity determined by this delay. For a more detailed discussion the reader is once more referred to the Ph.D. thesis of Boules (1987).

5.4 Propagation Velocity, Relativity, and Causality

Maxwell's equations predate the special theory of relativity by about 20 years, but they are invariant under a Lorentz transformation and thus satisfy the requirements of the special theory of relativity. The same holds true for the modified Maxwell equations of Eqs. (1.8-4) to (1.8-6) since they differ from Maxwell's equations by a duality transformation only. Hence, we are justified in applying our results about the propagation velocity of electromagnetic signals to the special theory of relativity.

Figure 5.4-1 shows a standard illustration of two points x_0, $-x_0$ in the same inertial system with an event at the same time $t = 0$ at both places. These events may cause other events at points in space–time that are located within the light cones defined by the dashed-dotted lines originating in points x_0 and $-x_0$. A signal would have to propagate with a velocity larger than c to reach a point outside the respective light cone. All points $x > 0$ will be reached first by the signal from x_0, all points $x < 0$ by the signal from $-x_0$.

Let us modify this illustration according to the results of Sections 5.2 and 5.3, assuming the electric field strength of a signal must exceed a certain threshold before it can be observed or cause an effect. For a certain distance the observed propagation velocity of the signal will be c; beyond this distance the velocity will be less. The points up to which the velocity is c are denoted P_c in Fig. 5.4-1. Beyond P_c the straight lines of the light cone boundary become curved, as shown by the dashed lines.

Let the electric field strength emitted from point $-x_0$ have half the amplitude of before, while the field strength emitted from x_0 remains unchanged. The points P_c on the light cone boundary of $-x_0$ are shifted to P_c', and the dashed lines of the curved boundary are replaced by the dotted lines. Previously the line $x = 0$ divided space–time into two regions reached first by simultaneous events in x_0 and $-x_0$, but this line is now shifted to $-x_c'$.

Let an event A be caused in a point x if a signal originating in x_0 is detected first, and an event B if a signal originating in $-x_0$ is detected first. If the amplitude of the electric field strength had no influence on the time of

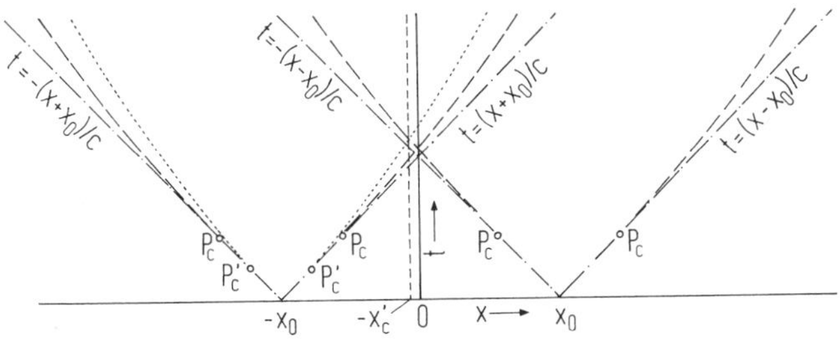

FIG. 5.4-1. Space–time diagram of two points x_0 and $-x_0$ from which signals are emitted. The dashed–dotted lines show the two light cones, the dashed lines, the modified light cones for equal signal amplitude, and the dotted lines a modified light cone due to reduced signal amplitude.

detection and thus the observed signal velocity, the event A would be caused in all points $x > 0$ and the event B in all points $x < 0$. However, in the example of Fig. 5.4-1 the dividing line for events A and B is no longer $x = 0$ but $x = x'_c$. Generally speaking, events would not only depend on the origin x_0, $t = 0$ in space–time of a signal but also on the amplitude (of its electric field strength). If the amplitudes of signals originating at x_0 and $-x_0$ vary in an unknown way, one would sometimes observe effect A and sometimes effect B close to $x = 0$, while at a greater distance from $x = 0$ one would always observe either effect A or effect B. Hence, it would appear that the causality law applies to certain points x in space but not to others, while the difference is in reality due to varying observed propagation velocities of signals.

Let us consider signals with superimposed noise. First we have to answer the question whether such noisy signals are relevant to the special theory of relativity. Since relativity relies heavily on the concept of "light flashes" as a means of signal transmission, we must accept electromagnetic waves not only as a possible but as the most distinguished implementation of signals.[1] Electromagnetic waves commonly come with superimposed noise, but this may not be so at the level of photons. Since photons are outside the scope of Maxwell's equations—modified or not—our results do not apply to them. Within Maxwell's theory, there does not seem to be an example of electromagnetic waves without noise.

Superimposed noise has in Fig. 5.4-1 very much the same effect as varying amplitudes of the signals originating in points x_0 and $-x_0$. The boundaries of the light cones, represented by either the dashed-dotted or the dashed lines, will hold on the average only. There may now be fluctuations to either side of these lines. In particular, the dividing line $x = 0$ for points reached first either by signals from x_0 or $-x_0$ will fluctuate around $x = 0$, and events A or B caused close to $x = 0$ will appear to be stochastic even though they would be deterministic in the absence of noise. Note that this is strictly due to the effect of noise on the observed propagation velocity of the signals.

---

[1] Other possible implementations of signals are neutrinos and gravitational waves. The generation and reception of neutrinos has been proven experimentally, but their use for signal transmission is still in the highly theoretical stage (Sáenz *et al.*, 1977, 1978; Überall *et al.*, 1979; Albers and Kotzner, 1978; Kelly *et al.*, 1979). The existence of gravitational waves has not yet been proven experimentally. In either case one would need transient solutions of quantum field theory and the general theory of relativity on the level derived here for the modified Maxwell equations, before one could discuss signal propagation intelligently. It may well turn out that such transient solutions would require modifications of quantum field theory and the general theory of relativity just as Maxwell's theory required it.

5.5 Finite-Periodic Pulses

Consider the pulse shown on top of Fig. 5.5-1, consisting of three periods of a square wave. We may represent it by a superposition of seven time-shifted step functions $S(\theta)$:

$$\begin{aligned} E_{E,3s}(0, \theta) = E_0[&S(\theta) - 2S(\theta - 6) + 2S(\theta - 12) \\ &-2S(\theta - 18) + 2S(\theta - 24) - 2S(\theta - 30) \\ &+ S(\theta - 36)] \end{aligned} \tag{1}$$

Let an electric excitation force with this time variation be applied to the plane $\xi = 0$ of a lossy medium. According to Eq. (2.1-80) and the results of Section 1.3 we get for the time variation of the magnitude of the electric field strength at the location ξ:

$$\begin{aligned} E_{E,3s}(\xi, \theta) = &E_E(\xi\ \theta) - 2E_E(\xi, \theta - 6) + 2E_E(\xi, \theta - 12) \\ &- 2E_E(\xi, \theta - 18) + 2E_E(\xi, \theta - 24) - 2E_E(\xi, \theta - 30) \\ &+ E_E(\xi, \theta - 36) \end{aligned} \tag{2}$$

We want to plot this function for $\xi = 4$. This requires first a plot of $E_E(\xi, \theta)$ for $\xi = 4$ for a somewhat larger range of θ then used in Fig. 5.2-1. Such a plot is shown in Fig. 5.5-2. Note that the jump at $\theta = 4$ is only about 2% of the

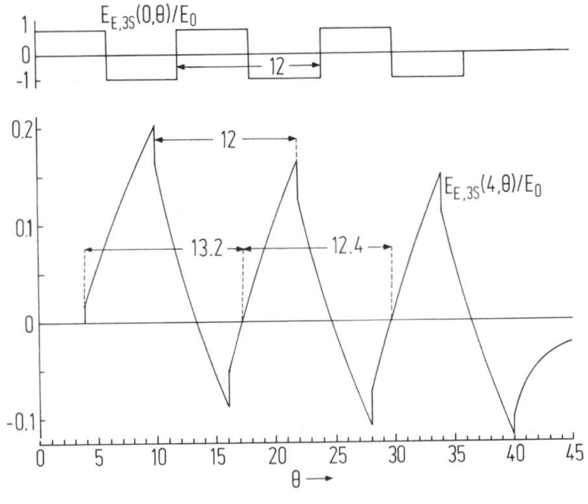

Fig. 5.5-1. A finite periodic pulse $E_{E,3s}(0, \theta)$ at the location $\xi = 0$ with three periods of a square wave, and the distorted pulse $E_{E,3s}(4, \theta)$ at the location $\xi = 4$. This illustration is based on computer plots by R. Boules, University of Alexandria, Egypt.

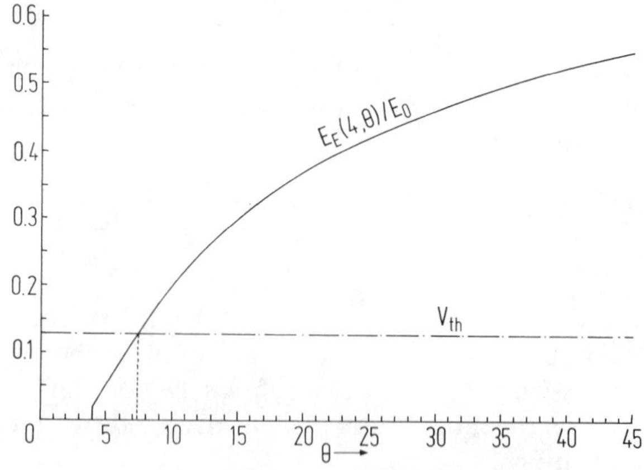

FIG. 5.5-2. The function $E_E(\xi, \theta)$ of Eq. (2.1-80) for $\xi = 4$ in the range $0 \leqq \theta \leqq 45$. This illustration is based on computer plots by R. Boules, University of Alexandria, Egypt.

maximum value of this function for $\theta \to \infty$. With this function one may readily plot the function $E_{E,3s}(4, \theta)$ in Fig. 5.5-1.

We are here interested in the change of the period of the original square wave. If we measure the duration of the period of $E_{E,3s}(4, \theta)$ between the discontinuities or jumps as shown, we get the same period $\Delta\theta = 12$ as for the square wave $E_{E,3s}(0, \theta)$ in Fig. 5.5-1. However, if we measure the period between zero crossings, we get the values $\Delta\theta = 13.2$ and 12.4 instead of 12. Hence, the periods have been stretched.

Let the three periods of the square wave $E_{E,3s}(0, \theta)$ be replaced by three periods of a sinusoidal wave. The discontinuities in Fig. 5.5-1 would then vanish and the distances between zero crossings would be distinguished for the definition of a period. Let the three periods in Fig. 5.5-1 be replaced by n periods. For large values of n the observed period of the zero crossings would approach the value $\Delta\theta = 12$. However, the average duration of a finite number n of periods would be larger than $\Delta\theta = 12$. We have here a *red-shift effect* for pulses consisting of a finite number of periods of a sinusoidal function. For periodic sinusoidal functions, which have an infinite number of periods, we can get such a red-shift by means of the Doppler[1] effect. Pulses with a finite number of periods—or finite-periodic pulses—may have the Doppler effect as well as the red-shift effect of Fig. 5.5-1. Only further study of sinusoidal pulses using the approximation by exponential ramp functions according to Figs. 1.3-9 and 1.3-10 can show whether this second red-shift effect is of interest in astrophysics or other fields.

[1] Christian Doppler, physicist, 1803–1853; born in Salzburg, Austria.

Figure 5.5-2 shows the same threshold voltage V_{th} as Fig. 5.2-1. The observed velocity of propagation at the crossing of this threshold voltage is now about $\xi/\theta = 4/7.2$ or $y/t = 4c/7.2$, while we obtained $y/t = 3c/4$ for $\xi = 3$ in Fig. 5.2-1. Hence, the observed velocity of propagation drops rapidly once the line V_{th} in Fig. 5.2-1 no longer intersects the jump of the function $E_E(\xi, \theta)$.

5.6 Magnetic Charges or Monopoles

We have shown that the range of application of Maxwell's equations can be extended to waves with non-negligible relative frequency bandwidth propagating in a medium with non-negligible losses by the addition of a magnetic current density. The transient solutions obtained from the modified Maxwell equations have shown a number of physical effects not known from the steady-state solutions of Maxwell's equations. Hence, our theory is more than a purely mathematical refinement of Maxwell's equations without any additional physical content.

Our theory does *not* prove the existence of magnetic charges or monopoles. Such a proof could come from experimental observation only. However, the theory strongly suggests the existence of magnetic currents. A similar situation existed half a century ago, when Dirac's[1] equations suggested the existence of a positron. If no magnetic currents should ever be found, it would be worth investigating why one has to add a magnetic current density to Maxwell's equations in order to get convergent transient solutions in lossy media. It may well be that magnetic currents exist only in lossy media during transients, which would explain why experiments failed that were designed to detect magnetic charges in the steady state in nonconducting media.

We make here a distinction between magnetic currents and magnetic charges, since our calculation suggests only the existence of magnetic currents but not of charges. A rest charge zero could yield a current, just as a rest mass zero can yield momentum and kinetic energy.

Whatever may come of magnetic currents, we cannot continue to use Maxwell's equations as the basis for the theory of electromagnetic waves since they permit no solution for the only type of wave known to exist: the wave with beginning and end propagating in a lossy medium. It is possible that another modification of Maxwell's equations can be found that will yield physically acceptable solutions without requiring magnetic currents.

[1] Paul Adrien Maurice Dirac, physicist, 1902–1984; born in Bristol, England.

This possibility is very definitely worth investigating. Here again we have a historical precedent, since Dirac's equations were not the only physically acceptable modification of the nonrelativistic Schrödinger equation. If neither such a modification of Maxwell's equations nor any magnetic currents can ever be found, we would have little choice but to conclude that the magnetic currents used in this book are a sign of a more fundamental problem that we cannot yet see.

6 Appendix

6.1 Spherical Wave in a Lossy Medium

All our calculations assumed planar wavefronts, which is usual in the study of reflection and transmission of waves. However, we concluded that Maxwell's equations have to be modified, and such a conclusion calls for more than usual care. Planar waves are an abstraction, and it could be that Maxwell's equations failed because of this abstraction. We will investigate the propagation of spherical waves in a medium with ohmic losses, since spherical waves are closer to reality.

We start from the modified Maxwell equations, Eqs. (2.1-2)–(2.1-4), and write them in spherical coordinates:

$$(r\sin\vartheta)^{-1}[H_\varphi\cos\vartheta + (\partial H_\varphi/\partial\vartheta)\sin\vartheta - \partial H_\vartheta/\partial\varphi] = \epsilon\,\partial E_r/\partial t + \sigma E_r \quad (1)$$

$$r^{-1}[(\partial H_r/\partial\varphi)\sin^{-1}\vartheta - H_\varphi - r\,\partial H_\varphi/\partial r] = \epsilon\,\partial E_\vartheta/\partial t + \sigma E_\vartheta \quad (2)$$

$$r^{-1}[H_\vartheta + r\,\partial H_\vartheta/\partial r - \partial H_r/\partial\vartheta] = \epsilon\,\partial E_\varphi/\partial t + \sigma E_\varphi \quad (3)$$

$$-(r\sin\vartheta)^{-1}[E_\varphi\cos\vartheta + (\partial E_\varphi/\partial\vartheta)\sin\vartheta - \partial E_\vartheta/\partial\varphi] = \mu\,\partial H_r/\partial t + sH_r \quad (4)$$

$$-r^{-1}[(\partial E_r/\partial\varphi)\sin^{-1}\vartheta - E_\varphi - r\,\partial E_\varphi/\partial r] = \mu\,\partial H_\vartheta/\partial t + sH_\vartheta \quad (5)$$

$$-r^{-1}(E_\vartheta + r\,\partial E_\vartheta/\partial r - \partial E_r/\partial\vartheta) = \mu\,\partial H_\varphi/\partial t + sH_\varphi \quad (6)$$

$$2r^{-1}E_r + \partial E_r/\partial r + (r\sin\vartheta)^{-1}[E_\vartheta\cos\vartheta + (\partial E_\vartheta/\partial\vartheta)\sin\vartheta + \partial E_\varphi/\partial\varphi] = 0 \quad (7)$$

$$2r^{-1}H_r + \partial H_r/\partial r + (r\sin\vartheta)^{-1}[H_\vartheta\cos\vartheta + (\partial H_\vartheta/\partial\vartheta)\sin\vartheta + \partial H_\vartheta/\partial\varphi] = 0 \quad (8)$$

Consider a TEM wave

$$E_r = H_r = 0 \quad (9)$$

and assume that the wave has circular symmetry for the angle φ:

$$\partial E_\varphi/\partial\varphi = \partial E_\vartheta/\partial\varphi = \partial H_\varphi/\partial\varphi = \partial H_\vartheta/\partial\varphi = 0 \quad (10)$$

We obtain from Eqs. (1)–(8):

$$H_\varphi\cos\vartheta + (\partial H_\varphi/\partial\vartheta)\sin\vartheta = 0 \quad (11)$$

$$-r^{-1}H_\varphi - \partial H_\varphi/\partial r = \epsilon\,\partial E_\vartheta/\partial t + \sigma E_\vartheta \quad (12)$$

$$r^{-1}H_\vartheta + \partial H_\vartheta/\partial r = \epsilon\, \partial E_\varphi/\partial t + \sigma E_\varphi \tag{13}$$

$$E_\varphi \cos\vartheta + (\partial E_\varphi/\partial\vartheta)\sin\vartheta = 0 \tag{14}$$

$$r^{-1}E_\varphi + \partial E_\varphi/\partial r = \mu\, \partial H_\vartheta/\partial t + sH_\vartheta \tag{15}$$

$$-r^{-1}E_\vartheta - \partial E_\vartheta/\partial r = \mu\, \partial H_\varphi/\partial t + sH_\varphi \tag{16}$$

$$E_\vartheta \cos\vartheta + (\partial E_\vartheta/\partial\vartheta)\sin\vartheta = 0 \tag{17}$$

$$H_\vartheta \cos\vartheta + (\partial H_\vartheta/\partial\vartheta)\sin\vartheta = 0 \tag{18}$$

According to Eq. (10), the functions H_φ, H_ϑ, E_φ, and E_ϑ depend on r and ϑ only. One may readily verify that the field strengths

$$H_\varphi(r, \vartheta) = H_\varphi(r)\sin^{-1}\vartheta \tag{19}$$

$$E_\varphi(r, \vartheta) = E_\varphi(r)\sin^{-1}\vartheta \tag{20}$$

$$E_\vartheta(r, \vartheta) = E_\vartheta(r)\sin^{-1}\vartheta \tag{21}$$

$$H_\vartheta(r, \vartheta) = H_\vartheta(r)\sin^{-1}\vartheta \tag{22}$$

satisfy Eqs. (11), (14), (17), and (18). With the substitutions

$$E' = E_\vartheta = E_\varphi, \qquad H' = -H_\vartheta = H_\varphi \tag{23}$$

one may rewrite the two pairs of Eqs. (12) and (16), as well as (13) and (15), as one pair:

$$-(r^{-1}H' + \partial H'/\partial r) = \epsilon\, \partial E'/\partial t + \sigma E' \tag{24}$$

$$-(r^{-1}E' + \partial E'/\partial r) = \mu\, \partial H'/\partial t + sH' \tag{25}$$

With the relation

$$r^{-1}\, \partial(rH')/\partial r = r^{-1}H' + \partial H'/\partial r \tag{26}$$

and the substitutions

$$rE' = E, \qquad rH' = H \tag{27}$$

we may rewrite Eqs. (25) and (24) as follows:

$$\partial E/\partial r + \mu\, \partial H/\partial t + sH = 0 \tag{28}$$

$$\partial H/\partial r + \epsilon\, \partial E/\partial t + \sigma E = 0 \tag{29}$$

These are the same equations as Eqs. (2.1-13) and (2.1-14), with y replaced by r. To obtain E and H in spherical coordinates we must thus use the modified Maxwell equations just as in the case of planar waves.

6.2 Notation and Formulas of Vector Analysis[1]

(1) Representation of a vector **a** by a set of orthogonal unit vectors, **i**, **j**, **k**:

$$\mathbf{a} = a_i\mathbf{i} + a_j\mathbf{j} + a_k\mathbf{k}$$

(2) Scalar product

$$\mathbf{a} \cdot \mathbf{b} = a_i b_i + a_j b_j + a_k b_k$$

(3) Vector product

$$\mathbf{a} \times \mathbf{b} = \begin{vmatrix} \mathbf{i} & \mathbf{j} & \mathbf{k} \\ a_i & a_j & a_k \\ b_i & b_j & b_k \end{vmatrix} = (a_j b_k - a_k b_j)\mathbf{i} + (a_k b_i - a_i b_k)\mathbf{j} + (a_i b_j - a_j b_i)\mathbf{k}$$

(4) $\mathbf{a} \times (\mathbf{b} \times \mathbf{c}) = \mathbf{b}(\mathbf{a} \cdot \mathbf{c}) - \mathbf{c}(\mathbf{a} \cdot \mathbf{b})$

(5) Vector operators

$$\text{grad}\,\phi = \nabla\phi, \qquad \text{div}\,\mathbf{f} = \nabla \cdot \text{f}, \qquad \text{curl}\,\mathbf{f} = \text{rot}\,\mathbf{f} = \nabla \times \mathbf{f}$$

The following relations hold in general coordinate systems.

(6) $\text{grad}(\phi + \psi) = \text{grad}\,\phi + \text{grad}\,\psi$

(7) $\text{grad}(\phi\psi) = \phi\,\text{grad}\,\psi + \psi\,\text{grad}\,\phi$

(8) $\text{grad}(\mathbf{f} \cdot \mathbf{g}) = (\mathbf{f} \cdot \text{grad})\mathbf{g} + (\mathbf{g} \cdot \text{grad})\mathbf{f} + \mathbf{f} \times \text{curl}\,\mathbf{g} + \mathbf{g} \times \text{curl}\,\mathbf{f}$

(9) $\text{div}(\phi\mathbf{f}) = \phi\,\text{div}\,\mathbf{f} + \mathbf{f} \cdot \text{grad}\,\phi$

(10) $\text{div}(\mathbf{f} \times \mathbf{g}) = \mathbf{g} \cdot \text{curl}\,\mathbf{f} - \mathbf{f} \cdot \text{curl}\,\mathbf{g}$

(11) $\text{curl}(\phi\mathbf{f}) = \phi\,\text{curl}\,\mathbf{f} + (\text{grad}\,\phi) \times \mathbf{f}$

(12) $\text{curl}(\mathbf{f} \times \mathbf{g}) = \mathbf{f}\,\text{div}\,\mathbf{g} - \mathbf{g}\,\text{div}\,\mathbf{f} + (\mathbf{g} \cdot \text{grad})\mathbf{f} - (\mathbf{f} \cdot \text{grad})\mathbf{g}$

(13) $\text{curl curl}\,\mathbf{f} = \text{grad div}\,\mathbf{f} - \nabla^2\mathbf{f}$

(14) $\text{curl curl}\,\phi \equiv 0$

(15) $\text{div curl}\,\mathbf{f} \equiv 0$

(16) $\nabla^2(\phi\psi) = \phi\,\nabla^2\psi + 2(\text{grad}\,\phi) \cdot (\text{grad}\,\psi) + \psi\,\nabla^2\phi$

(17) $\nabla^2 = \nabla \cdot \nabla = \text{div grad}$

In cartesian coordinate systems the orthogonal unit vectors, **i**, **j**, **k** are denoted $\mathbf{e}_x$, $\mathbf{e}_y$, $\mathbf{e}_z$.

(18) $\nabla = \mathbf{e}_x\,\partial/\partial x + \mathbf{e}_y\,\partial/\partial y + \mathbf{e}_z\,\partial/\partial z$

(19) $\text{grad}\,\phi = \nabla\phi = (\partial\phi/\partial x)\mathbf{e}_x + (\partial\phi/\partial y)\mathbf{e}_y + (\partial\phi/\partial z)\mathbf{e}_z$

(20) $\text{div}\,\mathbf{f} = \nabla \cdot \mathbf{f} = \partial f_x/\partial x + \partial f_y/\partial y + \partial f_z/\partial z$

(21) $\text{curl}\,\mathbf{f} = \text{rot}\,\mathbf{f} = \nabla \times \mathbf{f} = (\partial f_z/\partial y - \partial f_y/\partial z)\mathbf{e}_x + (\partial f_x/\partial z - \partial f_z/\partial x)\mathbf{e}_y + (\partial f_y/\partial x - \partial f_x/\partial y)\mathbf{e}_z$

---

[1] For a more extensive collection of formulas of vector analysis see, e.g., Gradshteyn and Ryzhik (1980).

or

$$\operatorname{curl} \mathbf{f} = \begin{vmatrix} \mathbf{e}_x & \mathbf{e}_y & \mathbf{e}_z \\ \partial/\partial x & \partial/\partial y & \partial/\partial z \\ f_x & f_y & f_z \end{vmatrix}$$

(22) $\mathbf{a} \cdot \operatorname{grad} = \mathbf{a} \cdot \nabla = a_x \, \partial/\partial x + a_y \, \partial/\partial y + a_z \, \partial/\partial z$

(23) $\nabla^2 = \nabla \cdot \nabla = \partial^2/\partial x^2 + \partial^2/\partial y^2 + \partial^2/\partial z^2$

In spherical coordinates the orthogonal unit vectors are denoted $\mathbf{e}_r, \mathbf{e}_\vartheta, \mathbf{e}_\varphi$.

(24) $\nabla = \mathbf{e}_r \, \partial/\partial r + \mathbf{e}_\vartheta r^{-1} \, \partial/\partial\vartheta + \mathbf{e}_\varphi (r \sin \vartheta)^{-1} \, \partial/\partial\varphi$

(25) $\operatorname{grad} \phi = (\partial\phi/\partial r)\mathbf{e}_r + r^{-1}(\partial\phi/\partial\vartheta)\mathbf{e}_\vartheta + (r \sin \vartheta)^{-1}(\partial\phi/\partial\varphi)\mathbf{e}_\varphi$

(26) $\operatorname{div} \mathbf{f} = r^{-2} \, \partial(r^2 f_r)/\partial r + (r \sin \vartheta)^{-1} \, \partial(f_\vartheta \sin \vartheta)/\partial\vartheta$
$+ (r \sin \vartheta)^{-1} \, \partial f_\varphi/\partial\varphi$

(27) $\operatorname{curl} \mathbf{f} = (r \sin \vartheta)^{-1}[\partial(f_\varphi \sin \vartheta)/\partial\vartheta - \partial f_\vartheta/\partial\varphi]\mathbf{e}_r$
$+ r^{-1}[(\partial f_r/\partial\varphi)\sin^{-1} \vartheta - \partial(r f_\varphi)/\partial r]\mathbf{e}_\vartheta$
$+ r^{-1}[\partial(r f_\vartheta)/\partial r - \partial f_r/\partial\vartheta]\mathbf{e}_\varphi$

(28) $\mathbf{a} \cdot \operatorname{grad} = a_r \, \partial/\partial r + a_\vartheta r^{-1} \, \partial/\partial\vartheta + a_\varphi (r \sin \vartheta)^{-1} \, \partial/\partial\varphi$

(29) $\nabla^2 = r^{-2} \, \partial(r^2 \, \partial/\partial r)/\partial r + (r^2 \sin \vartheta)^{-1} \, \partial(\sin \vartheta \, \partial/\partial\vartheta)/\partial\vartheta$
$+ (r \sin \vartheta)^{-2} \, \partial^2/\partial\varphi^2$

References and Bibliography

Some publications on nonsinusoidal waves and functions are listed even though they are not mentioned in the text.

Abraham, M. (1905). "Theorie der Elektrizität." Teubner, Leipzig.

Abraham, M., and Becker, R. (1932). "The Classical Theory of Electricity and Magnetism" (J. Dougall, trans., 8th Ed. of "Theorie der Elektrizität"). Hafner, New York.

Abraham, M., and Becker, R. (1950). "The Classical Theory of Electricity and Magnetism" (J. Dougall, trans., 14th Ed. of "Theorie der Elektrizität"). Hafner New York.

Adler, R. B., Chu, L. J., and Fano, R. M. (1960). "Electromagnetic Energy Transmission and Radiation." Wiley, New York.

Albers, J., and Kotzer, R. (1978). A neutrino communications system, *Conf. Proc. Neutrinos-78, Purdue University* C156–C167.

Alvarez, L. W., Eberhard, P. H., and Ross, R. R. (1970). Search for magnetic monopoles in the lunar sample. *Science* **167**, 701–703.

Amaldi, E. (1968). On the Dirac magnetic monopoles. *In* "Old and New Problems in Elementary Particles" (G. Puppi, ed.), pp. 1–61. Academic Press, New York.

Amin, M. B., and James, J. R. (1981a). Techniques for utilization of hexagonal ferrites in radar absorbers. Part I. Broadband planar coating. *Radio Electron. Eng. (London)* **51**, 209–218.

Amin, M. B., and James, J. R. (1981b). Techniques for utilization of hexagonal ferrites in radar absorbers. Part 2. Reduction of radar cross-section of h.f. and v.h.f. wire antennas. *Radio Electron Eng. (London)* **51**, 219–225.

Bates, R. H. T. (1971). A theorem for wide bandwidth echo-location systems. *J. Sound Vib.* **16,** 223–230.

Bebikh, M. V., and Saurin, A. A. (1985). A method of increasing the effectiveness of Walsh-transform devices (in Russian). *Izv. VUZ. Radioelektron.* **28,** 31–36.

Becker, R. (1957). "Theorie der Elektrizität" (revised by F. Sauter), 16th Ed., Vol.1. Teubner, Stuttgart.

Becker, R. (1964). "Electromagnetic Fields and Interactions" (trans. of "Theorie der Electrizität," 16th Ed., Vol. 1). Blaisdell, New York. Reprinted by Dover, New York, 1982.

Bernoulli, D. (1738). "Hydrodynamica" (in Latin). Basel.

Besslich, Ph. W. (1985). Spectral processing of switching functions using signal-flow transformations. *In* "Spectral Techniques and Fault Detection" (M. G. Karpovsky, ed.). Academic Press, New York.

Birks, J. B. (1948). Measurements of permeability of low conductivity ferrimagnetic materials at centimeter wavelength. *Proc. Phys. Soc. (London)* **60,** 282–292.

Borcherdt, G. T. (1961). Electromagnetic-radiation-absorptive article and method of manufacturing same. US Patent 2,992,426.

Boules, R. (1987). Propagation Velocity of Signals. PhD thesis Dept. Electrical Engineering, Catholic University of America, Washington, D. C.

Brillouin, L. (1914). Über die Fortpflanzung des Lichtes in dispergierenden Medien. *Ann Phys.* **44,** 203–240.

Brillouin, L. (1946). "Wave Propagation in Periodic Structures." McGraw Hill, New York.

Carlson, M. F., and Besslich, Ph. W. (1985). Adaptive selection of threshold matrix size for pseudogray rendition of images. *Opt. Eng.* **24,** 655–662.

Carrigan. R. A. (1965). Consequences of the existence of massive magnetic poles. *Nuovo Cimento* **38,** 638–641.

Chang Tong and Chen Wei-ren (1985). Generalized Rademacher functions (in Chinese, English Abstract). *Acta Math. Appl. Sin.* **8,** 257.

Chen, W. L. (1982). Walsh series analysis of multi-delay systems. *J. Franklin Inst.* **313,** 207–217.

Churchill, R. V., and Brown, J. W. (1978). "Fourier Series and Boundary Value Problems," 3rd Ed. McGraw-Hill, New York.

Close, C. M. (1966). "The Analysis of Linear Circuits." Harcourt, New York.

Cole, T. W. (1976). Polarizaton coding in a Hadamard-transform image scanner. *Opt. Commun.* **19,** 374–377.

Courant, R., Friedrichs, K., and Lewy, H. (1928). Über die partiellen Differentialgleichungen der mathematischen Physik. *Math. Ann.* **100,** 32–74.

Cuenod, M., and Durling, A. (1968). "A Discrete Time Approach for System Analysis." Academic Press, New York.

Dirac, P. A. M. (1931). Quantised singularities in the electromagnetic field. *Proc. R. Soc.* Ser. A **133,** 60–72.

Dirac, P. A. M. (1948). The theory of magnetic poles. *Phys Rev.* **74,** 817–830.

Downen, D. N., and Eichenberger, B.A. (1979). Reactive sheets. US Patent 4,162,496.

Emerson, W. H. (1973). Electromagnetic wave absorbers and anechoic chambers through the years. *IEEE Trans. Antennas Propag.* **AP–21,** 484–489.

Evans, S., and Kong, F. N. (1983). Gain and effective area of impulse antenna. Proc. *IEE Int. Conf. Antennas Propag. 3rd, London,* pp. 421–424.

Fleischer, R. L., Hart, H. R., Jacobs, I. S., and Price, P. B. (1969a).Search for magnetic monopoles in deep ocean deposits. *Phys. Rev.* **184,** 1393–1397.

Fleischer, R. L., Price, P. B., and Woods, R. T. (1969b). Search for tracks of massive, multiply charged magnetic poles. *Phys. Rev.* **184,** 1398–1401.

Forster, E. O., and Vanderbilt, B. M. (1977). Metal-filled plastic material. US Patent 4,024,318.

Geramita, A. V., and Seberry, J. (1979). "Orthogonal Designs: Quadratic Forms and Hadamard Matrices. " Dekker, New York.

Gómez, R., and Morente, J. A. (1985). Analysis of electric quadrupole radiation in the time domain: Applications to large-current radiators. *Int. J. Electron.* **58,** 921–931.

Gómez, R., Morente, J. A., and Olmedo, B. G. (1984). Some thoughts about the radiaton of antennas excited by nonsinusoidal currents. *Int. J. Electron.* **57,** 617–625.

Gómez, R., Morente, J. A., and Bretones, A. R. (1985). A new integro-differential equation for the time-domain analysis of thin-wire structures. *Proc. IEEE Int. Symp. Electromagn. Compat.* pp. 343–345.

Gradshteyn, I. S., and Ryzhik, I. M. (1980). "Table of Integrals, Series, and Products" (transl. from the 4th Russian Ed., by Scripta Technica, A. Jeffrey, ed.). Academic Press, New York.

Grimes, D. M. (1972). Ferrite radar absorbing material. US Patent 3,662,387.

Grimes, D. M., Raymond, W. W., Hach, R. J., and Walser, R. M. (1976). Magnetic absorbers. U.S. Patent 3,938,152.

Harmuth, H. F. (1981), "Nonsinusoidal Waves for Radar and Radio Communication." Academic Press, New York.

Harmuth, H. F. (1984). "Antennas and Waveguides for Nonsinusoidal Waves." Academic Press, New York.

Harmuth, H. F. (1985). Radiation of nonsinusoidal waves by a large-current radiator. *IEEE Trans. Electromagn. Compat.* **EMC–27,** 77–87.

Harmuth, H. F. (1987). "Information Theory Applied to Space–Time Physics" (in Russian, no English Edition due to the financial state of English language scientific publishing). MIR, Moscow, in press.

Hatakeyama, K., and Inui, T. (1984). Electromagnetic wave absorber using ferrite absorbing material dispersed with short metal fibers. *IEEE Trans. Magn.* **MAG-20,** 1261–1263.

Heitler, W. (1954). "The Quantum Theory of Radiation." Oxford Univ. Press, London.

Hussain, M. G. M. (1985a). An overview of the development in nonsinusidal wave technology. *Proc. IEEE Int. Radar Conf.* pp. 190–196.

Hussian, M. G. M. (1985b). Line-array beam-forming and monopulse techniques based on slope patterns of nonsinusoidal waveforms. *IEEE Trans. Electromagn. Compat.* **EMC-27,** 143–151.

Huygens, C. (1690). "Treatise on Light" (reprinted by Dover, New York, 1962).

Ishino, K., Watanabe, T., and Hashimoto, Y. (1977). Microwave absorber. U.S. Patent 4,003,840.

Ishino, K., Yamashita, H., Ono, N., and Hashimoto, Y. (1978). Electromagnetic wave-absorbing wall. U.S. Patent 4,118,704.

Jackson, J. D. (1975). "Classical Electrodynamics," 2nd Ed., Wiley, New York.

Johnstone, B. (1984). Absorbing metals catch Reagan's eye. *New Sci. (London)* **101,** (1399), 24.

Karanan, V. R., Frick, P. A., and Mohler, R. R. (1978). Bilinear system identification by Walsh functions. *IEEE Trans. Autom. Control* **AC-23,** 709–713.

Karbowiak, A. E. (1983). Walsh functions for electromagnetic field analysis. *URSI Symp. Electromagn. Theory, Santiago de Compostela, Spain.*

Kelly, F. J., Sáenz, A. W., and Überall, H. (1979). Telecommunication by high-energy neutrino beams. *In* "Long-Distance Neutrino Detection-1978" (*Am. Inst. Phys. Proc.*) (52), 113–128.

Kelso, J. M. (1964). "Radio Ray Propagation in the Ionoshpere." McGraw-Hill, New York.

Klingler, E. H. (1967). Tunable absorber. U.S. Patent 3,309,704.

Knott, E. F. (1979). The thickness criterion for single-layer radar absorbents. *IEEE Trans. Antennas Propag.* **AP-27,** 698–701.

Kraus, J. D. (1984). "Electromagnetics," 3rd Ed., McGraw-Hill, New York.

Kraus, J. D., and Carver, K. R. (1973). "Electromagnetics," 2nd Ed. McGraw-Hill, New York.

Kraus, J. D., and Carver, K. R. (1981). "Electromagnetics," Int. Student Ed. McGraw-Hill Kogakusha, Tokyo.

Krylov, A. N. (1929). On the transmission of currents in cables (in Russian). *J. Appl. Phys.* (Moscow) VI (2), 66.

Landau, L. D., and Lifshitz, E. M. (1971). "The Classical Theory of Fields," 3rd Ed. (M. Hamermesh, trans., from the Russian). Pergamon, Oxford.

Meinke, H. H., Ullrich, K., and Wesch, L. (1972). Wide band interference absorber and technique for electromagnetic radiation. U.S. Patent 3,680,107.

Miller, E. K., and Landt, J. A. (1980). Direct time-domain techniques for transient radiation and scatterng from wires. *Proc. IEEE* **68,** 1396–1423.

Morkoc, H., and Solomon, P. M. (1984). The HEMT: A superfast transistor. *IEEE Spectrum* **21,** (Feb.)28–35.

Naito, Y. (1973). Microwave absorbing wall element. U.S. Patent 3,720,951.

Neher, L. K. (1953). Nonreflecting background for testing microwave equipment. U.S. Patent 2,656,535.

Nielsen, W. G. (1981). Mat for multispectral camouflage of objects and permanent constructions. U.S. Patent 4,287,243.

Ormondroyd, R. F., and Alatsatianos, G. (1984). Compatibility of "carrier-free" electromagnetic Walsh waves with their sinusoidal counterparts in radio systems. *Proc. Int. Conf. Electromagn. Compat., Univ. Surrey* (Institution of Electronic and Radio Engineers Publ. No. 60), 69–77.

Palanisamy, K. R., and Prasada Rao, G. (1983). Optimal control of linear systems with delays in state and control via Walsh functions. *Proc. IEE* **130,** Part D, 300–312.

Paley, R. E. A. C., and Wiener, N. (1934). "Fourier transforms in the complex domain." *Am. Math. Soc. Coll. Publ.* **19.**

Pozar, D. M., McIntosh, R. E., and Walker, S. G. (1983). The optimum feed voltage for a dipole antenna for pulse radiation. *IEEE Trans. Antennas Propag.,* **AP–31,** 563–569.

Prasada Rao, G. and Palanisamy, K. R. (1983). Improved algorithm for parameter identification in continuous systems via Walsh functions. *Proc. IEE* 130, Pt. D, 9–16.

Prasada Rao, G., and Sivakumar, L. (1979). Identification of time-lag systems via Walsh functions. *IEEE Trans. Autom. Control* **AC–24,** 806–809

Pratt, B. C. (1961a). Nondirectional metal-backed, electromagnetic radiation-absorptive films. U.S. Patent 2,992,425.

Pratt, B. C. (1961b). Electromagnetic radiation absorptive article. U.S. Patent 2,996,710.

Reitz, J. R., Millford, F. J., and Christy, R. W. (1979). "Foundations of Electromagnetic Theory." Addison-Wesley, Reading, Massachusetts.

Richtmyer, R. D. (1957). "Difference Methods for Initial Value Problems." Wiley, New York.

Riemann, B. (1854). Über die Hypothesen, welche der Geometrie zu Grunde liegen. *In* "Gesammelte Mathematische Werke" (H. Weber, ed.), 272–287. Teubner, Leipzig, 1892.

Sáenz, A. W., Überall, H., Kelly, F. J., Padgett, D. W., and Seeman, N. (1977). Telecommunication with neutrino beams. *Science* **198,** 295–297.

Sáenz, A. W., Überall, H., and Kelly, F. J. (1978). High energy neutrino beams as a potential means of telecommunication. *Conf. Proc. Neutrino-78, Purdue Univ.* C168–C171.

Salisbury, W. W. (1952). Absorbent body for electromagnetic waves. U.S. Patent 2,599,944.

Sandars, P. G. H. (1966). Magnetic charge. *Contemp. Phys.* **7,** 419–429.

Sato, M., Iguchi, M., and Sato, R. (1984). Transient response of coupled linear dipole antennas. *IEEE Trans. Antennas Propag.* **AP–32,** 133–140.

Schmitt, H. J., Harrison, C. W., and Williams, C. S. (1966). Calculated and experimental response of thin cylindrical antennas to pulse excitation. *IEEE Trans. Antennas Propag.* **AP-14,** 120–127.

Schumann, W. O. (1948). "Elektrische Wellen." Carl Hanser, Munich.

Schwarzschild, K. (1903). Zur Elektrodynamik. II. Die elementare elektrodynamische Kraft. *Nachr. Akad. Wiss Göttingen, Math.-Phys. Kl.* 132–141.

Schwinger, J. (1969). A magnetic model of matter. *Science* **165,** 757–761.

Sinha, M. S. P., Rajamani, V. S., and Sinha, A. K. (1980). Identification of non-linear distributed system using Walsh functions. *Int. J. Control* **32,** 669–676.

Smirnov, V. I. (1964). "A Course of Higher Mathematics" (D. E. Brown, trans., from Russian). Pergamon, Oxford.

Sommerfeld, A. (1914). Über die Fortpflanzung des Lichtes in dispergierenden Median. *Ann. Phys.* **44,** 177–202.

Spetner, L. M. (1974). Radiation of arbitrary electromagnetic waveforms. *In* "Applications of Walsh Functions and Sequency Theory" (H. Schreiber and G. Sandy, eds.), pp. 249–274. IEEE, New York.

Stepanov, Y. G. (1968). "Antiradar Camouflage Techniques" (in Russian). Soviet Radio, Moscow.

Stratton, J. A. (1941). "Electromagnetic Theory." McGraw-Hill, New York.

Suetake, K. (1973a). Extremely thin, wave absorption wall. U.S. Patent 3,737,903.

Suetake, K. (1973b). Wide band flexible wave absorber. U.S. Patent 3,754,255.

Trifonov, A. P., and Senatorov, A. K. (1984). Noise immunity in pulsed signal reception with a nonsinusoidal carrier (in Russian). *Radiotek. Elektron.* **29,** 83–84.

Trifonov, A. P., Manelis, V. B., and Nechaev, E. P. (1985). The requirements imposed on the synchronization accuracy when using pulse-time modulation of signals with nonsinusoidal carrier (in Russian). *Izv. VUZ. Radioelektron* **28,** 39–43.

Tuinila, R. P., and Bayrd, R. O. (1969). Radio frequency absorber. U.S. Patent 3,441,933.

Tzafestas, S. G. (1983). Walsh transform theory and its application to systems analysis and control: An overview. *Math. Comput. Simul.* **25,** 214–225.

Tzafestas, S. G. (1985). Analysis and parameter computation of a delay coupled core nuclear reactor using Walsh series expansions. *In* "Systems Analysis and Simulation 1985, I" (A. Sydow, M. Thoma, and R. Vichnevetsky, eds.), pp. 105–113. Akademie-Verlag, Berlin, GDR.

Tzafestas, S. G., and Chrysochoides, N. (1977). Nuclear reactor control using Walsh variational synthesis. *Nucl. Sci. Eng.* **62,** 763–770.

Überall, H., Kelly, F. J., and Sáenz, A. W. (1979). Neutrino beams: A new concept in telecommunications. *J. Washington Acad. Sci.* **69,** (2) 48–54.

Wagner, K. W. (1953). "Elektromagnetische Wellen." Birkhäuser, Basel.

Wait, J. R. (1970). "Electromagnetic Waves in Stratfied Media," 2nd Ed. Pergamon, Oxford.

Wallin, E. W. (1977). Knitted camouflage material. U.S. Patent 4,064,305.

Wallman, H. (1948). Realizability of filters. *In* "Vacuum Tube Amplifiers" (G. E. Valley and H. Wallman, eds.), pp. 721–727. McGraw-Hill, New York.

Wartenberg, B. (1968). Measurement of the electromagnetic material constants μ and ϵ of ferrites in the mm-wave range. *Z. Angew. Phys.* **24,** 211–217.

Waters, W. E. (1983). "Electrical Induction from Distant Current Surges." Prentice-Hall, New York.

Wesch, L. (1967). Non-metallic packaging material with resonance absorption for electromagnetic waves. U.S. Patent 3,315,260.

Wesch, L. (1970). Resonance absorber for electromagnetic waves. U.S. Patent 3,526,896.

Wesch, L. (1973). Absorber for electromagnetic radiation. U.S. Patent 3,721,982.

Wesch, L., and Meinke, H. H. (1971). Electronmagnetic wave attenuating device. U.S. Patent 3,568,195.

Wesch, L., and Ullrich, K. (1969). Radar-proof and shell-proof building material. U.S. Patent 3,454,947.

Westphal, W. H. (1953). "Physik," 16th Ed. Springer-Verlag, Berlin and New York.

Wickenden, B. V. A., and Howell, W. G. (1978). Ferrite quarter wave type absorber. *Proc. Mil. Microwave Conf., Wembley, London* 310–317.

Wright, R. W. (1977a). Combined layers in a microwave radiation absorber. U.S. Patent 4,012,378.

Wright, R. W. (1977b). Magnetic ceramic absorber. U.S. Patent 4,023,174.

Wright, R. W., and Wright, J. W. (1975). Thin, lightweight electromagnetic absorber. U.S. Patent 3,887,920.

Index

All names in References and Bibliography are listed, except those of the primary authors.

E

F

G

H

I

J

K

L

M

N

O

P

Q

R

S

T

U

V

W

Y

Z